Reproductive Physiology of Mammals:

From Farm to Field and Beyond

Join us on the Web at

agriculture.delmar.com

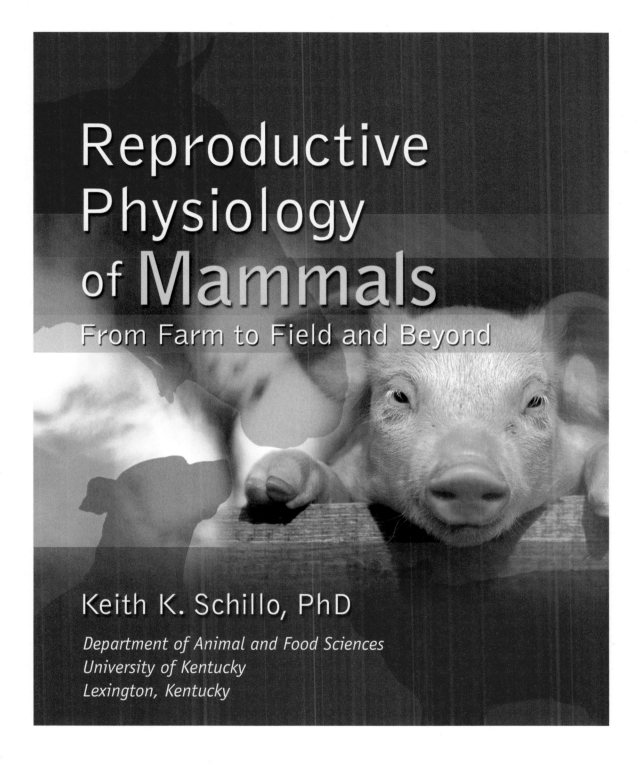

Reproductive Physiology of Mammals

From Farm to Field and Beyond

Keith K. Schillo, PhD

Department of Animal and Food Sciences
University of Kentucky
Lexington, Kentucky

DELMAR
CENGAGE Learning

Australia • Brazil • Japan • Korea • Mexico • Singapore • Spain • United Kingdom • United States

DELMAR
CENGAGE Learning

Reproductive Physiology of Mammals: From Farm to Field and Beyond
Keith K. Schillo

Vice President, Career and Professional Editorial: Dave Garza

Director of Learning Solutions: Matthew Kane

Acquisitions Editor: David Rosenbaum

Managing Editor: Marah Bellegarde

Product Manager: Christina Gifford

Editorial Assistant: Scott Royael

Vice President, Career and Professional Marketing: Jennifer McAvey

Marketing Director: Debbie Yarnell

Marketing Coordinator: Jonathan Sheehan

Production Director: Carolyn Miller

Production Manager: Andrew Crouth

Content Project Manager: Jeff Varecka

Art Director: David Arsenault

Technology Project Manager: Mary Colleen Liburdi

Production Technology Analyst: Thomas Stover

For product information and technology assistance, contact us at
Professional & Career Group Customer Support, 1-800-648-7450

For permission to use material from this text or product, submit all requests online at **cengage.com/permissions**. Further permissions questions can be e-mailed to **permissionrequest@cengage.com**.

Library of Congress Control Number: 2008923411

ISBN-13: 978-1-4180-3013-1

ISBN-10: 1-4180-3013-9

Delmar
5 Maxwell Drive
Clifton Park, NY 12065-2919
USA

Cengage Learning products are represented in Canada by Nelson Education, Ltd.

For your lifelong learning solutions, visit **delmar.cengage.com**

Visit our corporate website at **cengage.com**.

Notice to the Reader
Publisher does not warrant or guarantee any of the products described herein or perform any independent analysis in connection with any of the product information contained herein. Publisher does not assume, and expressly disclaims, any obligation to obtain and include information other than that provided to it by the manufacturer. The reader is expressly warned to consider and adopt all safety precautions that might be indicated by the activities described herein and to avoid all potential hazards. By following the instructions contained herein, the reader willingly assumes all risks in connection with such instructions. The publisher makes no representations or warranties of any kind, including but not limited to, the warranties of fitness for particular purpose or merchantability, nor are any such representations implied with respect to the material set forth herein, and the publisher takes no responsibility with respect to such material. The publisher shall not be liable for any special, consequential, or exemplary damages resulting, in whole or part, from the readers' use of, or reliance upon, this material.

Printed in the United States of America
1 2 3 4 5 6 7 12 11 10 09 08

As an educator, my goal is to help students develop a wisdom that will help them make informed judgments about what they can and should do as individuals living in a diverse society. This book is dedicated to that goal.

CONTENTS

CHAPTER 4   **SEXUAL DIFFERENTIATION**   **52**

CHAPTER 5   **FUNCTIONAL ANATOMY OF REPRODUCTIVE SYSTEMS: GENITAL
ORGANS**   **83**

CHAPTER 6   **FUNCTIONAL ANATOMY OF REPRODUCTIVE SYSTEMS:
NEUROENDOCRINE SYSTEMS**   **125**

PREFACE

This book is intended to serve as the primary text for an undergraduate-level course in reproductive physiology. Although the field of reproductive physiology spans the biological, medical, veterinary, and animal sciences, courses dealing specifically with this area of study are most typically offered through animal science departments. This text includes all of the major topics traditionally covered in this type of course. However, the overall perspective of the text deviates from conventional approaches in two important ways. Unlike most reproductive physiology books, this one develops a social context for the discipline in order to illuminate some controversial ethical issues associated with this field of study. In addition, the text takes more of a biological slant than more conventional books that focus primarily on agricultural perspectives of reproductive physiology. By presenting the science of reproductive physiology as a social institution, students can establish connections between their own lives and the subject matter. This helps them appreciate why they should study reproductive physiology and stimulates critical thought about pressing social issues related to this field. By emphasizing a biological perspective, students will be compelled to think more holistically and come to understand that physiologic processes exist as adaptations to an animal's environment. In addition to being personally enriching, this way of thinking will prove to be useful to students by helping them with careers in the life sciences.

WHY I WROTE THIS TEXT

I have been teaching reproductive physiology for 15 years. During this time I have tried to develop a course that illuminates the major scientific paradigms of the discipline while at the same time faces up to the social reality of our times. I have embraced two guiding principles in teaching this and other courses. First, I abide by the notion that undergraduate students should not be narrowly focused on their majors and judge the value of a course based solely on whether or not it will help them secure a job or gain admission to professional school. A richer understanding of the world is obtained when one assumes diverse perspectives. Not only does this approach make the world a much more fascinating place, it prepares students to successfully negotiate the pluralistic global culture in which they must live and work. My second guiding principle reflects the perspective of the late Richard Feynman, the renowned physicist. As a child, Feynman spent many summer weekends in the Catskill Mountains of New York. During one outing a boy asked Feynman to identify a bird. Feynman told the boy the name of the bird in English, Italian, Portuguese, Chinese, and Japanese, but added, "You can know the name of that bird in all the languages of the world, but when you're finished, you'll know absolutely nothing whatever about the bird. You'll only know about humans in different places and what they call the bird. So let's look at the bird and see what it's doing—that's what counts." In teaching reproductive physiology, I try to emphasize function over memorization of terminology and specific facts about the reproductive traits of particular species. As fascinated as I was with my first reproductive physiology course, my enthusiasm for the subject was dampened by being required to memorize things such as the penis lengths for a dozen species of mammals. There seemed to be no purpose for learning these facts other than to fill in a blank space on an examination. In my view much of this activity came at the expense of learning more about how these animals reproduce. Although I recognize that some memorization is required to establish a foundation of knowledge and to develop language that allows people to communicate within a discipline, I refrain from asking students to memorize tables of species-specific facts and instead emphasize universal concepts.

This text reflects the aforementioned principles that have guided my teaching activities. In order to provide students with a broad and diverse education in reproductive physiology, discussions offer a comparative perspective of mammalian reproduction and seek to explain not only how mammals reproduce, but why species display particular patterns of reproduction. This approach accommodates the conventional approaches of teaching livestock reproduction so students can learn the specialized knowledge required for their major, and also provides continuity with their previous courses in biology. The text also broadens the scope of reproductive physiology in a way that some may view as a radical departure from conventional texts on this subject. Students often ask why they are required to take certain classes. In an attempt to address this question, the text begins with a discussion of the relevance of reproductive physiology in human societies and ends with an analysis of the two most controversial issues associated with reproductive physiology research. When I teach this subject I am conscious of two

elephants looming in the classroom: the use of nonhuman animals by humans and the abortion of human fetuses. Students are keenly aware of these issues and have strong opinions about them. However, they often lack experience discussing contentious topics in a civil and informative manner. A course in reproductive physiology provides the unique opportunity to engage students in discussions about these important social issues. In light of the fact that our current generation of students is largely disengaged from civic activities, it seems irresponsible for educators to squander any opportunity to engage students in discussions of issues that are likely to influence how they live. For instructors who are brave enough to address these issues in their classrooms, I offer analyses of major arguments concerning human use of animals and abortion of human fetuses. My hope is that this will provide a resource to motivate students to take more active roles in civic life.

One of the greatest challenges teachers face is deciding what they should teach. The explosion of knowledge in reproductive biology requires that teachers make judicious decisions about what material they will cover in class. As noted earlier, the text includes all of the topics generally considered as the core knowledge of reproductive physiology. However, this text places greater emphasis on fundamental biological principles and provides less information on reproductive management of farm animals than more conventional texts.

Some readers may accuse me of abandoning Feynman's principle because my extensive use of scientific terminology in the early chapters, especially those dealing with anatomy. My decision to do this is based on the observation that most animal science students lack basic skills in developmental biology, anatomy, and endocrinology. Most of the students who enroll in my class find this material to be extremely useful, especially the 80 percent who plan to apply to veterinary school. In keeping with my preference for function, I have tried to emphasize major concepts common to most mammals and have kept the inclusion of species-specific data to a minimum. For those who have an interest in providing extensive quantitative descriptions of the reproductive patterns of mammals I recommend they consult *Asdell's Patterns of Mammalian Reproduction: A Compendium of Species-Specific Data.*

ORGANIZATION OF THE TEXT

The text consists of four major sections. The first two chapters make up the first section and attempt to establish a context for studying reproductive physiology. Chapter 1 provides a general description of the discipline and discusses why it is important to human societies. Chapter 2 outlines the focus of the book and introduces students to fundamental concepts that govern sexual reproduction.

The second section of the book consists of Chapters 3 through 7. These chapters deal with background information that is a prerequisite for understanding the major reproductive processes considered in subsequent chapters. This section begins with the organization of reproductive systems and is followed by more detailed accounts of how the sex of individuals is established, the anatomy of reproductive organs, and an overview of endocrine principles. Chapters on reproductive anatomy depart from what has been the conventional approach. Most, if not all, reproductive physiology texts describe the anatomy of male and female reproductive tracts in separate chapters. I have chosen to describe them in the same chapter and organize discussions based on the major components of the tracts (e.g., gonads, genital ducts, external genitalia) as opposed to sex. This approach emphasizes the fact that the reproductive tissues of adult males and females are comparable

and arise from the same embryonic tissues. This reinforces concepts developed in the chapter dealing with sexual differentiation.

The core of any reproductive physiology course consists of descriptions of the major reproductive processes. These are included in Chapters 8 through 17. The arrangement of topics is similar to that of other texts. However, my coverage of puberty, sexual behavior, and the effects of environment on reproduction is more substantive than what is typically included in texts oriented toward animal science.

The last section consists of a single chapter that provides detailed analyses of two important ethical issues pertaining to the science of reproductive physiology. Chapter 18 begins by exploring the relationship between science and ethics and then provides several of the most significant ethical arguments concerning the moral status of nonhuman animals and the human conceptus.

The pedagogical approach of this text is based on the assumption that most students can comfortably manage four to five major concepts for each major topic of a course. The chapters of this text are organized around this principle. Each begins with a list of teaching objectives that introduce the major concepts. The subdivisions of the text elaborate on these concepts. At the conclusion of each chapter, I provide a summary of concepts in language that students should be able to remember. Discussion questions at the end of each chapter emphasize applications of the major concepts. These questions can be used to generate classroom discussions, as homework assignments, or as exam questions.

In an attempt to create a readable text, I have refrained from using a purely scientific style of writing. This means I provide a narrative account of physiology as opposed to the reporting of research results with references cited in the text. While writing this book, I was constantly reminded of the familiar adage: you can't please all of the people all of the time; you can only please some of the people some of the time. This became even more apparent when I read the insightful comments of the individuals who reviewed early drafts this work. The beauty of higher education is that there is remarkable diversity in the way instructors teach a particular subject. In animal science, there is a schism between those who favor an applied approach, emphasizing management of livestock and those who embrace a basic biology approach, emphasizing fundamental knowledge about animals. I have tried to respect each of these camps and therefore run the risk of pleasing no one. Nevertheless, my approach is justified by the demographic characteristics of the contemporary animal science student as well as the current research climate in reproductive physiology. Most students have little or no experience with livestock and are more interested in companion animals, horses, and nondomesticated animals. Moreover, the overall importance of reproductive physiology in animal science has diminished greatly during the past 20 years. The vast majority of journal articles in this field are related to issues in human medicine, involving clinical research or basic research with laboratory animals. Although it is important for students to understand livestock reproduction, such knowledge represents only a small fraction of what they will have to understand about reproductive physiology in order to be successful in their chosen careers. Whether we approve or disapprove of this state of affairs seems irrelevant at this point. As educators, we have an obligation to train students in ways that correspond to the social reality in which they must live.

Keith K. Schillo

ACKNOWLEDGMENTS

I should like to acknowledge my colleague Dr. William Silvia for his assistance with preparing and photographing various specimens for text figures in the anatomy chapters and Dr. Tom Curry for letting us use his microscopy laboratory for preparing photographs of tissue sections. I also thank Dr. Sandra Legan, Dr. Peter Hansen, Dr. Cheryl Sisk, Dr. Heather Figueira, Mr. Michael Meyer, and Ms. Remi Tagawa for providing me with photographs. My wife Annabel deserves special thanks for her support, editorial assistance, and ideas about how the text should be organized. Finally, I thank the following reviewers for their thoughtful comments and helpful suggestions:

Peter Hansen, Ph.D.
University of Florida
Mark Diekman, Ph.D.
Purdue University
David Fernandez, Ph.D.
California State University at Pomona
Anita Oberbauer, Ph.D.
University of California at Davis
Henry (Buz) R. Bireline
Santa Fe Community College
Gary Anderson, Ph.D.
University of California at Davis
Niki Whitley, Ph.D.
University of Maryland

ABOUT THE AUTHOR

Dr. Schillo is a native of New York State. He was born in Buffalo and spent his teenage years living in Camillus, a suburb of Syracuse. In 1975, he completed a B.S. degree in Animal Science from Cornell University. He went on to earn a M.S. degree in Animal Science from Purdue University in 1977 and a Ph.D. in Endocrinology-Reproductive Physiology from the University of Wisconsin in 1981. After his graduate education, Dr. Schillo pursued postdoctoral training in neuroendocrinology at the University of Illinois before joining the faculty of the Department of Animal and Food Sciences at the University of Kentucky in 1984. During his career at the University of Kentucky, Dr. Schillo has directed a research program dealing with the neuroendocrine control of reproduction in ruminants. He is the author of over 50 scientific articles and book chapters covering topics ranging from the effects of environment on reproduction in female sheep and cattle to regulation of sexual behavior in bulls. For the past 15 years, Dr. Schillo has devoted most of his time to teaching. He currently teaches endocrinology, animal physiology, environmental and agricultural ethics, a senior capstone course in animal science, and reproductive physiology. Dr. Schillo is the 2007 recipient of the Gamma Sigma Delta Master Teacher Award and the University of Kentucky President's Award for Diversity, honors that recognize his dedication to helping students become responsible citizens in a culturally diverse society. He currently lives in Winchester, Kentucky with his wife, Dr. Annabel Kellam, who is a fiber artist.

HOW TO USE THIS BOOK

Learning a particular subject requires more than remembering pertinent facts and understanding major concepts. Developing knowledge also requires familiarity with the structure and boundaries of the subject. Concept maps can facilitate this process. In addition to providing an overview of concepts a field of study encompasses, it provides a template for sorting the multitude of information one encounters when learning a new subject. The following concept map reflects the perspective of reproductive physiology upon which this book is based. The relationship between a concept map and a person planning to study a subject is analogous to the relationship between a map of a foreign country and a person planning a trip to that country. When traveling it is useful to learn something about the geography of a country before you travel through it. Likewise, when embarking on a new area of study, it is useful to know how the subject is structured.

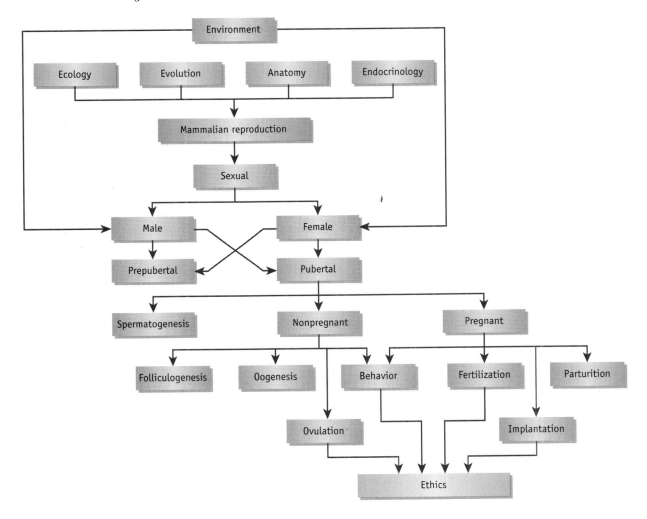

The central focus of the book is mammalian reproduction (highlighted in red). The core concepts that are necessary for developing a thorough understanding of this subject are highlighted in yellow. Concepts highlighted with green are concepts are necessary for developing a comprehensive understanding the core concepts; that is, background or prerequisite concepts. Ethics (highlighted in blue) is neither a core nor prerequisite concept. However, it is a concept that places the field of reproductive physiology in social context. What we know about mammalian reproduction, how we came to know these things, and how we apply the knowledge we develop have ethical implications that impact everyone. The lines between the boxes portray relationships between concepts. The natures of these relationships become clearer as one gains a deeper understanding of the subject matter. In addition, as students become more educated in a subject they may envision more relationships among concepts, and build even more complex maps.

The chapters included in this text reflect the structure of this concept map. The prerequisite concepts (green) are addressed by Chapters 1 through 7. Chapter 18 addresses ethical implications (blue). The remaining chapters address the core concepts. The order of topics presented in this book is consistent with that of previous texts on reproductive physiology and reflects a popular approach to teaching this subject. The organization of chapters is based on the assumption that a student can accommodate only about four or five major concepts for any given topic. In light of this idea, each chapter begins with a list of no more than five objectives that correspond to major chapter headings, major concepts summarized at the end of each chapter, and several discussion questions provided at the end of each chapter. The discussion questions can be used in one of several ways. First, they can be assigned as homework. Second, they can serve as questions for quizzes and exams. Finally they can be used in class to introduce material and(or) stimulate discussions. For the most part the questions require students to apply concepts, to solve problems, and to provoke critical questioning of subject matter.

Most chapters include a text box titled "Focus on Fertility" that describes a classic experiment, an application, or implication of reproductive physiology research. These are intended to be both entertaining and informative. They are intentionally brief and provide only an overview of the topic. These topics could be explored in more detail by the instructor in the form of supplemental material or by students in the form short research assignments (group projects, term papers, or oral presentations).

Becoming educated in a particular discipline requires an appropriate vocabulary. The vocabulary of reproductive physiology is particularly rich and complicated because the discipline encompasses a broad array of topics. This book includes an extensive glossary to facilitate vocabulary development. The terms defined in the glossary appear in bold letters in the text. With respect to testing, it may be preferable to ask students to use terminology appropriately in written answers or apply terms in objective test questions, rather than require them to memorize and regurgitate definition of terms.

Introduction

CHAPTER
OBJECTIVES

- Describe the focus of this book.

- Provide a brief overview of the field of reproductive physiology.

- Establish a context for studying the reproductive physiology of mammals.

FOCUS OF THE BOOK

This book is about the reproductive physiology of mammals. Reproduction refers to the process of self-replication. Physiology is the branch of biology that deals with the functions and activities of organisms. Therefore, reproductive physiology can be defined as a branch of biology that deals with how organisms replicate. It is important to realize that an understanding of physiology requires some appreciation for anatomy, or the structure of organisms. Obviously it would be difficult, if not impossible, to explain how something works without knowing its parts and how they are related to each other. For example, one couldn't understand how an automobile moves without being aware of parts such as the engine, gears, drive train, and wheels, as well as knowledge of how these parts interact with each other. Likewise, it is impossible to understand how dogs reproduce without first learning something about the structure of their reproductive organs.

We will be concerned with the reproductive physiology of mammals. In general mammals are endothermic, homeothermic animals that are hair covered, and feed their young with milk produced by the female's mammary glands. Other defining characteristics include the presence of a diaphragm; three middle-ear ossicles; heterodont dentation (differentiated teeth); sweat, sebaceous and scent glands; a four-chambered heart; and a large cerebral cortex. A more precise account of what it means to be a mammal requires an understanding of how these animals fit into the overall classification system of animals. There is disagreement concerning the classification of animals. Figure 1-1 shows one widely accepted system. Mammals make up the *Class Mammalia,* which belongs to the *Phylum Chordata.* Mammals and other animals with backbones (fishes, amphibians, reptiles, and birds) comprise the *Subphylum Vertebrata. Class Mammalia* can be divided into two subclasses: *Prototheria* and *Theria.* The only prototherian

1

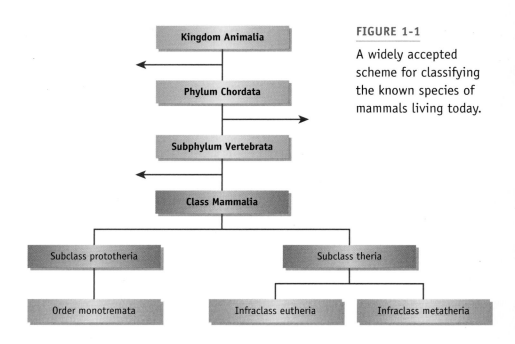

FIGURE 1-1

A widely accepted scheme for classifying the known species of mammals living today.

mammals that exist today are members of the *Order Monotremata*. This order consists of two families of mammals that lay heavily yolked eggs with leathery shells. In addition, females of this subclass nourish their young with milk that is produced by mammary glands and secreted through pores that align the belly (e.g., duck-billed platypus, spiny anteater, and echidna). The other sub-class of mammals *Theria* includes two infraclasses: *Eutheria* and *Metatheria*. There are approximately 270 species of metatherian mammals. These animals are also known as the marsupials. Females of these species give birth to extremely immature offspring, which then migrate from the birth canal to an abdominal pouch where they nurse until they become independent (e.g., opossums, kangaroos, wombats and koalas).

Most of the information presented in this book pertains to the eutherian mammals. This group is typically referred to as the placental mammals, but this terminology can lead to confusion because marsupials have placentas, albeit ones different from those of eutherian mammals. The placentas of eutherian mammals consist of two extra-embryonic membranes; that is, the allantois and the chorion. The placentas of marsupials consist only of tissue from the yolk sac, a structure similar to the yolks of avian eggs.

We will concern ourselves only with the few orders of eutherian mammals with which humans deal most often: *Rodentia* (e.g., rats and mice), *Primata* (e.g., monkeys and apes), *Lagomorpha* (e.g., rabbits and hares), *Carnivora* (e.g., dogs and cats), *Perissodactlyla* (e.g., horses), and *Artiodactyla* (e.g. cattle and sheep). It is important to bear in mind that our focus on reproductive

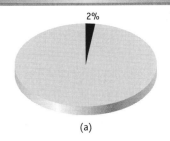

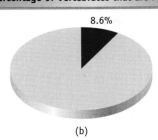

FIGURE 1-2

The percentage of known animal species that are vertebrates (a) and the percentage of known vertebrate species that are mammals (b).

physiology is an extremely narrow one. Vertebrate species account for only 2 percent of the known animal species living today, whereas mammalian species account for slightly more than 8 percent of the known vertebrate species (Figure 1-2). Thus mammals make up less than 0.2 percent of the known species of animals, and we will consider only a small fraction of these animals.

Much of the information presented in this book concerns domestic mammals, particularly those used in agriculture. However, when appropriate, discussions will be extended to include wild mammals since interest in managing the reproductive physiology of these animals has grown in recent years. We will also examine some of the more important principles of human reproduction. There are two major reasons for emphasizing livestock species. First, we know much more about the reproductive physiologies of livestock species than those of other animals (except laboratory rodents). Second, reproductive technologies are used extensively in livestock production. This provides opportunities to link fundamental scientific knowledge with practical applications of this knowledge. The primary goal of this book is to develop an understanding of the basic mechanisms of reproduction, but we will also consider how these mechanisms can be manipulated to enhance or reduce birth rates, and explore some of the ethical issues associated with these practices.

REPRODUCTIVE PHYSIOLOGY AS A SCIENTIFIC DISCIPLINE

Science can be thought of as a system for developing knowledge. This involves asking questions and constructing hypothetical answers that are either supported or rejected by experimentally-derived data. Scientists consider two types of questions: ultimate and proximate. Ultimate questions deal with "why?" These are the so-called big questions. Why does a robin fly? Why does a cow chew its cud? Answers to these types of questions require theoretical assumptions about the world. Biologists typically ground answers to why questions in evolution theory. According to this theory, robins fly and cows chew

1

their cud because of natural selection. In other words, these traits either made these animals more fit (enhanced their ability to survive and reproduce), or didn't make them less fit. We won't concern ourselves very much with these types of questions. Rather, we will focus on proximate questions, or questions of "how?" How is the sex of an organism determined? How does a sheep become sexually mature? How does pregnancy occur?

In studying this subject, you will have to deal with three types of facts. First there are scientific *names* for cell types, reproductive organs, and the various chemical signals that regulate reproduction. The only way to learn this type of material is to memorize it. Think of this as a necessary means to build a vocabulary that will allow you to speak the language of reproductive physiology. The second type of fact you will encounter deals with *measurements*; for example, duration of pregnancy, length of an estrous cycle, dimensions of various reproductive organs, and so on. These types of facts help you achieve a frame of reference. For example, in planning a trip, it's crucial to know how far you will have to travel. Likewise, if you are planning to manage a herd of cows to produce beef, it is necessary to know the length of pregnancy, and so on. Finally, there are *concepts*. These are scientific claims (hypotheses) about how reproductive processes work. For example, you will read that estrogen causes a female to come into heat. This may seem like a well-accepted fact that virtually everyone accepts as truth, but it is really a hypothesis that is supported by abundant experimental data. A hypothesis is assumed to be true until there is good evidence to reject it. Although it is important to understand the scientific basis of this type of fact, it is beyond the scope of this book to provide detailed discussions of the scientific research supporting all the concepts presented in this book. Nevertheless, some experiments are particularly worthy of attention as instructional tools, and will be discussed where appropriate.

In their pursuit of answers to questions about how animals reproduce, reproductive physiologists use a variety of techniques and experimental approaches. When a particular species is first studied, much of the scientific work involves *description*; for example, a careful and thorough documentation of various reproductive characteristics (sexual behaviors, breeding patterns, number of offspring, and so on). Once the reproductive traits have been characterized, research typically turns to the study of the mechanisms controlling reproduction. These types of studies are *experimental* in nature. In other words, they are designed to test specific hypotheses about reproductive processes. For example, once descriptive information about the pattern of reproduction in sheep became available, it was possible to perform experiments that tested specific hypotheses regarding the mechanisms controlling this pattern of reproduction. Of particular importance are experiments that tested the hypothesis that day length determines the time of year that sheep engage in reproductive activity. These types of experiments can involve a variety of approaches, including the following:

- Anatomic (studies describing the gross and microscopic structures of reproductive tissues)

- Physiologic (surgical, pharmacologic, and hormonal studies of how reproductive tissues function)

- Biochemical (studies of chemical activities of cells within the reproductive system)

- Molecular Genetics (studies of how genes regulate activities of reproductive cells)

A CONCEPTUAL FRAMEWORK FOR REPRODUCTIVE PHYSIOLOGY

Every discipline of science is based on a set of basic assumptions about the world; that is, a conceptual framework. Such assumptions are law-like in the sense that they are rarely, if ever, challenged. When they are challenged it is because they no longer help us make sense of the world. In the biological sciences, the theory of evolution provides a set of foundational assumptions about why living things are the way they are. The basic assumption of evolution theory is that the distinguishing features of particular organisms originate in preexisting organisms, and that differences among organisms are the result of natural selection over successive generations. It is useful to establish some fundamental concepts concerning evolution and reproduction:

- Different species express different reproductive strategies.
- A reproductive strategy is an expression of an individual organism's genotype.
- Species express a particular reproductive strategy because individuals that expressed this strategy have greater reproductive fitness than others.
- Reproductive fitness of an individual is a function of how many of its offspring survive to reproduce; that is, how much of an individual's genes are spread through the population.

Keep these concepts in mind as you learn about the diverse reproductive traits discussed in this book. They will help provide insight into why mammals reproduce in so many different ways.

A GLOBAL CONTEXT FOR REPRODUCTIVE PHYSIOLOGY

Students frequently ask why they should study a particular subject. This is a fair question because we are more inclined to learn something when we find it relevant to our lives. Why should anyone study reproductive physiology? One reason is that this subject is concerned with a basic inclination of all living things, including humans. All one has to do is watch television, surf the Internet, or read a popular magazine to be reminded that humans are interested in reproduction and a wide array of subjects related to it. Reproduction has fascinated people for millennia. Another reason for understanding reproductive physiology is more practical; that is, knowledge of reproductive physiology can have a direct impact on the quality of our lives. Historically, humans have employed their understanding of reproduction to achieve two goals: 1) to limit growth of human populations and 2) to enhance food production. This has been the challenge of primitive societies as well as modern ones. Sustaining a human population with adequate food remains an important challenge and will continue to be as long as we humans inhabit this planet.

Agriculture and Human Population

According to paleontologists, humans have inhabited the earth for approximately 3 million years. For most of this time, humans accounted for what amounts to only a miniscule portion of the earth's ecology. However, during the past 10,000 years the human population has increased in an exponential manner (Figure 1-3). In other words, as the human population increases, the amount of time required for the population to double is decreasing. Today over 6 billion people inhabit the earth and they have had an impact on every square centimeter of the planet. What has been responsible for this change?

One of the fundamental laws of ecology is that the population of a particular species varies directly with available food supply. During the past 10,000 years, production of food increased due to agriculture. For every advance in humanity's ability to produce food, there has been a corresponding increase in population. For example, the introduction and use of high-yielding varieties of rice and other crops in the 1960s (the so-called Green Revolution) is largely responsible for the doubling of the world's population during the past 40 years. Geographer Jared Diamond explains how the development of agriculture stimulated population growth in early human societies. Most of the living matter on this planet is useless as food because it is indigestible, poisonous, low in nutritional value, tedious to prepare, or difficult or dangerous to hunt or gather. Typically only a small amount (0.1 percent) of biomass on a particular parcel of land is available as human food. By cultivating

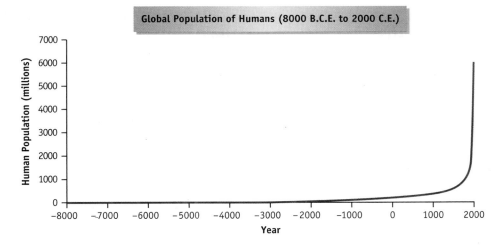

Global Population of Humans (8000 B.C.E. to 2000 C.E.)

FIGURE 1-3

Human population growth between 8,000 B.C.E. (before common era) and 2,000 C.E. (common era). The exponential increase during the past 1,000 years is attributed in part to advances in agriculture. Data for the period between 0 and 2,000 C.E. were taken from Cohen (1995). Population estimates for the period between 8,000 B.C.E. and 0 were generated via extrapolation.

selected species of plants and animals, humans can make as much as 90 percent of the biomass available as food. As we shall see in later discussions, the major determinant of reproductive rate in healthy organisms is level of nutrition. Agriculture increases human populations by increasing the amount of edible food per unit of land.

Use of domestic animals contributes to the expansion of human populations via direct and indirect effects. Use of livestock directly affects population by providing sources of high-quality nutrients (particularly protein). The amount of food provided in the lifetime of a cow, ewe, or sow is much greater than that of wild animals because domestic animals are used as breeding stock to provide offspring that are killed annually for food. Domestic animals also affect human populations by enhancing production of crops. This effect is brought about by providing manure (fertilizer), as well as power for agricultural technologies (e.g., plows) that facilitate planting and harvesting. In addition to these direct effects, the use of domestic animals enhances human populations indirectly. Raising livestock and crops contribute to the development of a "settled existence," meaning that agriculturalists do not re-locate as frequently as hunter-gatherers or pastoralists. This less-mobile lifestyle leads to a shortened birthing interval, which results in greater population density.

The success of agrarian societies is widely evident. Aside from a few remaining pastoral or hunter-gatherer societies, the vast majority of humans living today depend on agriculture for survival. In fact, the dominance of western civilization is largely attributed to the development of agriculture. For better or worse, development of agriculture in the West facilitated the development of advanced weaponry (steel swords, guns, and so on), ocean-going ships, political organization, writing, and epidemic diseases, developments which allowed

Western peoples to conquer other peoples and spread their culture to almost every part of the globe.

Although agriculture permitted the development of technologically advanced societies, it also resulted in significant unintended consequences such as: 1) loss of natural habitat; 2) reduction in wild foods (primarily fish and shellfish); 3) loss of genetic diversity in wild species; 4) erosion of soil; 5) reduction in fossil fuel reserves; 6) pollution of air and water; 7) destruction of native species by invasions of alien species; and 8) atmospheric changes (ozone depletion, and greenhouse gases). Each of these problems can be directly attributed to increasing human populations along with an increased environmental impact of each individual. As agrarian populations expand they seek to increase food production by one or more of the following practices: securing more land, using more resources, and employing new technologies to increase per capita production of food. Unfortunately, these practices often enhance environmental degradation. Societies with expanding populations, limited resources, and minimal access to technologies that can boost food production are prone to such environmental problems. For example, the extensive deforestation of the Caribbean nation of Haiti is in part attributed to the desperate attempts of an impoverished people to meet their growing demands for fuel and food. It is important to emphasize that such problems are not confined to poor nations. All nations struggle with environmental problems stemming from increasing populations and/or increasing resource consumption.

Carrying Capacity of Earth

The preceding discussion sets the stage for explaining how a knowledge of reproductive physiology might be useful, if not essential, for stabilizing population growth and resource use. Before this can be made clear, it is necessary to discuss what it means to stabilize population growth and resource use. The concept of **carrying capacity** is helpful in this regard (Figure 1-4).

Charles Darwin's (1809–1882) insights into evolution were based in part by the ideas of economist Thomas Malthus (1766–1834). Both men came to realize that populations expand exponentially when resources (e.g., food) are not limited. They also realized that resources are never unlimited and that populations tend to increase more rapidly than resources. When an organism lives in an environment where there is a surplus of resources, population growth increases exponentially. As resources become more limiting (due to diminishing surpluses of resources), population growth decelerates. Eventually, the rate of resource consumption becomes equal to rate of resource

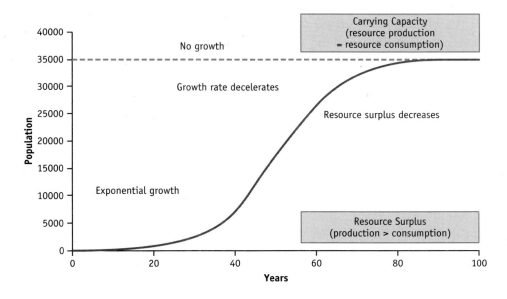

FIGURE 1-4

1

Relationship between the population of a particular species and the availability of vital resources (e.g., food and water) within a well-defined ecosystem. Note that when the rate of resource production is equivalent to rate of resource consumption the population size stabilizes, which is the carrying capacity of the ecosystem.

replenishment, and the population stabilizes. The point at which this occurs is known as carrying capacity.

The carrying capacity concept may seem simple, but it can be quite difficult to determine, especially when it comes to estimating the earth's carrying capacity for humans. With respect to humans, carrying capacity is a function not only of available resources, but also of how such resources are consumed. The manner in which humans use resources is determined by population size, use of technology, and cultural attitudes. First, consider how human numbers affect carrying capacity. The carrying capacity of the earth today may be less than it was 10,000 years ago because a significant amount of some resources have been used and there are considerably more humans using the resources. On the other hand, because of certain technologies, we are able to use resources today that were not available to earlier human cultures (e.g., petroleum and electricity). This brings us to the second variable affecting carrying capacity; the ability to develop and employ new technologies. Although technologies can expand the resource base, they can also lead to more rapid depletion of resources. This depends to some extent on cultural attitudes, the third variable affecting carrying capacity. The types of technologies developed by a culture as well as the ways in which such technologies are used are related to its values. For example, cultures that value forests as sacred places are less likely to develop and use forest-clearing technologies than are cultures that value forests purely as resources to support human development. So, taking all of these variables into account, what then is the carrying capacity of the earth?

1

Hopfenberg calculated the earth's carrying capacity for humans between 1960 and 2000 taking into account food production and patterns of food consumption, and compared these estimates to those based entirely on food production. What is striking about these data is that the carrying capacities based solely on food production are two to three times greater than those that take into account actual food production and food consumption patterns. For example, in the year 2000, the estimate of carrying capacity based on food production only is over 20 billion humans, whereas the estimate that takes into account how food is actually distributed and used is only 6 billion, the population of the Earth in 2000. The discrepancy between how many people can be adequately fed and how many are actually well fed is due primarily to food distribution patterns. Major problems include breakdown of normal production, disruption of transportation, and distribution due to wars or natural disasters (flooding and drought), and insufficient access by certain individuals and/or groups due to social, political, and economic conditions. For example, each year in the United States (a nation that produces more than enough food to meet the nutritional needs of its people) approximately 12 million children and 8 million adults suffer from a chronic shortage of nutrients needed for growth and good health.

The fact that the earth could support many more people than actually inhabit the planet may seem re-assuring. However, it is important to keep in mind that there is a limit to how many resources the earth can provide (Figure 1-5). Populations can exceed carrying capacity. When this occurs, the rate of resource consumption exceeds the rate of resource production, and shortages ensue. A decline in per capita consumption of resources can lead to a rapid drop (crash) in population size due to decreased reproductive rates and increased mortality rates. Resource shortages, along with increased competition for limited resources, leads to social and political problems, which tend to lower living standards. The drop in population will continue until it reaches a level that can be sustained by available resources. In cases where the population surge results in depletion of resources the population will continue to fall and eventually reach extinction.

To illustrate this concept, consider what life would be like in a world inhabited by 10 billion people, the projected global population in the year 2050 according to the World Bank and the United Nations. In order to meet the nutritional requirements of this many humans, Smil (1994) estimates that agricultural production of food energy would have to increase by 60 percent. This could be achieved by improving production efficiencies, reducing

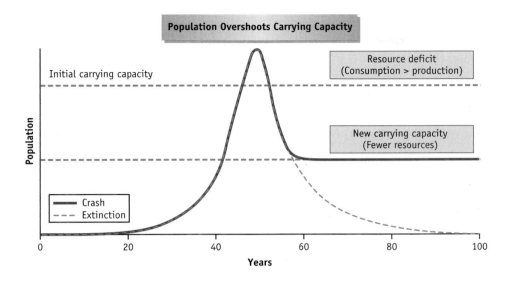

FIGURE 1-5

1

Relationship between the population of a particular species and availability of vital resources when the population surges and exceeds carrying capacity. Note that when consumption of resources exceeds production, a crash in population size ensues. If the overconsumption of resources is limited a new carrying capacity is established. When overconsumption is severe, leading to exhaustion of resources, the species can become extinct.

waste, and lowering the amount of fat in our diets. This is no simple task, and the implications may not be too appealing, especially to those of us living in wealthy societies. Currently, the wealthiest nations, with only 20 percent of the world's population, consume approximately 67 percent of all fossil fuels. Much of the technology responsible for enhanced food production during the past century is dependent on fossil fuels (e.g., heavy agricultural machinery, and inorganic nitrogen fertilizer). Unless there is a more equitable distribution of energy, it is unlikely that poor nations can adopt the petroleum-driven technologies required to meet growing demands for food. Moreover, a failure of rich nations to stabilize their energy use is likely to increase the economic gap between rich and poor nations, thereby exacerbating social and political unrest and leading to regional conflicts as well as refugee and immigration problems.

Even if energy distribution were more equitable and food production more efficient, these factors alone will not resolve problems related to population pressures. What happens if the population grows beyond 10 billion people? As noted previously, there is a limit to our ability to increase resources. At some point we reach the point of diminishing return; that is, the costs of increasing resources become greater than the gains derived from the resources. For example, even though the earth's reserves of petroleum seem vast, once 50 percent of the reserves are depleted (sometime between the years 2005 and 2010 according to some estimates), it becomes increasingly more costly to extract it. Thus the amount of oil that is actually available for our use is much less than that which exists in the ground.

As illustrated in Figure 1-5, once population size expands to a point where resource consumption exceeds resource replenishment, severe shortages ensue resulting in a crash in population size. No one knows when, or even if, this will occur on a global basis. However, over the millennia, such catastrophic changes have occurred on a regional basis with devastating consequences for some human societies; for example, the builders of the stone statues of Easter Island in the South Pacific Ocean, the Anaszi of the southwestern United States, and the Maya of Central America. Such crises are not unknown in more recent times. In 1994, the African nation of Rwanda experienced the third largest genocide since 1950 (the killing of 800,000 Tutsi's or 11 percent of the Rwandan population by the Hutu majority). The immediate or proximate cause of this violence was ethnic hatred instigated by politicians who were attempting to stay in power. However, the ultimate cause of this genocide may have been population pressure and scarcity of resources. In the early 1990s, Rwandan society was experiencing tremendous population expansion (3 percent annually), shrinking farm sizes, degradation of farm land, and disputes over access to scarce agricultural resources (land and water). Areas that experienced the most intensive violence were those where agricultural resources were most limited and food shortages most severe. This case should not be interpreted to mean that population pressure *always* leads to genocide or that other factors didn't contribute to this catastrophe. Certainly societies can and do choose different approaches to alleviate population pressures. Nevertheless, the events in Rwanda illustrate what can occur when available resources fail to sustain a population.

Challenges for Reproductive Physiology

The previous discussion illuminates two challenges for the field of reproductive physiology: 1) increasing efficiency of food production and 2) reducing birth rates of humans. Reproductive physiologists who study livestock reproduction have sought to increase production efficiencies of dairy and meat-producing animals. During the previous century, research with livestock has lead to the development of various technologies that enhance reproductive efficiency; for example, artificial insemination and estrous synchronization. Such practices have increased the amount of food a particular animal produces during its lifetime as well as the amount of animal-derived products a farmer can produce. Therefore, fewer livestock (and fewer farmers) are required to sustain a particular level of production. Meanwhile, reproductive physiologists interested in human reproduction have developed birth control technologies that can be used to lower birth rates. Together, these technologies can be used to stabilize population growth and resource use.

As noted earlier, livestock have always played important roles in agriculture. However, in recent years, some people have questioned whether production of animal-derived foods is an effective means to enhance the nutrition of humans on a global basis. Meat and dairy products are luxury foods that only the wealthiest nations can afford to include in daily diets. David Pimental notes that on a global basis, only 30 percent of protein in human diets is derived from animals. In wealthy nations such as the United States, 70 percent of the protein consumed by humans is from animal-derived foods. On the average, 4 kg of plant-derived protein is required to produce 1 kg of animal protein. Thus, a diet that is rich in animal protein requires large amounts of crops to sustain livestock production. In the United States, the annual per capita consumption of meat is 120 kg. In order to support this diet, 91 percent of crops suitable for human consumption (cereal, legume, and vegetable protein) are fed to livestock. The earth may not have enough resources to sustain this type of diet on a global basis. In order to supply this amount of animal protein to the current population, all of the world's grain harvest would have to be fed to livestock. Moreover, such a heavy dependence on animal protein would result in a 30 percent reduction in total amount of protein available for human consumption (due to the inefficiency of converting plant to animal protein).

Clearly, a diet rich in foods derived from grain-fed livestock is impractical and highly unsustainable. However, this does not mean that livestock should be excluded from agriculture. Animal protein is of high quality and can be produced in ways that are less wasteful than those currently employed in industrialized nations. For example, a more moderate use of global grain supplies could support a global per capital consumption of 40 kg of meat per year, an amount that would vastly improve human diets in many regions of the world. Whatever system of livestock production is adopted, knowledge of reproductive physiology will be essential for the efficient management of resources required to sustain production of animal-derived foods.

As mentioned earlier, human invention can have a positive effect on carrying capacity. Knowledge of reproductive physiology can help bring about equilibrium between human populations and resource availability. However, having the knowledge and means to accomplish these goals doesn't guarantee success. Cultural attitudes determine how technology is used (or whether or not it is used at all). People have to be willing and able to use a particular technology. For example, the lower birth rates of industrialized nations has been attributed to not only birth control technologies, but also policies that empower women to the extent that they have access to and are free to use such technologies.

All societies struggle with issues concerning human reproduction and resource use. Over the centuries many societies have collapsed because of an inability to cope with population pressure and resource use. Jared Diamond reminds us that the success of a society (whether or not it survives) depends largely on how it chooses to cope with prevailing environmental conditions. Values play an important role in this regard. The colonies of Norse Vikings failed in Greenland after several hundred years because the colonists retained European values, which proved to be incompatible with the ecology of Greenland (e.g., devoting great effort to raising cattle without adequate forages). In contrast, the small (1.8 square mile) island of Tikopia, located in the Southwest Pacific Ocean, has been occupied continuously for 3,000 years, largely because its inhabitants adopted practices that limit population growth (e.g., abstinence, abortion, and ritual suicide) and restricted their use of essential resources (e.g., invoking food taboos that prevent over-fishing). This is not to say that changing values is the only key to survival. The survival of a society also depends on its ability to recognize and retain values that have been responsible for its success. The path society should take is not always clear. Sometimes, people don't foresee the negative consequences of their lifestyles, or they realize the consequences too late to circumvent disaster. Moreover, conflicts between those who want to retain traditional values (conservatives) and those who advocate changing values (progressives) are common in all societies and give rise to contentious issues that can prevent or delay decisive action. Finally, it should be noted that in spite of having awareness of environmental problems, some societies may lack the political power to choose how they will live. For example, an indigenous society that occupies a rainforest may not have the political power to resist the logging and development that destroys the habitats on which they rely for survival.

The previous discussion highlights the fact that the science of reproduction takes place in a social context. In other words, the values of society influence research on mammalian reproduction and the knowledge that we gain from these studies can have an impact on society. Therefore, it should not be surprising that there are numerous social issues concerning how this research should be conducted as well as how research results should be applied. Ultimately these are ethical issues. In the concluding chapter of this book we will examine ethical aspects of some of the more controversial issues associated with reproductive physiology.

SUMMARY OF MAJOR CONCEPTS

- Reproductive physiology is the study of the reproductive functions and activities of organisms.

- As a branch of biology, reproductive physiology is based on the assumption that a species expresses a particular reproductive strategy because of natural selection.

- Research in reproductive physiology focuses on describing patterns of reproduction as well as characterizing the mechanisms responsible for these patterns.

- Reproductive physiology can serve societies by developing technologies that help sustain the carrying capacity of Earth.

DISCUSSION

1. Imagine that you are a wildlife biologist supervising a project to establish a population of wolves on an island. The island has viable populations of other animals on which wolf packs can prey. Describe expected changes in the size of the wolf population over the next 30 years, assuming that no other animals are brought to the island after the initial stocking of wolves. Explain these changes.

2. Some people argue that new advances in agriculture (e.g., biotechnology) should be pursued in order to keep pace with a growing human population. Others argue that this should not be done because agriculture cannot keep up with population expansion. What are some of the strengths and weaknesses of these opposing views?

3. Consider the following questions: a) Why do humans have sex without reproducing? b) How do humans have sex without reproducing? c) Should humans have sex without reproducing? How do these questions differ?

REFERENCES

Cohen, J.E. Population growth and Earth's human carrying capacity. *Science* 269:341–346.

Diamond, J. 2005. *Collapse.* New York: Viking Penguin.

Diamond, J. 1998. *Guns, Germs, and Steel.* New York: W.W. Norton and Co.

1

Hopfenberg, R. Human carrying capacity is determined by food availability. *Population and Environment* 25:109–117.

Klare, M.T. 2001. *Resource Wars.* New York: Henry Holt and Co.

Pimental, D. W. Dritschilo, J. Krummel, J. Kutzman. 1975. Energy and land constraints in food protein production. *Science* 190:754–761.

Smil, V. 1994. How many people can the Earth feed? *Population and Development Review* 20:255–292.

United Nations. 1992. *Long-Range World Population Projections: Two Centuries of Population Growth 1950–2150.* New York: United Nations.

World Bank. 1992. *World Development Report.* New York: Oxford University Press.

CHAPTER

2

Life, Reproduction, and Sex

CHAPTER OBJECTIVES

- Define sex.

- Describe asexual and sexual reproduction.

- Review mitosis and meiosis.

- Discuss why sexual reproduction occurs.

- Characterize the major reproductive strategies of mammals.

WHAT IS SEX?

This first stanza from the poem "Sex," by Arthur Guiterman (1871–1943) raises some important fundamental questions about reproduction. Guiterman is describing a common type of reproduction—cell division. But is this **sex**? There can be much confusion regarding the meaning of sex and how it is related to reproduction. This is because we use the word in several different ways. Sex and reproduction seem to be linked, but when people speak of having sex, they are not necessarily referring to engaging in a reproductive activity. We also identify individual humans according to their sex (male or female), but are we always making reference to their roles in the reproductive process? This seems unlikely. For example, when we identify a human as male or female, we frequently rely on characteristics (hair style, type of clothing, and so on) that reflect social norms, not biological traits such as type of genitalia. What exactly do we mean when we speak of sex? Before we can engage in a study of reproductive physiology, it is important to clear up any confusion regarding sex and reproduction.

As noted in the first chapter, reproduction can be defined as a means by which an organism replicates itself. There are two major modes of reproducing; **sexual** and **asexual** (meaning without sex). One way to develop a deeper understanding of these terms is to explore how sex is related to reproduction. Specifically, we will consider whether sex is a necessary and/or sufficient condition for reproduction. A necessary condition means that something (e.g., sex) is a prerequisite for something else (e.g., reproduction). A sufficient condition means that something is all that is required for something else. We will establish such relationships between sex and reproduction as we examine various definitions of sex.

There are several ways of conceptualizing sex. In the broadest, biological sense of the term, sex is the process whereby a new individual arises from the recombination of genes from separate sources. Using this definition, it is possible to identify three different types of

FIGURE 2-1

Three types of sex: a. exchange of DNA in bacteria (transgenic sex), b. merging of two organisms (hypersex), and c. merging of gametes (meiotic sex).

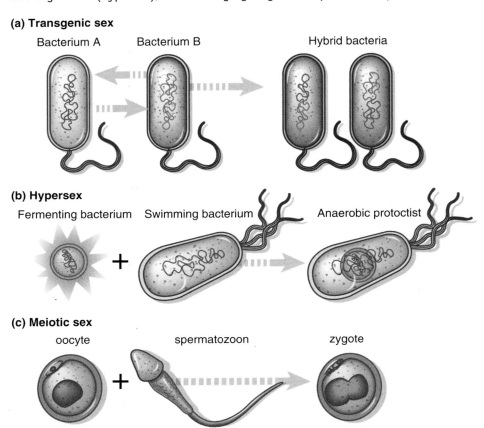

(a) Transgenic sex

Bacterium A   Bacterium B   Hybrid bacteria

(b) Hypersex

Fermenting bacterium   Swimming bacterium   Anaerobic protoctist

(c) Meiotic sex

oocyte   spermatozoon   zygote

sex (Figure 2-1). The first is **transgenic sex** (also known as conjugation) practiced by bacteria. Bacteria frequently receive genes from other bacteria or viruses. For example, the multicellular bacterium, *Streptomyces griseus,* can receive genes from *Escherichia coli,* a vastly different single-celled bacterium.

The second type of sex is **hypersex,** characteristic of the protoctists; microorganisms that comprise one of the five kingdoms of living organisms (i.e., protoctista; plantae, animalia, fungi, and bacteria). The amebas, to which Guiterman refers, are members of this kingdom. Hypersex occurs when different organisms merge and develop a permanent symbiotic relationship. You are undoubtedly acquainted with at least one example of this phenomenon. The familiar mitochondria, oxygen metabolizing organelles found in almost all eukaryotic cells, were once bacteria that merged with anaerobic bacteria (those that can not metabolize oxygen) to create aerobic organisms that can metabolize oxygen.

The third type of sex is perhaps the one with which you are most familiar; **meiotic sex.** This is the type of sex practiced by animals, fungi, and plants, and involves specialized sex cells (**gametes**) that possess only half the number of chromosomes as other body cells (somatic cells). Gametes are produced via **meiosis,** a type of cell division that reduces the number of chromosomes in a cell (also known as reduction division). During meiotic sex, two different types of gametes, each with a **haploid** (half the full complement) number of chromosomes, undergo **syngamy** (fusion to form one cell) to form a **zygote,** with the **diploid** (full complement) number of chromosomes.

The important lesson from this brief survey of sex is that in the community of living organisms, *sex is not necessary for reproduction.* Only in the case of meiotic sex is sex a necessary condition for reproduction. Although bacteria can exchange genetic material and unicellular protoctists can merge to form larger cells, in neither of these cases is the exchange of genes necessary for reproduction to occur. Bacteria reproduce via a process known as binary fission; that is, simple cell division. Unicellular protoctists will fuse with each other, but they can also reproduce by dividing. This discussion illustrates a second way of defining sex; that is, as a mode of reproduction. In sexual reproduction, production of new individuals requires the recombination genes from two parents. The other major approach to reproducing is asexual reproduction. Organisms that reproduce asexually do so by **mitosis** or cell division; that is, one individual cell divides into two identical copies.

The close association between sex and reproduction has significant implications for organisms that rely solely on this mode of reproduction. One of the most important is that in order to engage in sexual reproduction, individuals must have the ability to recognize members of their own species as well as members of the so-called opposite sex. This requirement permeates almost every aspect of animal life. The anatomic, behavioral, and physiologic differences between males and females of a particular species serve the purpose of ensuring that the two types of gametes (**spermatozoa** and **oocyte**) encounter each other and merge. Thus, in the case of sexual reproduction, sex can also refer to the reproductive role (i.e., providing male or female gametes) of an individual. In this sense of the term, to say that someone is male or female is analogous to saying that the individual is a scientist, or actor; that is, to make reference to something that individual does in a particular context.

There is a tendency for humans to over-extrapolate the aforementioned implications of sexual reproduction. Although it is true that humans reproduce sexually (recombination of genes via fusion of gametes), it is also well-known that humans frequently practice sex (assume reproductive roles)

2

but without reproducing. Unless the individuals practicing sex are **fertile** (i.e., have the ability to conceive), reproduction will not occur. **Infertility** (temporary inability to reproduce) can occur spontaneously (e.g., due to disease) or can be induced intentionally (e.g., use of birth control). For centuries, humans have practiced various means of birth control in order to sever the link between having sex and reproducing. The implication of this discussion is that sex can be thought of in yet another way; that is, as one's **sexuality.** Sexuality refers to practices in which a person engages to achieve sexual gratification, not necessarily for the purpose of reproducing. There are a variety of practices that we label as sex that have little, if anything, to do with reproduction; for example, oral sex, anal sex, cyber sex, phone sex, and so on. In these cases, the sexual practices of individuals do not coincide with their reproductive roles in meiotic sex. Whether or not they should coincide is an ethical matter that has been the subject to considerable debate for centuries.

In light of the previous discussion, we can conclude that among all species sex is not a necessary condition for reproduction. Many organisms reproduce without sex (asexual reproduction), and some of them engage in sex without reproducing. In sexually reproducing species, sex is usually a prerequisite for reproduction. However, recent success with cloning in some animals raises the possibility that sex is not a necessary condition for reproduction even in sexually reproducing species. Is sex a sufficient condition for reproduction? Clearly not in species that reproduce asexually. In sexually reproducing species, sex is a sufficient condition for producing a zygote. However, reproduction also depends on that zygote developing into an individual that can reproduce. In mammals, a male and female might produce gametes, mate, and produce a zygote, but reproduction can still fail due to a variety of reasons, including embryonic and neonatal death (either spontaneous or induced).

In summary, sex has several different meanings. With respect to mammalian reproduction, the subject of this book, we will use the term to make reference to a mode of reproduction that requires the recombination of genes via fusion of gametes. With respect to individuals engaging in sexual reproduction we will use the terms male and female to refer to their roles in this reproductive process.

MITOSIS AND MEIOSIS

The hallmark of sexual reproduction is the production of gametes (gametogenesis) via meiosis. However, the production of sperm cells and oocytes also involves mitosis. Early in development, a series of mitotic divisions produces an

abundance of diploid cells that are the precursors of cells that eventually undergo meiosis to form gametes. In somatic tissues, mitosis contributes to growth
by increasing cell numbers. You should become familiar with the details of
mitosis and meiosis because these cellular mechanisms are of fundamental
importance in reproductive physiology. Discussions of **spermatogenesis** (production of spermatozoa) and **oogenesis** (production of oocytes) in later chapters are based on the assumption that you understand these processes.

Figure 2-2 illustrates the process of mitosis in a cell from an organism for
which the diploid number of chromosomes is four. As you may recall from
elementary biology, chromosomes exist in homologous pairs. Thus in this
case, the organism has two homologous pairs of chromosomes. In mitosis,

FIGURE 2-2

A cell containing two homologous pairs of chromosomes undergoes mitosis to produce two identical
offspring cells.

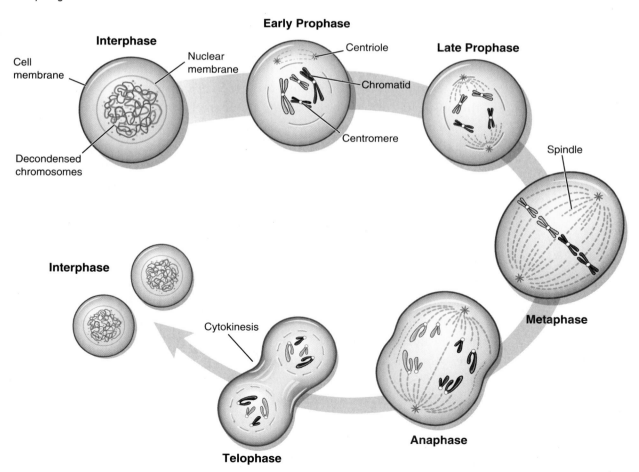

one parent cell will divide and give rise to two identical offspring cells (also known as daughter cells). Our example begins with the **interphase** portion of a cell cycle; that is, the phase when the cell is not dividing. Interphase chromosomes are decondensed; that is, long and thin. By the end of this phase, chromosomes begin to contract and condense; that is, shorten and thicken. In addition, the chromosomes replicate such that each chromosome consists of two identical **chromatids** joined at the **centromere.** During early mitotic prophase, the **centrioles** begin to migrate laterally and the nuclear membrane begins to break down. By late **prophase,** the **spindle** forms; that is, a network of microtubules along which chromosomes migrate. At this time, the chromosomes migrate towards the equatorial region of the cell. **Metaphase** is the time when the centrioles are located at opposite ends of the cell, joined by the spindle. The distinguishing feature of metaphase is the alignment of chromosomes along the equator of the cell. During **anaphase,** the centromere of each chromosome splits and each chromatid migrates to opposite poles of the spindle. Finally, during **telophase,** the spindle degrades, chromosomes decondense, two nuclear membranes form around each set of chromosomes, and **cytokinesis** (division of the cytoplasm) occurs, resulting in two identical offspring cells.

Meiosis (Figure 2-3) differs from mitosis in three important ways. First, one parent cell gives rise to four offspring cells (gametes). Second, each of the offspring cells has only half the number of chromosomes as the parent cell; that is, a haploid number of chromosomes. Third, there are two cell divisions involved in meiosis.

As with mitosis, meiotic interphase is the stage preceding division. By the end of this phase chromosomes have condensed and have replicated to form chromatids and centromeres. Meiotic prophase is much longer than mitotic prophase, and can be divided into four steps. During the **leptotene** stage, chromosomes begin to thicken. By the time of the **zygotene** stage the chromosomes are fully condensed and the homologous pairs begin to line up parallel to each other. Throughout early prophase, the centrioles migrate laterally and eventually occupy opposite sides of the cell. During the third, or **pachytene,** stage of meiotic prophase, chromosomes are shortened and thickened and lie in pairs. By the time the cell enters the diplotene/diakinesis stage the chromosomes shorten further and microtubules begin to radiate from the centrioles. During this later part of prophase, homologous pairs of chromosomes are tightly linked and their chromatids overlap. The point at which this crossing over occurs is known as the **chiasmata.** This permits the exchange of pieces of chromatids between homologous pairs of chromosomes. Meiotic metaphase 1 is characterized by the disappearance of the nuclear membrane, a well-defined spindle, and homologous pairs of chromosomes aligned at the equatorial

FIGURE 2-3

A cell containing two homologous pairs of chromosomes undergoes meiosis to produce four gametes, each containing half as many chromosomes as the parent cell.

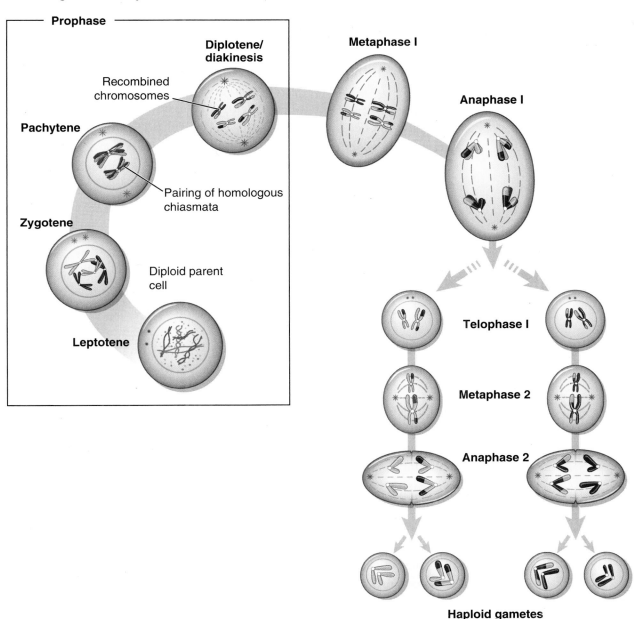

region of the cell. During anaphase 1, homologous chromosomes move in opposite directions along the spindle. By late telophase 1, nuclear membranes have formed and cytokinesis is complete, yielding two offspring cells, each containing one member of each homologous pair of chromosomes. During the remaining stages of meiosis, the offspring cells undergo another division

2

similar to mitosis. Each chromosome, with its chromatids, moves to the cell equator along the spindle during metaphase 2. During anaphase 2, the chromosomes separate and each half migrates to opposite ends of the cell. In telophase 2, a second division is completed giving rise to two gametes; that is, cells with half as many chromosomes as the original parent cell.

One of the most important features of meiosis is the fact that each of the four gametes arising from the parent cell is genetically unique. This is due to the random segregation of homologous chromosomes and chromatid exchange among homologues during the first meiotic prophase. This feature of meiosis enhances genetic variation among offspring. As discussed in the next section, genetic variation is important in evolution. A species with greater genetic variation is more likely to adapt to environmental changes, and therefore is more likely to survive than a species with less variation.

WHY SEXUAL REPRODUCTION?

One can't help but marvel at the intricate mechanisms governing sexual reproduction. Research aimed at understanding how these mechanisms work has been compared to peeling an onion. Once one level of understanding is achieved there is a deeper layer waiting to be peeled back. Whether you are a student beginning your study of this subject, or an experienced scientist seeking to understand molecular mechanisms regulating fertility, a daunting question will linger in the back of your mind. Why is there sexual reproduction in the first place? As noted in the first chapter, this is an ultimate question requiring an answer based largely on theory. You might think it's a waste of time to theorize about such things, but before you reject this type of inquiry, consider what cartoonist and author James Thurber once wrote; "Sometimes it is better to know some of the questions than all of the answers." Science is all about knowing which questions to ask. In fact, scientists are reluctant to say that they actually know the answer to a scientific question. The best they can do is to provide a hypothesis, which can be thought of as a tentative answer to a question. Much can be learned from questioning and developing theoretical answers. Asking why there is sexual reproduction provides the opportunity for us to put our best scientific theories to use. If such theories provide a satisfactory explanation, then we enhance our understanding of the world we live in. Such understanding can help us live more skillfully— at least until a better explanation comes along!

We will address why sexual reproduction exists in two steps. First, we will consider why reproduction can exist in the world. Second, we will consider why sexual reproduction exists, and why it is maintained.

Reproduction, Life, and Thermodynamics

What is the most fundamental trait of all living beings? Any organism that we consider to be alive has the ability to maintain itself in the presence of a continually changing environment. Biologist Lynn Margulis calls this property is **autopoiesis,** meaning self-maintenance. An autopoietic entity is self-bounded (has a membrane or skin), self-generating (the boundary is produced by the entity), and self-perpetuating (they use energy continuously to maintain themselves, even when they are not growing or reproducing). A more familiar way of understanding autopoeisis is to say that each living being is determined by its own metabolism; that is, its own internal set of biochemical processes that provide energy for vital processes and activities. Reproduction is one vital process supported by metabolism. When an organism fails to shunt energy to the biochemical reactions supporting reproductive activity, it fails to reproduce.

Autopoietic entities are open, thermodynamic systems. In other words, they are systems through which energy flows. Some of the energy is captured to run life-sustaining processes, and the rest is lost as heat and wastes (urinary, fecal, and gaseous). Living beings capture energy in different ways. Plants and other photosynthetic organisms capture solar energy in the form of various hydrocarbons (carbohydrates, lipids, and proteins), which they store and use to fuel their metabolic processes. Animals capture energy from plants, either directly by consuming the plants (herbivorous animals), or indirectly by consuming other animals that consume plants (carnivorous animals), or both (omnivorous animals). Figure 2-4 shows the energy flow through a cow, which is an herbivore. Energy from plants is consumed, digested, absorbed, and transported to various tissues where it is metabolized or stored for later use. In mammals, much of this dietary energy is used for basal metabolism, voluntary movements, thermoregulation, lactation, growth, and, of course, reproduction. This scheme illuminates the important relationship between reproduction and energy metabolism. When there isn't enough dietary energy to fuel all of these vital processes some processes, cease. As it turns out, reproduction is among the lowest priorities; that is, it is usually the first process to be eliminated during times of restricted energy intake.

The aforementioned scheme outlining the partitioning of energy in living beings illustrates one of the most important fundamental laws of physics; the so-called second law of thermodynamics. In any system, whether it is an inanimate object such as a television or an autopoietic entity such as a cow, a portion of the energy consumed by the system will be lost as **entropy.** Entropy is the energy that is not available to the system to power its work. Not all the

FIGURE 2-4

The cow as an open thermodynamic system. Some of the dietary energy from plants is captured and metabolized to support various vital processes, including reproduction. However, a good deal of this energy is lost as waste (entropy).

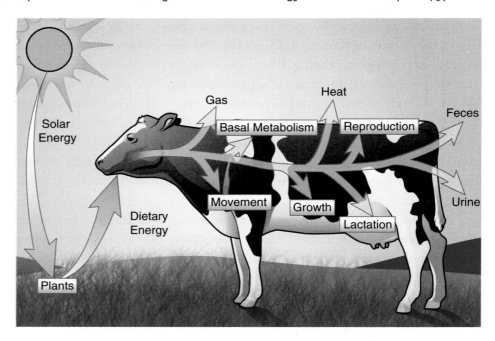

electrical energy flowing into your television is used to provide pictures and sound. A good portion of that energy can't be used and is given off as heat. The same is true for living beings. Using our cow as an example, the energy dissipated in the form of heat and various wastes is not available to the animal. Another way of looking at this phenomenon is to consider how the different forms of energy behave. The energy of the plants that serve as the cow's feed, is potential energy; energy based on a particular chemical structure of hydrocarbon nutrients. When the cow metabolizes these chemicals, energy is liberated (kinetic energy). Some of it is used to power various biochemical pathways, whereas the rest is distributed among molecules that make up waste products. Wastes represent kinetic energy that is too randomly distributed to be of use to the animal. From a purely energetic perspective, the cow (or any living being) is a system that facilitates the movement of energy from a highly ordered state to a more random state. This insight provides the basis for an explanation of why reproduction can exist in the first place.

As noted earlier, living beings are open systems; they are open to incoming energy. However, they are all part of a very large closed system known as

BOX 2-1 Focus on Fertility: Entropy and Human Lifestyles

The entropy law of thermodynamics is fundamentally important, yet most of us are either unaware of it, or ignore it in our day-to-day activities. Nevertheless, the implications of this law are of practical significance and influence all aspects of our lives. Living beings metabolize dietary energy to support various vital processes. Because this process isn't 100 percent efficient, metabolism produces various wastes which are forms of entropy. With the exception of humans, all living beings rely solely on solar energy, either directly (photosynthetic organisms) or indirectly (organisms that consume photosynthetic organisms or organisms that consume organisms that consume photosynthetic organisms). Humans, on the other hand, employ other fuels to produce, process, and distribute their foods. This means that humans generate considerably more entropy than other life forms, especially humans who live in industrialized societies. Jeremy Rifkin illustrates this by analyzing the entropy associated with procuring the loaf of bread you routinely purchase from your nearest grocery store. The following energy-requiring steps are required to make this bread available to the consumer: 1) wheat is planted, cultivated, and harvested; 2) wheat is transported by trucks to a bakery plant; 3) wheat is processed (refined and bleached) into flour; 4) flour is enriched with vitamins and minerals; 5) preservatives and dough conditioners are added to the flour; 6) dough is formed into loaves and baked; 7) bread loaves are packaged in plastic wrappers and boxed; 8) loaves are trucked to grocery stores; 9) bread is housed in climate-controlled grocery stores until purchased; 10) consumer transports bread (via automobile) to her home; and 11) bread is consumed. Each of these steps requires energy and generates entropy; that is, waste. The energy costs of human reproduction in an industrialized world can also be quite high. Finding a mate and establishing a family typically requires energy expenditures for transportation, clothing, entertainment, and so on. Couples with fertility problems may expend even more energy by using assisted reproductive technologies (artificial insemination, in vitro fertilization, embryo transfer, and so on). Other aspects of human life are no less complicated. Consider the energy-dependent steps involved with going to work, taking a vacation, watching a movie or downloading music to your MP3 player. When we examine our lives from a thermodynamic perspective, it becomes clear that we humans consume an astounding amount of energy and generate staggering amounts of entropy in the form of solid wastes, greenhouse gases, and chemical pollutants.

the universe. According to thermodynamic theory, the total amount of energy in a closed system remains constant, but energy can exist in different forms (potential and kinetic). When an energy gradient exists within a closed system (i.e., when energy is distributed between potential and kinetic forms) the system moves toward a state of equilibrium, which is the most disordered or most homogeneous state. This explains why heat flows from a warm body to a cool one, or why molecules diffuse from a high concentration to a lower one. At equilibrium both bodies become the same intermediate temperature or the two solutions end up with the same intermediate concentration. Astronomers believe that our universe is moving towards a state of equilibrium; that is, the highly ordered potential energy of the stars, including our sun, is slowly being converted to disordered (kinetic) energy. Life on earth contributes to this process; that is, living beings capture the potential energy of the sun and dissipate it to various forms of kinetic energy. Reproduction is one of the life sustaining processes that plays a role in this process. Therefore, one can say that life (and reproduction) exists because of energy gradients in the universe. Life is only one of countless mechanisms that help move the universe to a state of equilibrium.

Why Sexual Reproduction?

This question can be broken down to two parts. First, why did sexual reproduction occur in the first place? Second, why is sexual reproduction maintained? These questions arise from the so-called paradox of sex; that is, sex is widespread, but seems too costly to be beneficial to an organism. The costs associated with this type of reproduction seem to be considerably higher than those for asexual reproduction and include: 1) costs associated with mating or conjugation, 2) costs associated with producing offspring, and 3) costs associated with the risk of producing offspring via randomly mixing genes from two individuals. If this is indeed true, then why do so many species practice sexual reproduction? Why didn't asexual organisms out-compete the sexual organisms?

With respect to the cost of mating, considerable time and energy are required to find and secure mates. In some sexually reproducing species, mate selection involves investment of energy in particular body shapes, coloration schemes, behavioral displays, and so on. Moreover, searching for and mating with a partner makes sexual reproduction much slower than asexual reproduction. The costs associated with producing offspring stem from the fact that the reproductive unit in sexual reproduction is the mating couple. In asexual reproduction, one individual can give rise to two individuals, whereas in sexual reproduction two individuals are required to produce one individual. Thus the net reproductive rates for the two types of reproduction are 1 and 0.5, respectively. In order for a sexually reproducing pair to achieve the same reproductive rate as an asexually reproducing individual, they have to invest in producing twice as many offspring. Finally, there are risks associated with mixing genes from different individuals. In general, recombination of genes destroys advantageous gene combinations faster than it creates new ones.

With these costs in mind, it is difficult to understand how sexual reproduction can exist in so many species. It may be easier to explain why sex arose in the first place than to explain why it persisted, so let's take on the simpler question first. One sound explanation is that sex originated as a nonlethal infection in bacteria. The genes that allow a bacterium to be copied and transferred to other individuals might have spread throughout a population because the rate of infection by these genes was much faster than the onset of any lethal or fitness-reducing consequences. Although this may explain how sex got its start, it doesn't explain why sex persisted as a mode of reproduction.

Many biologists argue that sexual reproduction should have disappeared soon after its appearance. They assert that asexually reproducing species would have out-competed sexually reproducing species, due to the high costs of sexual reproduction, *unless* sexual reproduction offered some other advantages the outweighed these costs. The conventional explanation for why sexual reproduction is maintained is that it increases variation in offspring, and that this variation allows sexual organisms to adapt to lethal environmental changes faster than asexual organisms. Although this is a standard textbook explanation, there are serious problems with this idea. First, sex doesn't necessarily increase genetic variation in a species. Second, natural selection need not promote genetic variation. Third, evolution need not favor increased genetic exchange. It is beyond the scope of this book to explore these criticisms in any detail. However, the basis of these criticisms can be summarized in the following manner. Sexual reproduction is only one of many mechanisms whereby genetic variations can arise. Mutations, symbiogenesis, and various forms of stress also generate variations, both in sexual and asexual organisms. Any attempt to make generalizations regarding the relative benefits of sexual reproduction is complicated by the fact that the type of reproduction expressed by a particular species is determined by the specific context in which it evolved; that is, its natural history. As noted in a previous section, organisms have engaged in different types of sex for billions of years, irrespective of how they reproduce. Sexual reproduction may not be as much a trait that has been directly selected for as much as it is an evolutionary path by which sexually-reproducing organisms came to be. This mode of reproduction may very well have been an improvisational aspect of evolution as suggested by reproductive physiologist Irving Rothchild. In other words, sex may not have provided any advantage to early organisms who practiced it (e.g., bacteria), but descendents made use of it in different types of environments. In other words, sex may not have been advantageous until environmental conditions made it so. What this means is that an understanding of why a certain species reproduces the way it does requires a detailed understanding of not only the

2

genetic changes that occurred in its evolution, but also the environmental changes that may have accompanied such changes.

SEXUAL REPRODUCTION IN MAMMALS

The evolutionary path from bacterial transgenic sex to the meiotic sex of animals is complex and has been the subject of considerable research and speculation in biology. For our own purposes, we will have to ignore the past 3.5 billion years of evolution and focus our attention on animals that have been around for only the last 200 million years; that is, mammals.

As we study mammalian reproduction, it will become clear that the cost of reproduction in these species is very large. A significant amount of energy is required for mammals to find and select mating partners, not to mention the costs of pregnancy, lactation, and rearing offspring. This is true whether we are speaking of wild mammals or domestic mammals. The high cost of reproduction in mammals is readily apparent in the livestock industry. Producers of cattle, sheep, pigs, and horses spend a tremendous amount of time and money on reproductive management. If the cost of reproduction in mammals is so high, then why have mammals endured over the past 200 million years? We can address this question in the same way we addressed the question about sexual reproduction. Mammals have been successful because their traits allow them to thrive in the environments they inhabit. There really is no good way to understand why an animal reproduces the way it does without understanding its environment and how it interacts with it.

SUMMARY OF MAJOR CONCEPTS

- In biology, sex can refer to the creation of a new individual via recombination of genes from separate sources, or a mode of reproduction (fusion of gametes which are produced via meiosis).

- Sex is neither a necessary nor sufficient condition for reproduction.

- Reproduction is one of several life-sustaining processes that are dependent on the ability of living beings to capture and metabolize energy.

- The biochemical reactions that sustain life obey the laws of thermodynamics; that is, they capture part of the potential energy of the sun to support metabolism and dissipate the remaining energy as waste or entropy.

2

- Although sexual reproduction requires more energy than asexual reproduction, it has proven to be a highly successful strategy for numerous species living under a wide variety of circumstances.

DISCUSSION

1. How many times during the past 24 hours has the word sex entered your thoughts and/or conversations? How was this word used? In other words, what definitions did you employ? Give some examples of how you might use the term differently.

2. In basic arithmetic, you learned that $1+1=2$. However, with respect to sexual organisms $1+1=1$. Explain this discrepancy.

3. Women and men who engage in heavy athletic training (e.g., long-distance running, swimming, bicycling, and so on) experience infertility; that is, they fail to produce sufficient gametes to reproduce. Explain this in terms energy metabolism and thermodynamics.

4. Is sex a necessary and/or a sufficient condition for reproduction in humans? Explain.

REFERENCES

Margulis, L. and D. Sagan. 1997. *What is Sex?* New York: Simon and Schuster.

Otto, S.P. and T. Lenormand. 2002. Resolving the paradox of sex and recombination. *Nature Reviews Genetics* 3:252–261.

Rothchild, I. 2003. The yolkless egg and the evolution of eutherian viviparity. *Biology of Reproduction* 68:337–357.

Rifkin, J. 1980. *Entropy: A New World View.* New York: The Viking Press.

3

Organization and Structure of Mammalian Reproductive Systems

CHAPTER OBJECTIVES

- Describe the major anatomic components of the reproductive system.

- Discuss how the major components of the reproductive system are related functionally.

- Review basic anatomic terminology.

- Review the basic principles of gross and microscopic anatomy that are pertinent to reproductive physiology.

THE REPRODUCTIVE SYSTEMS OF MAMMALS

The reproductive systems of mammals can be analyzed into five major components:

- the **gonads,**
- the **genital ducts,**
- the **external genitalia,**
- the **pituitary gland,** and
- the **hypothalamus.**

The gonads are the gamete-producing organs; that is, **testes** in males and **ovaries** in females. The genital ducts are involved with the transport of gametes and/or the development of offspring. The genital ducts of males include the **epididymis** and the **ductus deferens,** whereas the female genital ducts include the **oviducts, uterus,** and innermost portion of the **vagina.** The external genitalia are the organs of copulation (sexual connection between a male and female).

The **scrotum** and **penis** make up the male external genitalia, whereas the **vulva** (consisting of the major and minor labia), **clitoris,** and outermost portion of the vagina (**vestibule**) make up the female external genitalia. The pituitary gland is located at the ventral surface of the forebrain and consists of anterior and posterior lobes. The hypothalamus is a small region of forebrain located above the pituitary gland. A more detailed description of the sexual anatomy of males and females is provided in subsequent chapters. Figure 3-1 provides a schematic representation of the major components of the mammalian reproductive system as well as the functional relationships among these tissues. A few general concepts concerning each of the major subdivisions are included in the following sections.

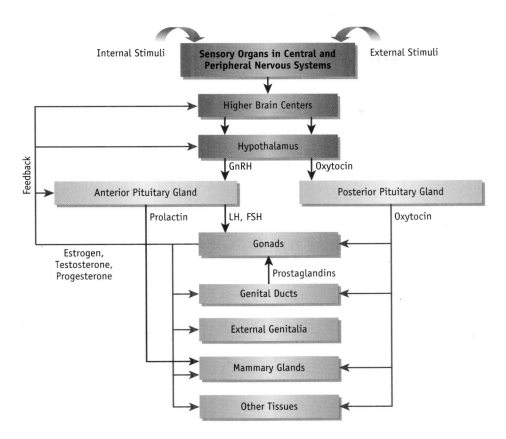

Internal Stimuli — **Sensory Organs in Central and Peripheral Nervous Systems** — External Stimuli

Higher Brain Centers

Hypothalamus

GnRH · Oxytocin

Anterior Pituitary Gland — Posterior Pituitary Gland

Prolactin · LH, FSH · Oxytocin

Gonads

Prostaglandins

Genital Ducts

External Genitalia

Mammary Glands

Other Tissues

Feedback

Estrogen, Testosterone, Progesterone

FIGURE 3-1

Organization of the reproductive system and major routes of communication among tissues. Various reproductive hormones serve as the means of communication among tissues. Note that the information flows in an efferent direction from the central nervous system to the pituitary, and finally to the genital organs, and that afferent information from the gonads feeds back on the central nervous system and pituitary gland.

The Hypothalamic-Pituitary Axis

It is wellknown that various environmental stimuli influence reproduction in animals. For example, the reproductive activity of many animals is restricted to a particular time of year and/or set of environmental conditions. Thus it is readily apparent that animals have the ability to sense environmental conditions, such as day length, and convert this information into internal signals that affect the activity of reproductive tissues. Animals also have the ability to sense changes in their internal environments (e.g., concentrations of gases, ions, and metabolites in extracellular fluids), and these too can affect reproductive activity. It is obvious that the central nervous system (brain and spinal cord) provides the ability to sense and respond to such environmental changes. Various sense organs (e.g., the eyes, olfactory bulb, chemical receptors, and so on) respond to changes in the environment and convey this information to the brain via afferent (flowing into the central nervous system) neuronal pathways. Diverse neuronal inputs are integrated (blended) and transformed into humoral (blood-borne) signals within the hypothalamic-pituitary axis.

Numerous afferent neurons converge on the hypothalamus and impinge upon neurosecretory cells that respond to these inputs by releasing hypothalamic hormones into the blood. Some of these hormones travel directly to the anterior pituitary gland to induce the release of other hormones that enter the general circulation and regulate other organs. **Gonadotropin-releasing hormone (GnRH)** is a hypothalamic hormone that plays a pivotal role in regulating reproductive activity. GnRH stimulates the release of two **gonadotrophic hormones (gonadotropins)** from the anterior pituitary gland: **luteinizing hormone (LH)** and **follicle-stimulating hormone (FSH).** The gonadotropins support gametogenesis in the testes and ovaries and stimulate production and secretion of gonadal hormones, which regulate other reproductive tissues, including the brain. With respect to the latter action, hormones produced by the gonads feed back on the brain and pituitary gland to regulate release of hypothalamic and pituitary hormones. **Prolactin** is another anterior pituitary hormone that plays a role in reproduction. This hormone is best known for its role in promoting synthesis of milk in the mammary glands. Prolactin also influences ovarian function in some species (e.g., rodents). The anterior pituitary gland is involved with much more than reproduction. It also produces several hormones that regulate a variety of physiologic processes including growth, thyroid activity, and adrenal activity.

The posterior pituitary gland is also involved with the regulation of reproduction. Unlike the anterior pituitary gland, this lobe does not produce hormones. Rather, it stores hormones that are produced by neurons located in the hypothalamus. One of these hormones is **oxytocin.** It acts on the female genital ducts and mammary glands to regulate the birthing process and milk secretion.

Gonads, Gonadal Hormones, and Genital Ducts

The testes and ovaries have two major functions: 1) production of gametes (spermatozoa and oocytes) and 2) production of gonadal hormones. As noted earlier, these processes are dependent on gonadotropins. Thus any disruption in release of GnRH by the hypothalamus and/or release of LH and FSH by the anterior pituitary gland results in lowered fertility and reduced production of gonadal hormones.

Gonadal hormones act on the genital ducts, external genitalia, mammary glands, brain, anterior pituitary gland, and other tissues not directly involved with reproduction. In the embryo, gonadal hormones regulate development of the genital ducts and external genitalia. At sexual maturity these hormones stimulate development of **secondary sex traits;** that is, specific characteristics that distinguish males from females, but are not directly involved with reproduction.

In adults, gonadal hormones are important in maintaining proper function of the genital ducts as well as in regulating gonadotropin secretion.

Physiologic relationships between the ovaries and female reproductive tract are of particular importance. Hormones produced by the ovaries (e.g., estrogen, testosterone, and progesterone) promote changes in the vagina, uterus, and oviducts to support copulation and pregnancy. Moreover, the uterus produces hormones (e.g., prostaglandins) that affect ovarian function.

Gonadal hormones act on the hypothalamus and anterior pituitary gland to regulate release of LH and FSH. Hormonal relationships between the gonads and the hypothalamic-pituitary unit can be negative or positive. For example, estradiol, progesterone, and testosterone can feed back on the hypothalamus and pituitary gland to suppress release of LH and FSH. This is commonly known as a **negative feedback** loop. Negative feedback loops serve the purpose of keeping hormone levels within a narrow range. On the other hand, gonadal hormones can also stimulate release of GnRH, LH, and FSH. Such **positive feedback** loops are not as common as negative feedback loops and serve the purpose of inducing rapid increases in concentrations of a hormone. We will examine these types of feedback relationships in later chapters dealing with the hormonal control reproduction.

REVIEW OF GROSS ANATOMY

Before you can begin learning how a physiologic system functions it is necessary to develop an appreciation for the structures of tissues that make up the system. The following sections provide a brief review of general anatomic principles which will prepare you for chapters that provide anatomic descriptions of the male and female reproductive organs.

Directional Terms and Planes

An object is identified by its relationships with other objects. Anatomy requires the ability to visualize and describe such relationships in three dimensions. Thus it is necessary to have terms to establish a frame of reference. Figure 3-2 illustrates terms that are commonly used to describe locations of anatomic structures in four-footed animals. These are particularly useful in studying gross (visible to the naked eye) anatomy.

The first task is to establish direction, like the points on a compass. **Cranial** or anterior refers to the front, or toward the head, whereas **caudal** or posterior refers to the rear, or toward the tail. Within the head, forward is called **rostral** (toward the nose). The upper (back) part of the body is referred to as **dorsal,**

FIGURE 3-2

Directional terms and body planes in four-legged animals.

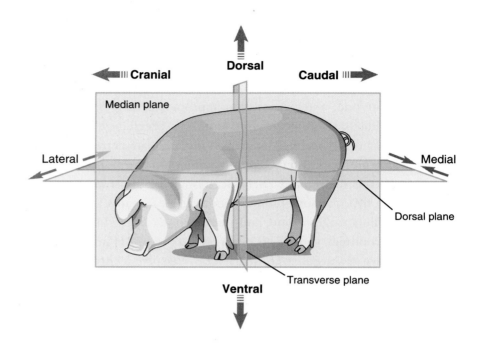

whereas the undersurface is referred to as **ventral.** These terms alone would be sufficient if we lived in a two-dimensional world. However, our world consists of three dimensions, so we require terminology to divide the body into different planes.

Differentiating between left and right allows us to describe anatomy in three dimensions. Two terms are important in this regard. The **median** plane passes through the body from head to tail and divides it into equal left and right halves. A **sagittal** plane is any plane parallel to the median plane; that is, dividing the body into unequal left and right halves. You may run across the term "mid-sagittal plane," which is another term for median plane.

The body can be further divided using terms to describe additional planes. A **transverse** plane occurs at a right angle to the median plane and divides the body into front and rear halves. Finally, a **dorsal** plane is at right angles to the median and transverse planes and divides the body into upper and lower halves.

Meaningful anatomic descriptions require more than identifying location by plane and direction. Although it is accurate to state that the uterus is caudal to the transverse plane and dorsal to the dorsal plane, this doesn't tell us very much because many other structures (kidneys, bladder, intestines, and so on) are also located in this general location. A more precise description of the uterus' location requires terminology that can describe spatial relationships among structures. The term **medial** refers to the middle or center. To say that the uterus is medial to something is to say it is in the center of it; for example, the uterus is medial to the ovaries. **Lateral,** on the other hand, means away from the center; that is, the ovaries lie laterally to the uterus. **Deep** and **superficial** refer to depth relative to the body surface. For example, the skin is superficial to muscle, whereas the kidney is deep relative to the skin. Finally, **proximal** and **distal** refer to the relative distance between structures. For example, the oviducts are proximal to the ovaries, whereas the uterus is distal to the ovaries.

Body Cavities

Additional precision in locating anatomic structures can be gained by dividing the body into major cavities. The dorsal cavity contains the brain in its cranial cavity and the spinal cord which lies in the vertebral cavity. The ventral cavity includes the thoracic cavity, located cranially, and the abdominal and pelvic cavities (or abdominopelvic cavity), located caudally. The thoracic and abdominopelvic cavities are separated by the diaphragm.

Much of our attention will focus on the abdominopelvic cavity; the location of the reproductive tracts of males and females. Therefore it is worthwhile to elaborate on the anatomy this region. Figure 3-3 is a schematic illustration of the abdominopelvic cavity showing the position of the female reproductive tract relative to some other familiar organs. This cavity is surrounded by the body wall, which consists of an outer layer of skin followed by a double layer of fascia (sheet of fibrous tissue), a musculoskeletal layer, and an inner layer of fascia. The entire abdominopelvic cavity is lined by a thin,

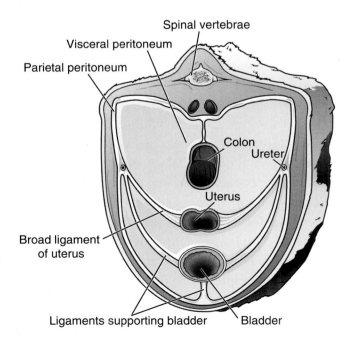

Spinal vertebrae
Visceral peritoneum
Parietal peritoneum
Colon
Ureter
Uterus
Broad ligament of uterus
Ligaments supporting bladder
Bladder

FIGURE 3-3

Transverse section through the abdominopelvic cavity of a cow, showing the peritoneum and suspensory ligaments supporting the colon, uterus, and bladder.

translucent membrane; that is, the **peritoneum.** The peritoneum forms a sac that encases the organs located in this region. During development, abdominal-pelvic organs form outside the peritoneum, near the body wall. As they develop they move into the cavity carrying the peritoneum with them. Thus, each organ is embedded in a fold of peritoneum, which acts to suspend the organs from the body wall. These folds are called **omenta** (a fold passing from the stomach to other viscera), **mesentery** (a fold that attaches to the intestines to the dorsal wall of the abdominal cavity), and **ligaments** (folds that connects viscera other than the digestive organs to the abdominal wall. The reproductive organs are suspended in the abdominopelvic cavity by various ligaments, the names of which we will encounter in a later chapter. The peritoneum that surrounds an organ is called the **visceral** peritoneum, whereas the peritoneum that lines the body cavity is called the **parietal** peritoneum.

MICROSCOPIC ANATOMY

In order to fully appreciate how reproductive tissues interact, it is necessary to develop an understanding of how these tissues are organized at the cellular level. Body tissues fall into one of four categories: **epithelial tissue, connective tissue, nervous tissue,** and **muscle tissue.** Each of these tissue types plays an important role in the structure and function of the organs involved with reproduction.

Origin of Major Tissue Types

The major body tissues develop from one of three layers of germ cells in the embryo. We will study embryogenesis later in the book. At this point, it is only necessary that you understand the organization of the early embryo as it relates to development of body tissues.

Soon after its formation, the zygote undergoes a series of mitotic divisions giving rise to numerous identical cells. Within 5 to 12 days, the embryo becomes a **blastocyst** (Figure 3-4), which consists of an **inner cell mass,** a group of cells clumped together at one end of the embryo, and the **trophectoderm,** a layer of thin cells (**trophoblasts**) forming a cyst-like cavity called the **blastocoele.** The inner cell mass will develop into the body of the embryo, whereas the trophectoderm will form extra-embryonic membranes that will ultimately form the fetal part of the placenta. Three distinct layers of cells will develop from the inner cell mass. The cells closest to the trophoblast become the **endoderm,** which will eventually give rise to the linings of the gastrointestinal and respiratory tracts, as well as endocrine glands. The outermost layer (**ectoderm**) will give rise to the nervous system, skin, and hair. The layer of cells between

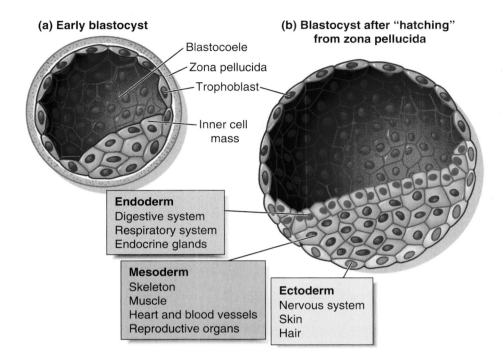

(a) Early blastocyst

- Blastocoele
- Zona pelucida
- Trophoblast
- Inner cell mass

(b) Blastocyst after "hatching" from zona pellucida

Endoderm
Digestive system
Respiratory system
Endocrine glands

Mesoderm
Skeleton
Muscle
Heart and blood vessels
Reproductive organs

Ectoderm
Nervous system
Skin
Hair

FIGURE 3-4

A blastocyst sectioned medially before (a) and after (b) it "hatches" from a membrane called the zona pellucida. Cells of the inner cell mast differentiate to give rise to germ layers, which eventually form major types of tissues.

3

the inner and outer layers is called the **mesoderm** and provides the precursors for the muscular, skeletal, cardiovascular, and reproductive systems.

As the embryo develops it takes on the physical characteristics of the species of which it is a member and the various organ systems, including the reproductive organs, assume their familiar structures. In general, organs are composed of several different types of tissues. The general anatomic features of the major tissue types found in reproductive tissues are described in the following sections.

Epithelial Tissue

Epithelial tissue (**epithelium**) accounts for a large portion of the mass of reproductive organs. Epithelial cells exist in aggregates and typically form sheets that serve as coverings, linings, or organ surfaces. The cells of epithelial tissues are in close association with each other. Therefore, the intercellular space is small and not penetrated by blood vessels. In almost every case, these aggregates of cells lie on a supporting **basement membrane,** a noncellular layer of connective tissue that is produced and secreted by epithelial cells.

Epithelial tissue is classified based on the shape and arrangement of cells (Figure 3-5). With respect to shape, epithelial cells can be squamous (thin and flat), cuboidal (equal height and width), or columnar (height is greater than width). Simple epithelium consists of only one layer of cells, whereas

FIGURE 3-5

Major types of epithelial tissue found in reproductive organs: simple squamous (a), simple cuboidal (b), simple columnar (c), stratified squamous (d), stratified cuboidal (e), stratified columnar (f), transitional (g), and pseudostratified (h).

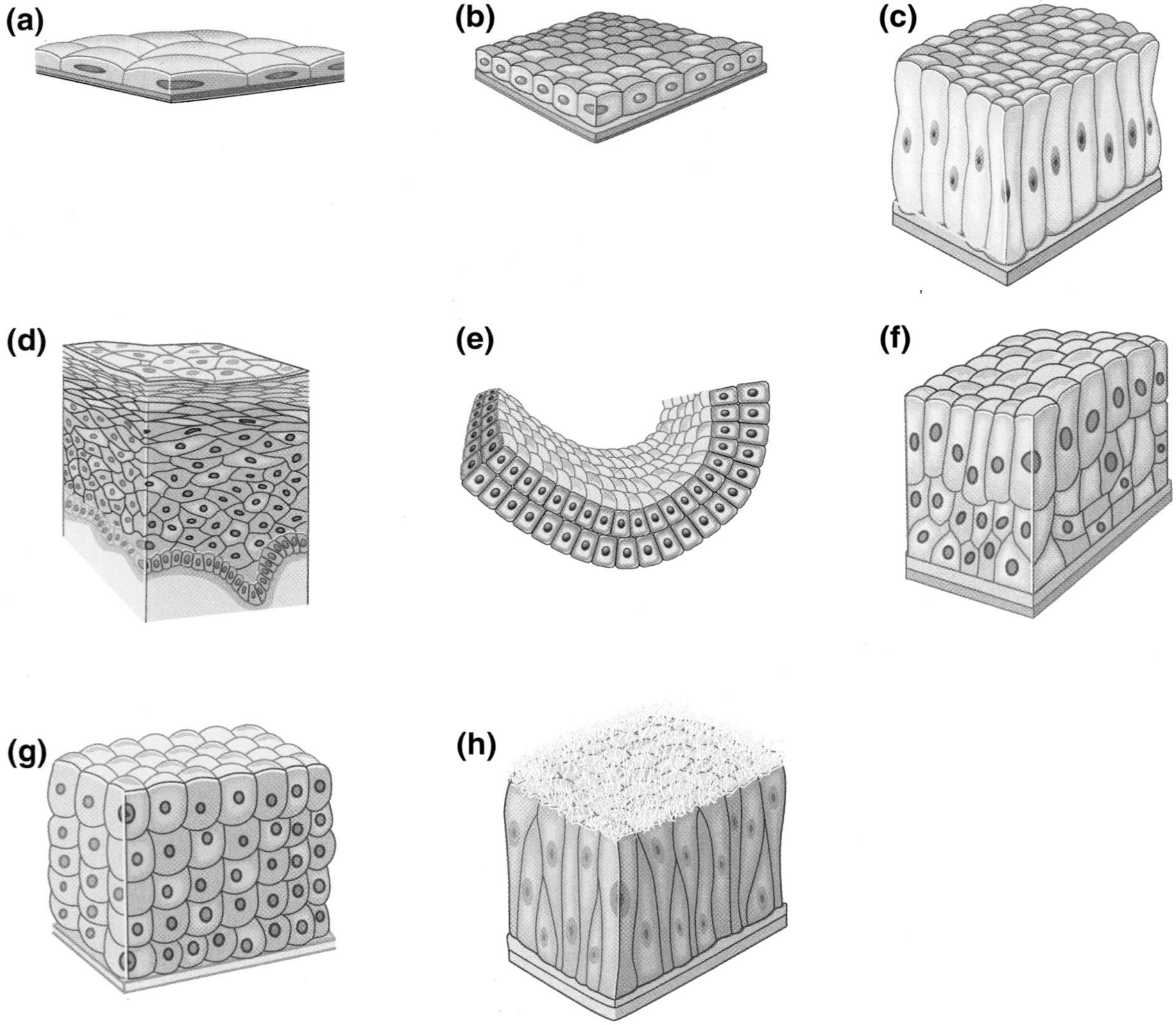

3

(a) **(b)** **(c)**

(d) **(e)** **(f)**

(g) **(h)**

stratified epithelium consists of two or more layers of cells. Some common types of epithelial tissue include: simple squamous, stratified squamous, simple cuboidal, stratified cuboidal, simple columnar, stratified columnar, pseudostratified, and transitional.

Simple squamous epithelium forms thin sheets of tissue. Examples of this type of epithelial tissue are the **endothelium** (inner lining of the heart and blood and lymph vessels), the **mesothelium** (inner lining of the body cavities), and the **mesenchyme** (embryonic connective tissue). One of the most common types of epithelium is the simple cuboidal. This type of tissue covers the surface of ovary. Simple columnar epithelium forms the linings of the digestive tract as well as other tubular organs such as the uterus and cranial vagina. The cells of this tissue may have **cilia,** projections that extend into the **lumen,** or cavity, of the tubular organ.

Pseudostratified epithelium consists of long and short cells that overlap, giving the impression of a stratified organization. Closer examination reveals that this tissue is made up of only a single layer. This type of arrangement is seen in the trachea of the respiratory system. Transitional epithelium is made up of layers of cells of varying shapes. The urinary bladder provides an excellent example of this type of tissue. Areas of the body that require protection typically have stratified squamous epithelium. In this type of tissue, only the outermost layers of cells are usually squamous. The layer resting on the basement membrane is frequently columnar. The skin and tubular organs that are prone to trauma have this type of epithelial tissue; for example, caudal vagina.

Epithelial cells form various types of glands. These can be either **exocrine** or **endocrine** glands (Figure 3-6). In both cases epithelial cells have secretory functions. In the case of exocrine glands, cells secrete their products into ducts which carry secretions to a free surface (either internal or external). In contrast, endocrine glands are ductless; cells secrete their products into the surrounding extracellular space, adjacent to capillaries. Both types of glands are found in reproductive tissues.

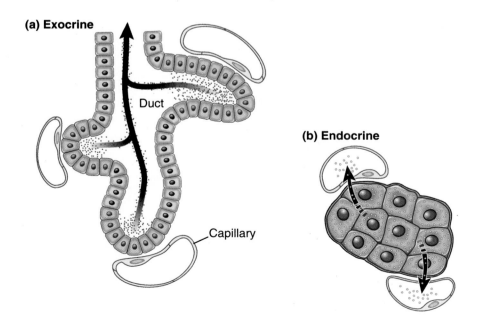

(a) Exocrine

Duct

Capillary

(b) Endocrine

FIGURE 3-6

Two types of multicellular glands consisting of epithelial: (a) exocrine gland and (b) endocrine gland.

3

Glands can be unicellular or multicellular. In the former case, specialized cells are scattered within an epithelial lining; for example, endocrine cells in the epithelial lining of the digestive tract. Multicellular glands consist of sheets of specialized epithelial cells that, together, serve some secretory function. Multicellular glands are organized in different ways. Simple, multicellular glands consist of one or more secretory portions and an unbranched excretory duct. The secretory end may take on the shape of a straight tube, a coiled tube, or alveolus (pear shape). Examples of simple, multicellular glands can be found in the intestine and skin (sweat glands). Multicellular glands with a compound structure consist of secretory units organized around branches of a duct, each of which drains into the main excretory duct. The shape of the secretory unit can be tubular, alveolar, or tubuloalveolar (Figure 3-7). Most of these glands are lobulated; that is, subdivided into smaller sections called lobules. The testis is an excellent example of this type of gland. Within each lobule there is a network of small ducts that drain into lobular ducts which drain into several main excretory ducts.

Connective Tissue

Connective tissue is ubiquitous and exists in a variety of structures. All types of connective tissues are derived from a portion of the embryonic mesoderm called the **mesenchyme.** Mesenchymal cells are stem cells that give rise to each type of connective tissue found in adults. The major adult connective tissues

FIGURE 3-7

The epithelial cells of exocrine glands are organized around branches of excretory ducts. The cells can be arranged in tubular, alveolar, or tubuloalveolar shapes.

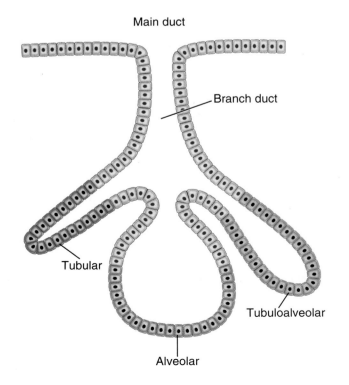

include the blood (cells and plasma), supportive connective tissues (cartilage and bone), and proper connective tissue. Our major concern will be with proper connective tissues, the type that connects organs to one another and suspends organs from the body wall. This type of tissue contains different types of cells that are separated by an extracellular material called intercellular substance. This is composed of both fibrous and amorphous components. The latter component is called ground substance and consists of mucopolysaccharides. One of the more common types of cells seen in connective tissue is the **fibroblast.**

There are two types of proper connective tissue; loose and dense. Loose connective tissue (Figure 3-8) is distributed throughout the body forming the superficial fascia; filling spaces between organs and binding organs together. The intercellular substance of loose connective tissue is produced by long, flat fibroblasts and consists of collagenous and elastic fibers embedded in the ground substance. Dense connective tissue contains the same types of fibers as loose connective tissue but the fibers are arranged in tight, parallel bundles to form tendons and ligaments, or in tightly interwoven networks to form dense matting (e.g., the dermis of the skin).

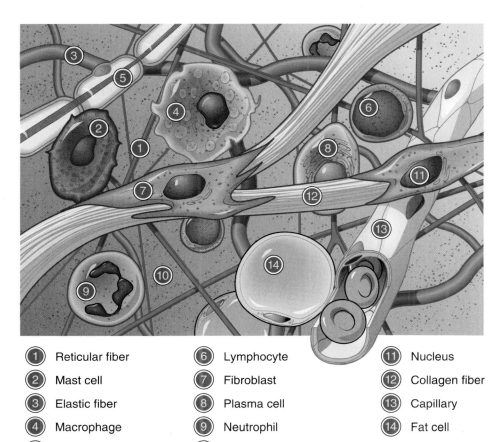

FIGURE 3-8

Loose connective tissue consists of various cell types and blood vessels supported by a matrix consisting of ground substance and fibers.

①	Reticular fiber	⑥	Lymphocyte	⑪	Nucleus
②	Mast cell	⑦	Fibroblast	⑫	Collagen fiber
③	Elastic fiber	⑧	Plasma cell	⑬	Capillary
④	Macrophage	⑨	Neutrophil	⑭	Fat cell
⑤	Nerve fiber	⑩	Ground substance		

3

Circulatory Tissue

The circulatory system provides the primary means of communication among the reproductive tissues. In order to fully appreciate how this communication system works, it is necessary to review some basic principles of circulatory anatomy. Many of the reproductive organs contain endocrine cells that produce various regulatory chemicals in response to particular blood-borne signals. These regulatory substances are secreted into the extracellular fluids and then diffuse locally to affect adjacent cells and/or enter the blood to affect other, more distant cell types. Endocrine tissues are richly supplied with blood. Arterial blood flows into this area via an arteriole (small artery), which supplies a capillary plexus or network (Figure 3-9). A capillary is an extremely small and delicate blood vessel. The wall of capillary is formed by a single layer of endothelial cells and associated pericytes (undifferentiated cells that can transform into cells such as fibroblasts and smooth muscle cells). Both cell types are embedded in a basement membrane. Small slits (pores) form at the borders between adjacent endothelial cells. This allows substances of low molecular size to pass between the

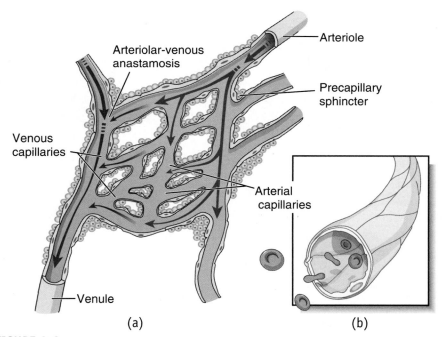

(a)   (b)

FIGURE 3-9

Typical capillary plexus supplying blood to an organ (a). Blood flows into the plexus via an arteriole and exits via a venule. Blood flow through the capillary network is regulated by sphincters that contract to constrict the capillary and occlude the vessel. In some tissues, blood can flow directly from the arteriole to the venule through anastomoses. Blood vessels are formed by endothelial cells arranged to form a cylinder (b).

blood and surrounding extracellular fluid. There are five types of capillaries. Continuous capillaries consist of endothelial cells that lack pores. This prevents exchange of fluids between the blood and surrounding extracellular fluid. This type of capillary is commonly found in nervous tissues and forms the so-called blood-brain barrier. In contrast, the endothelial cells of fenestrated capillaries contain pores through which small amounts of fluid can be transported. These exist primarily in endocrine glands, intestines, and kidneys. There are areas in the central nervous system with this type of capillaries. These regions seem to be important in the transduction of humoral signals into neuronal signals. For examples, certain regions detect blood concentrations of sodium in order to help the brain regulate circulating concentrations of this ion. These areas might also be involved with mediating the effects of metabolic state on reproductive activity. The third type capillary is the sinusoidal capillary. These capillaries are larger and more irregular in shape than the former types and have an inconspicuous basement membrane. They are found in endocrine organs as well as the carotid and aortic bodies. Sinusoids are larger than sinusoidal capillaries and they frequently lack a basement membrane. Venous sinuses are the largest type of capillary and consist of only endothelial cells and a discontinuous basement membrane. These provide the means for blood to drain from the brain. Regardless of the type of capillary, branches of the capillary plexus converge to form a venule (small vein), which returns blood to the venous circulation. Blood flow into capillary beds is regulated by smooth muscle cells called precapillary sphincters.

Although the transition between arterial and venous blood typically occurs in a capillary bed, direct connections (**arteriovenous anastomoses**) between a small artery and vein permit a short-circuiting of capillary beds (Figure 3-9). These arrangements allow blood to be shunted away from tissues during times of low activity (e.g., the digestive tract between meals), or to assist with temperature regulation (shunting blood to chilled areas or shunting blood to an appendage and/or surface to promote cooling). Arteriovenous anastomoses also permit a counter current exchange of substance; that is, direct transfer of material between blood vessels. These structures are not necessarily required for counter current exchange. For example, heat or molecules that will diffuse across the walls of blood vessels can be transferred between arteries and veins that lie in very close proximity to each other. You will encounter this latter type of vascular arrangement in both the male and female reproductive tracts.

Erectile tissue is a unique type of vascular tissue found primarily in the reproductive system (e.g., penis and clitoris), but also exists in nasal tissue. Such tissue is characterized by numerous, tightly packed spaces which are lined by endothelial cells. These areas are supplied by arterioles which are innervated by neurons which keep vessels constricted (closed). During periods of stimulation, the arterioles relax and allow the spaces to engorge with blood.

Nervous Tissue

Nerve cells provide another means for communication among reproductive tissues. As noted earlier, the activity of the reproductive system proper (i.e., the gonads, genital ducts, and external genitalia) is regulated by the pituitary gland, which is regulated by the brain. A good deal of our later discussions will focus on how the hypothalamus regulates pituitary function as well as how reproductive organs communicate with the nervous system. Therefore, it is worthwhile to review some basic information about how information is transmitted within the nervous system. A schematic drawing of a nerve cell, also called a neuron, is shown in Figure 3-10. The neuron consists of a cell body, dendrites (processes emanating from the cell body), an axon, and terminal branches of the axon. Neurons transmit information in the following way. Upon stimulation, the cell body becomes depolarized. This initiates a wave of depolarization along the axon away from the cell body and into the

FIGURE 3-10

Drawing of the type of neuron commonly observed in the central nervous system (a). Interaction between neurons occurs at synapses, which are specialized points of contact between adjacent nerve cells (b).

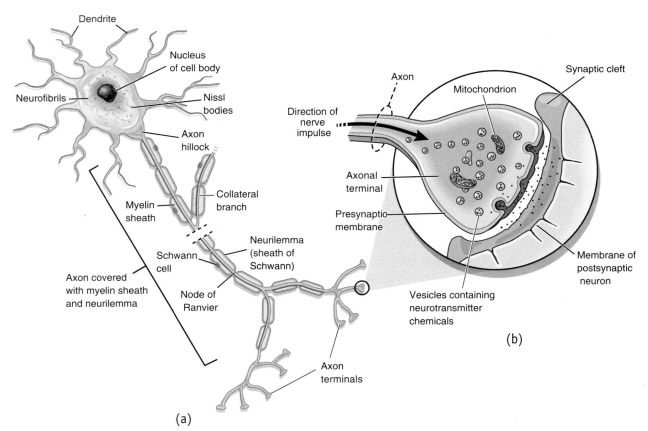

axon terminals. Depolarization of the axon terminals induces the release of chemical messengers. In cases where the axon terminals end in close proximity to another cell, the chemical messengers are called neurotransmitters. The area in which an axon terminal makes contact with another cells is called a synapse (Figure 3-10). Neurotransmitters released from the presynaptic neuron can either stimulate or inhibit the postsynaptic cell. When the chemical messenger is released into the extracellular space and enters a nearby capillary, the chemical is called a **neurohormone.** Neurons that conduct impulses to other cells typically have myelin sheaths, which enhance speed of transmission. Nerve cells that secrete neurohormones are not myelinated.

Finally, it is important to clarify some terminology regarding the organization of neurons. A **nerve fiber** is another name for axon. A bundle of parallel nerve fibers within the central nervous system is called a **tract.** A similar bundle of fibers in the peripheral nervous system is called a **nerve.** A grouping of nerve cell bodies is called a **nucleus** in the central nervous system, and a **ganglion** outside the brain and spinal cord.

Muscle Tissue

Both smooth muscle and skeletal muscle play important roles in regulation of reproductive activity. Obviously, the skeletal muscles that permit voluntary movement play are essential in sexual behaviors. You are undoubtedly familiar with the basic structure and function of skeletal muscle. The structure and function of smooth muscle (Figure 3-11) may be less familiar.

Nucleus of smooth
muscle cell

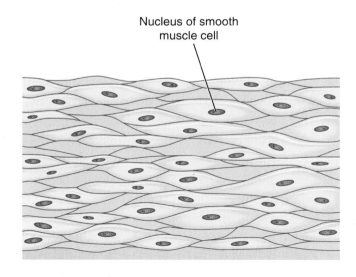

FIGURE 3-11

Longitudinal section of the small intestine of the sheep showing a layer of smooth muscle. Smooth muscle cells typically exist in bundles, which are held together by reticular fibers. These cells lack the banding patterns characteristic of skeletal muscle cells.

3

The genital ducts of males and females are lined with a layer of smooth muscle, which is primarily concerned with transport. Rhythmic contractions of smooth muscle in these tissues facilitate movement of fluids and gametes along the lumens of these organs. The smooth muscle cells of these tissues are arranged in two ways; longitudinally (along the length of the duct) and circular (as a concentric ring around the lumen of the duct). Smooth muscle cells are identified by their spindle shape and a centrally-located nucleus. They lack visible striations due to the lack of an ordered arrangement of contractile fibers. These cells are regulated by the autonomic nervous system which evokes either contraction or relaxation of smooth muscle cells. Microscopic examination reveals that axons of autonomic nerve cells terminate near smooth muscle cells, but do not form the distinctive neuromuscular junctions characteristic of skeletal muscle tissue. In the genital ducts, contraction of the longitudinal layer causes a shortening of the tube, whereas contraction of the circular layer causes the diameter of the tube to shrink. Humoral factors also affect activity of smooth muscle cells.

ORGANIZATION OF ORGANS

There are three major types of organs: tubular, solid, and membranous. In each case, two general types of tissues can be distinguished. The cell types that give the organ the ability to perform a particular function make up the **parenchyma.** These are usually epithelial cells. For example, the gametes and cells that support gamete development make up the parenchyma of the testis and ovary. Likewise, neurons are the parenchymal cells of the central nervous system. The remaining tissues (i.e., connective tissues) are known as the **stroma.**

Structure of Tubular Organs

Tubular organs are made up of concentric layers of tissues (tunics). Organs such as those found in the digestive tract have four tunics: tunica **mucosa,** tela **submucosa,** tunica **muscularis,** and tunica **adventitia** or tunica **serosa** (Figure 3-12). There is significant variation in the arrangement of these layers. Depending on the organ, particular layers may be missing or altered.

The tunica serosa, or outermost layer, faces the body cavity and consists of a surface layer of mesothelium reinforced by irregular fibroblastic connective tissue. Its major function is to provide support as well as the tissue through which nerves, blood vessels, and lymphatics enter an organ. Tubular organs not closely associated with the celomic cavities (e.g., urethra) have a different type of outer covering; tunica adventitia. This is a loose connective tissue devoid of mesothelium. The tunica muscularis lies beneath the serosa. In its

FIGURE 3-12

Drawing of a section of small intestine showing transverse and longitudinal perspectives of concentric layers of tissues. In this example, all of the possible layers are represented. Depending on the organ, one or more layers may be absent.

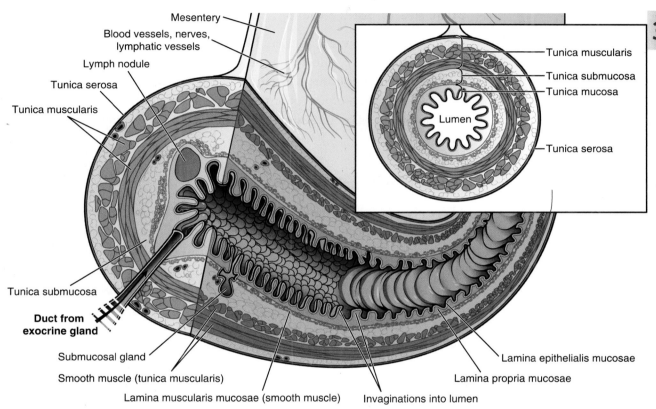

most organized state, this layer consists of two layers of smooth muscle; an outer longitudinal layer (running the length of the tube) and an inner circular layer (concentric to the circumference of the tube). These layers provide the means for tubular organs to contract. The next layer of tissue is the tela submucosa, a loosely organized layer of connective tissue containing large blood vessels, nerves, nerve plexes, and autonomic ganglia. The nerves in this layer regulate motility of the tunica muscularis. The tunica mucosa is the innermost layer of tissue, bordering on the lumen of the organ. Three sublayers can be distinguished within this layer. The lamina epithelialis mucosa consists of a layer of one or more types of epithelial cells. This outer layer is supported by the lamina propria mucosa, a connective tissue layer that contains small blood vessels and nerves that regulate and nourish the epithelial cells. In some cases, a lamina muscularis mucosa underlies the lamina propria mucosae. This thin layer of smooth muscle is less developed than the tunica mucosa, sometimes consisting of only a single layer of cells. It is not uncommon for the mucosal epithelium to evaginate into the lumen to form villous projections **(villi).** This layer might also invaginate into the lamina propria mucosa to form mucosal glands.

Structure of Solid Organs

Solid organs (e.g., testis, liver, and kidney) are enveloped by a **capsule,** or a dense collagenous connective tissue. Strands of looser connective tissue project from the capsule into the parenchyma and divide it into smaller units called **lobules.** These projections are called **trabeculae.** The capsule and trabeculae provide structural support for blood vessels and nerves to enter and leave the organ.

Structure of Membranous Organs

There are two types of membranes: **mucous** and **serous.** We have already encountered the former type; that is, the tunica mucosa of tubular organs is a mucous membrane. Mucous membranes consist of a layer of moist epithelial cells that rests on a layer of connective tissue. Secretions from the epithelial cells of the lamina epithelialis mucosa and/or glands within the lamina propria mucosa keep the outer lining moist.

Serous membranes are made up of a layer of mesothelium and surrounding connective tissue. These membranes line spaces within the body cavities and are moistened with the serous fluids found in these areas. The peritoneum is one familiar example of this organ type.

SUMMARY OF MAJOR CONCEPTS

- The reproductive system includes portions of the central nervous system (i.e., hypothalamus), pituitary gland, gonads, genital ducts, and external genitalia.
- Communication among reproductive organs occurs via humoral and neuronal mechanisms.
- Anatomic descriptions are facilitated by terminology that identifies direction and major body cavities.
- Each of the major tissue types (epithelial, connective, nervous, and muscle) is involved with reproductive function.
- Organs are identified by the types of cells making up their parenchymal tissue as well as the organization of the parenchyma and stroma.

DISCUSSION

1. Imagine that you are consulted for an opinion regarding the infertility of a 1-year-old dog. The semen from dog does not contain sperm cells and the testes appear atrophied. Overall, the dog's appearance resembles that of a puppy; that is, the animal doesn't have the masculine traits of normal adult males of a similar age. On hunch, you decide to give the dog daily injections of GnRH. Within a month, the dog's testes enlarge and start producing sperm cells. Moreover, the dog looks more like an adult. The owner is pleased and asks you what sort of miracle you performed. How would you explain your treatment? What would you conclude if your treatment failed to produce noticeable effects?

2. What organs would you expect to find in the following locations: dorsal cavity, caudal portion of the abdominal cavity, thoracic cavity, lateral to the heart, superficial to the abdominal muscles?

3. Why wouldn't you expect to see stratified squamous epithelium forming the small intestine, or simple cuboidal epithelium forming the external genitalia?

4. Suppose you take simultaneous blood samples from an artery supplying an organ and a vein that drains the organ. Also assume that this organ produces and secretes a hormone that exerts its effect somewhere else in the body. How would the concentrations of this hormone compare in the artery and vein from which you sampled? Explain your answer.

5. In general terms explain why some spinal injuries result in erectile dysfunction in males.

REFERENCES

Banks, W.J. 1993. *Applied Veterinary Histology (Third Edition)*. St. Louis: Mosby Year Book.

Cochran, P.E. 2004. *Laboratory Manual for Comparative Veterinary Anatomy and Physiology*. Clifton Park, New York: Thomson Delmar Learning.

Frandson, R.D., W.L. Wilke, and A.D. Fails. 2003. *Anatomy and Physiology of Farm Animals*, 6th edition. Philadelphia: Lippincott, Williams & Wilkins.

Sexual Differentiation

- Describe the sexual organization of mammalian bodies.

- Describe how sex is determined.

- Describe the development and differentiation of the gonads, genital tract, external genitalia, and secondary sex traits.

- Describe major anomalies in sexual differentiation.

OVERVIEW OF SEXUAL DIFFERENTIATION

The males and females of most animals can be distinguished by various anatomic features that are said to be sexually differentiated. A characteristic that differs in shape and function between males and females is said to be a sexually dimorphic trait. These characteristics are related to the reproductive roles of each sex. Such traits play either direct or indirect roles in sexual reproduction. Those that play direct roles include the gonads, genital ducts, and external genitalia.

The so-called **secondary sex traits** play indirect roles in the reproductive process. In other words, they are sexually differentiated traits that may facilitate sexual contact, but are not required for the production, transport, or fusion of gametes. Examples of secondary sex traits include sex-based differences in mammary gland development, body size, body shape, body composition, pattern of coloration, pattern of hair growth, and sexual behavior. You may recall that sexual reproduction requires the ability of individuals to recognize members of their own species as well as members of the other sex. It is likely that secondary sex traits allow males and females of a particular species to recognize each other in order to form mating pairs.

The sexual differentiation of animals involves several developmental steps that begin during embryogenesis (development of the embryo) and end with **puberty** (sexual maturation). In viviparous animals (i.e., most mammals) embryogenesis occurs in the uterus. In oviparous animals (birds, reptiles, amphibians, and monotreme mammals) embryogenesis occurs in the egg. Your understanding of sexual differentiation will be facilitated if you remember three important concepts.

- The sexual organization of an individual occurs in several steps.

- Each of these steps depends on the previous one.

- Almost all sexual characteristics arise from indifferent (neither male nor female) precursors.

Figure 4-1 summarizes the major steps in sexual differentiation. Briefly, **chromosomal sex** is determined at syngamy and reflects the genetic composition of the father's sperm cell and the mother's oocyte. The chromosomal sex determines what type of gonad develops in the embryo. This is commonly referred to as **sex determination.** Subsequent steps are referred to as **sexual differentiation.** In the first step of sexual differentiation, the differentiated gonad directs development of the genital ducts and external genitalia. By the time of birth, newborn mammals have fully differentiated genital ducts and external genitalia, but males and females are otherwise indistinguishable. However, during sexual maturation, the gonads begin producing large amounts of hormones that induce development of secondary sex traits. These characteristics are the most visible manifestations of sexual dimorphisms, and are the ones we typically rely on to identify a particular sex in humans and other animals.

A word of caution seems appropriate before we engage in a deeper discussion of sexual differentiation. Descriptions of this process typically emphasize anatomic and physiologic *differences* between males and females. This can promote the notion that the sexes are opposites; that is, that the sexes are somehow set against each other. Although this view has been popularized in our culture, it isn't consistent with what we know about the biology of sexual reproduction. Sexual reproduction requires cooperation between the sexes. Although it is true that the reproductive traits of males and females are clearly

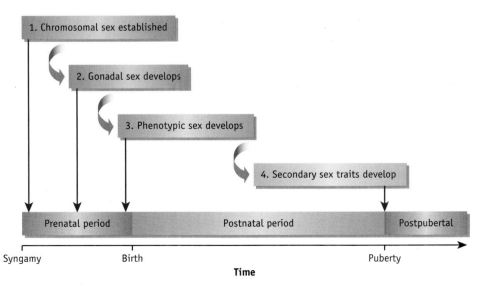

1. Chromosomal sex established

2. Gonadal sex develops

3. Phenotypic sex develops

4. Secondary sex traits develop

| Prenatal period | Postnatal period | Postpubertal |

Syngamy   Birth   Puberty

Time

FIGURE 4-1

Overview of sexual differentiation in mammals. Note that sexual differentiation occurs as a sequence of steps, each one serving as a prerequisite for the next and that the full process spans the time between conception and sexual maturation.

distinguishable, they promote complex interactions between a male and female, not antagonism. Moreover, sexually dimorphic traits usually serve similar biological functions in each sex and develop from the same embryonic structures. For example, the ovaries and testes of adult females and males serve similar physiologic functions and consist of analogous cell types that originate from the same embryonic precursor cells.

CHROMOSOMAL SEX

In some animals (e.g., some reptiles and fish), sex is determined by environmental conditions such as the ambient temperature that prevails during incubation of eggs. Sex determination in mammals is a more stable process; that is, sex ratios are the same regardless of environmental conditions. The sex of an individual mammal is determined by its genotype. More specifically, the type of primary and secondary sex traits expressed by an individual is due to particular genes that control sexual differentiation. A gene, the functional unit of heredity, consists of particular segment of DNA that makes up part of a chromosome. Chromosomes exist as homologous pairs; that is, two chromosomes that have similar size and shape. Chromosomes are usually studied by making a karyotype. A karyotype is produced by photographing the metaphase chromosomes of a single cell and then matching the chromosomes according to their sizes and shapes. Finally, the chromosomes are ordered according to size. Matching chromosomes is facilitated by using various imaging techniques that reveal similarities in DNA sequences. Homologous pairs of chromosomes have analogous gene sequences as well as similar sizes and shapes.

Sexually reproducing organisms have two major types of chromosomes; that is, **sex chromosomes** and **autosomes.** Sex chromosomes contain genes that are particularly important in determining whether an individual will develop ovaries or testes. There are two types of sex chromosomes in mammals: X and Y. Although the sex chromosomes are considered to be a homologous pair, there is typically a marked difference in size between the two types. The X chromosome is a medium-size chromosome with a submedial (almost in the middle) centromere. The Y chromosome resembles short, acrocentric autosomes (i.e., having the centromere near one end). The short arms (chromatids) of the X and Y chromosomes have homologous segments that pair during meiosis. This is the site at which the chiasmata form, permitting the exchange of DNA between these regions.

The standard way of making reference to an individual's karyotype is to note the total number of chromosomes followed by the type of sex

TABLE 4-1 Karyotypes and gonadal sexes of various anomalies in human sexual differentiation

Autosomes	Sex Chromosomes	Gonad	Syndrome Name
44	X0	Ovaries	Turner's
44	XX	Ovaries	Normal female
44	XXX	Ovaries	Superfemale
44	XY	Testes	Normal male
44	XXY	Testes	Klinfelter's
44	XYY	Testes	Supermale
66	XXX	Ovaries	Triploid (lethal)
66	XXY	Testes	Triploid (lethal)
44	XX^{sxr}	Testes	Sex reversal[1]

[1]In this case a small piece of the Y chromosome is translocated to an X chromosome.

4

chromosomes. For example, normal human males have the 46, XY karyotype, whereas normal human females have the 46, XX karyotype. The importance of the sex chromosomes in determining an individual's sex is revealed by studying various anomalies of sexual differentiation in humans. Table 4-1 lists the karyotypes and gonadal sexes of normal males and females as well as those of individuals with some of the most common anomalies. Each of these anomalies can be classified as aneuploidy; that is, the total number of chromosomes differs from that which is characteristic for a particular species. In these particular cases the discrepancy is due to variations in number of sex chromosomes. Aneuploidy can result from errors in mitosis in the zygote, or from errors in meiosis in development of the oocyte; that is, failure of chromatids or homologous pairs of chromosomes to separate.

After studying this table, it becomes clear that development of testes is associated with presence of a Y chromosome. Of particular importance is the observation that the presence of only one X chromosome results in development of ovaries. Based on this observation it is clear that a Y chromosome plays a pivotal role in testicular organogenesis. Interestingly, the ovaries of X0 individuals are not fully functional, suggesting that two X chromosomes are required for normal ovarian development. Moreover, the X chromosome contains genes that regulate vital processes; that is, embryos that lack an X chromosome (e.g., 45, 0Y) do not survive.

In somatic cells of 46, XX females, one of the X chromosomes is inactivated. The inactivation of an X chromosome is random. Gonadal cells that are precursors for gametes have only one active X chromosome, but both are active during meiosis and development of the oocyte.

Studies of human karyotypes and corresponding sex traits provide the basis for the hypothesis that the Y chromosome contains putative genes that induce development of the male gonad. Advances in molecular genetics research lead to identification of such a gene; that is, **the sex-determining region of the Y chromosome (SRY).** This area consists of a 35 kb region of DNA located on the short arm of the Y chromosome. Most scientists believe that this gene acts as a switch that initiates expression of other genes that direct development of the male gonad. It is likely that the genes that are activated by SRY reside on the Y chromosome as well as on autosomes. The actual gene products controlling testicular development are various peptide growth factors, some of which have been fully characterized.

The extent to which the X chromosome is involved in gonadogenesis is less clear. Apparently, the X chromosome of males is not necessary for normal testicular function. However, males with 44, XXY karyotypes have impaired spermatogenesis (production of sperm cells). In females, two X chromosomes are required for normal ovarian development. Based on these observations, it seems likely that there are particular genes that direct ovarian organogenesis, but such genes may be inhibited in the presence of genes expressed by the Y chromosome.

OVERVIEW OF EMBRYONIC DEVELOPMENT: THE PIG AS A MODEL

As noted earlier, phenotypic sex depends on **gonadal sex.** In order to understand how this occurs, you must become familiar with some of the major aspects of embryonic development. The pig serves as a useful model to study this process because embryogenesis in this species has been thoroughly characterized. Although the embryonic ages at which specific developmental events occur varies among mammals, the order in which they occur is similar across species.

Embryogenesis in the pig occurs between 12 days of gestation and birth (114 days). It is useful to divide this process into two periods: embryonic and fetal. The embryonic period refers to the time when the **embryo** prevails and begins with establishment of the body axis (i.e., distinguishable head and tail regions) and ends when males and females can be distinguished by their external genitalia; that is, between 12 and 36 days of gestation. The fetal period refers to the time after which the fetus develops. In a **fetus** the structures of major organ systems have differentiated and the species of the **conceptus** can be identified; for example, a pig embryo takes on the shape of a pig. Embryology can be an extremely difficult area of study, due to the fact that all of the anatomic systems are developing simultaneously. Our examination of embryogenesis will be simplified by the fact that we are concerned only with development of the urogenital system.

Day 17 Through 18

Figure 4-2 illustrates the major anatomic features of the pig embryo at 17 days of age. At this stage of development the embryo has rudimentary nervous, cardiovascular, urinary, and digestive systems and is intimately associated with the so-called **extra-embryonic membranes; chorion, amnion, allantois,** and **yolk sac.** Figure 4-2 shows only a portion of these membranes. A protective amniotic membrane directly surrounds the embryo, encasing it in an amniotic space. A prominent yolk sac (vitellus) protrudes from the ventral surface of the embryo and is important in providing nutrients. At this stage, an

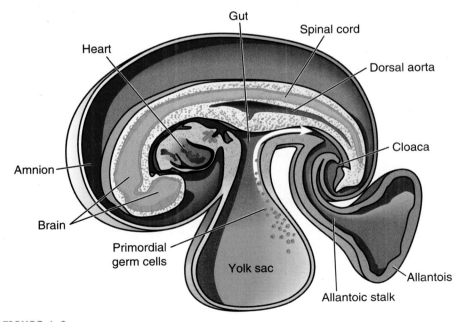

FIGURE 4-2

Mid-sagittal section through a 17 day-old pig embryo showing rudiments of the major organ systems and developing extra-embryonic membranes. At this stage of development, primordial germ cells migrate from the lining of the yolk sac to the gonadal ridges, along the mesentery of the developing hindgut.

allantois is developing from the hindgut. Both the embryo and the aforementioned membranes are surrounded by the chorion (not shown). The allantois eventually grows to surround the amnion and fuses with the chorion to form the placenta in eutherian mammals.

Development of the reproductive system is closely related to development of the urinary system (Figure 4-3). At 17 days of age, the **mesonephros,** an early embryonic kidney, is a large organ consisting of a tight mass of convoluted tubules. One end of each tubule is blind and is richly supplied with capillaries that receive blood from the dorsal aorta. The other ends of these tubules drain into bilateral (one on each side) **mesonephric ducts** that empty into the cloaca, also known as the **urogenital sinus.** Two **uretic buds** can be seen immediately above the points where the mesonephric ducts enter the cloaca . These will develop into the **metanephric kidneys** (metanephroi) that become the permanent kidneys later in development. The **metanephric ducts** exit from these kidneys and enter the bladder. In adults these ducts are known as the ureters.

At the stage of development described in the previous paragraph, both male and female embryos have identical (undifferentiated) gonads; that is,

FIGURE 4-3

Details of the caudal region of the pig embryo showing major structures of the developing urogenital system (shaded in red). As the gonads and genital ducts differentiate the mesonephros (blue) regresses.

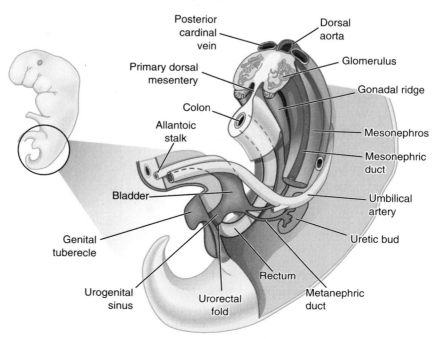

the **gonadal ridges** (Figure 4-3). These appear as knots in the connective tissue located on either side of the central, dorsal aorta, above the hindgut in the lower thoracic (between neck and abdomen), and upper lumbar (lower back between the ribs and pelvis) region of the embryo. A layer of columnar-shaped cells covers the ventral surface of each genital ridge. This tissue layer is the **germinal epithelium.** The genital ridge is located superficial to and medial to the developing mesonephros. At this period of development, the gonads are devoid of cells that will become gametes. **Primordial germ cells,** the progenitors of gamete cells, originate in the inner lining of the yolk sac near the developing allantois (Figure 4-2). They migrate from the yolk sac to the genital ridges via the connective tissue of the hindgut and the mesentery which supports the hindgut. Primordial germ cells have well-defined pseudopodia (temporary protoplasmic processes) similar to amoeba, and move via amoeboid action.

The 18-day-old pig embryo has a well-defined **genital tubercle** (Figure 4-3), a swelling on the ventral surface between the umbilical cord and opening of the urogenital sinus. This will eventually form the male or female external genitalia.

Day 28

Figure 4-4a depicts major features of the genital system at day 28. Well-defined gonads appear along the medial face of each mesonephros. These are elongated structures and have a germinal epithelium that has been invaded by primordial germ cells. Microscopic examination reveals that the gonads have initiated differentiation. Two ducts can be seen along each mesonephros. The mesonephric **(Wolffian)** ducts drain the mesonephroi and empty into the urogenital sinus. On each side of the embryo, a smaller paramesonephric **(Müllerian)** duct appears between the mesonephros and mesonephric duct. At this point these ducts have blind, growing points directed toward the urogenital sinus. Finally, the genital tubercle enlarges and develops a furrow along its median (center) axis; that is, the urogenital slit (not shown).

Day 36

By day 36, the end of the embryonic period, the gonads have differentiated into either testes or ovaries (Figure 4-4b). In addition, two sets of genital ducts (Wolffian and Müllerian) are readily apparent and extend from the mesonephros to the urogenital sinus. The mesonephric kidneys have begun

FIGURE 4-4

Development of the urogenital system in the pig embryo between days 28 (a) and 36 (b). At this stage of development, the gonad is beginning to differentiate, but the genitals are indifferent.

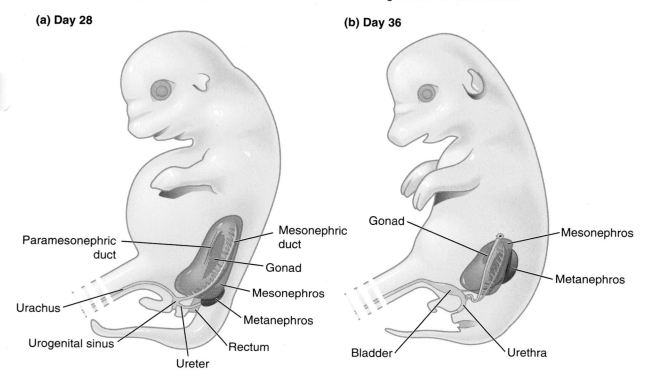

(a) Day 28

Paramesonephric duct

Mesonephric duct

Gonad

Mesonephros

Urachus

Metanephros

Urogenital sinus

Rectum

Ureter

(b) Day 36

Gonad

Mesonephros

Metanephros

Bladder

Urethra

to retract and metanephric kidneys have enlarged and have migrated cranially (forward). A ureter can be seen draining each metanephros and emptying into the urogenital sinus. The urogenital sinus has elongated to form a tubular urethra that connects to the bladder. The external genitalia have just begun to differentiate.

Day 55

By 55 days of age (Figure 4-5) the mesonephros has regressed and the metanephros has begun producing urine. Each kidney is drained by a ureter that connects with the bladder. Urine is voided either through the **urachus** to the allantois, or through the urethra into the amniotic cavity. In the 55-day-old male fetus, (Figure 4-5a) the testes are located in the abdominal region near the lateral surface of the caudal (toward the rear) pole of the metanephroi. The paramesonephric ducts have regressed leaving only the mesonephric ducts that become connected with the testes via

FIGURE 4-5

Development of male (a) and female (b) reproductive tracts in the 55 day-old pig embryo. At this stage of development the gonads, genital ducts, and external genitalia have become sexually differentiated. Notice that in the male, the testes has not yet descended from the abdomen into the scrotum.

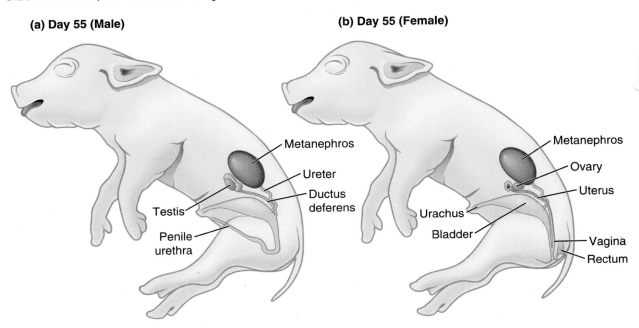

(a) Day 55 (Male)   **(b) Day 55 (Female)**

a few remaining tubules of the mesonephros. The mesonephric duct will become the epididymis and ductus deferens, the tubular system that allows sperm cells to be ejaculated into the urethra in adults. The genital tubercle of males has elongated and moved cranially toward the umbilical cord to form the penis and the ventral opening to the urogenital sinus has closed to form the penile urethra. The entire penis is embedded in the muscle of the ventral body wall.

At 55 days of age the ovaries of the female fetus are closely associated with the paramesonephric ducts (Figure 4-5b). The mesonephric ducts no longer exist. At the innermost region, the two paramesonephric ducts remain separate and form the left and right oviducts and uterine horns. In contrast, the outermost portions of the two paramesonephric ducts fuse to form the body of the uterus and part of the vagina. Changes in the genital tubercle of the female fetus is less dramatic that those of the male. The clitoris does not become closely associated with the ventral body wall, and the urogenital slit does not close. This allows formation of a vaginal opening, which is protected by the vulva.

4

FIGURE 4-6

Location of the testes in the male pig embryo between day 55 and 110 of gestation. By the end of pregnancy the testes occupy the scrotum which protrudes noticeably beyond the rump.

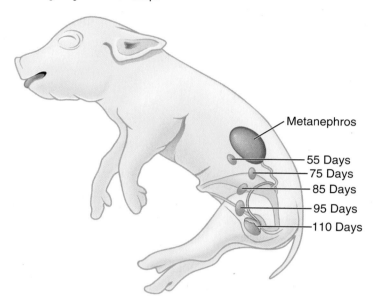

Metanephros

55 Days
75 Days
85 Days
95 Days
110 Days

Days 55 Through 114

By 55 days, the female pig fetus has assumed most of the particular form that is characteristic of its species. Although the male fetus is also recognizable as a pig, the shape of the male fetus changes dramatically during the last 20 to 25 days of pregnancy. This change is due to descent of the testes from the abdomen to the scrotum. Figure 4-6 traces the migration of the testes during this time period. Note that by day 80, the testes have moved to the bottom of the abdominal cavity in the area of the groin. A few days before birth, they have entered the scrotal swellings, which, in the pig, protrude noticeably beyond the buttocks. The details of this process will be considered later in this chapter.

DETAILS OF SEXUAL DIFFERENTIATION

The previous discussion provides a brief overview of the critical events in the sexual differentiation of mammals. Its purpose is to help you understand the temporal relationships among these developmental changes. In the next several sections, we will examine these changes in greater detail so you can understand the physiologic mechanisms that regulate them. However, before delving into these details it is helpful to review the overall regulation of sexual differentiation of the genital organs (Figure 4-7). As noted in an earlier section, differentiation of the gonads is determined by the chromosomal sex. In embryos that are genetically male, SRY induces development of testes, whereas ovaries develop in the absence of SRY. The presence of testes induces masculinization of the genital ducts and external genitalia. These effects are mediated by two testicular hormones. Testosterone, a steroid, stimulates development of the Wolffian ducts leading to formation of the epididymis and ductus deferens, the duct system that drains the testes. **Antimüllerian hormone (AMH)** is a peptide that induces regression of the Müllerian ducts. Testosterone also induces masculinization of the external genitalia, but this effect depends on its conversion to dihydrotestosterone (DHT). Development of the female genitalia occurs in the absence of testosterone and AMH. Without AMH, the Müllerian ducts develop into the oviducts, uterus, and cranial vagina, and the external genitalia form the vulva, clitoris, and caudal vagina (vestibule). The lack of testosterone causes the Wolffian ducts to regress.

FIGURE 4-7

Overview of mechanisms controlling sexual differentiation of the gonads, genital ducts, and external genitalia.

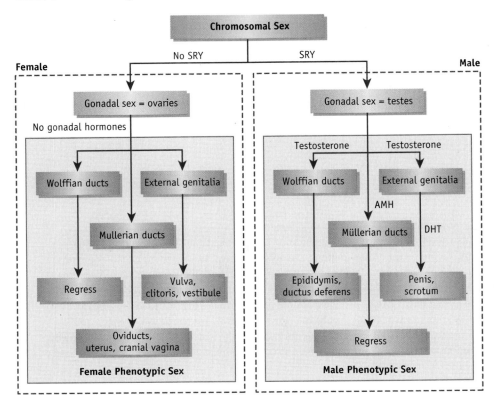

BOX 4-1   Focus on Fertility: The Jost Paradigm

Much of our current understanding of sexual differentiation in mammals can be attributed to the work of the French scientist, A. Jost. His work in the late 1940s forms the basis of what is now known as the "Jost Paradigm" of sexual differentiation. More specifically, Jost and his coworkers demonstrated that the embryonic gonads influenced development of the genital ducts via a local, humoral (relating to a body fluid) mechanism. Figure 4-8 summarizes Jost's experiments with rabbit embryos. As noted previously, two indifferent genital ducts are present in the non-differentiated embryo (Figure 4-8a). In males (Figure 4-8b), the Wolffian ducts develop

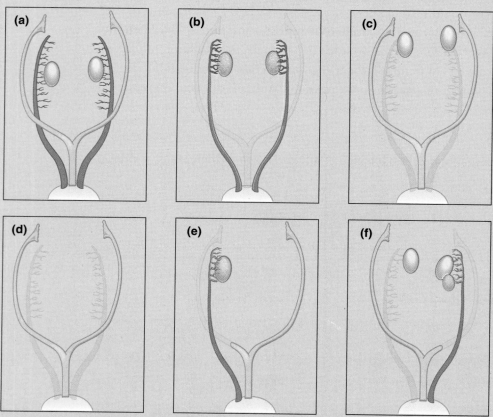

Figure 4-8 Results of experiments concerning the sexual differentiation of rabbit embryos first conducted by Jost during the mid-twentieth century. Before sexual differentiation (a) embryos possess an indifferent gonad and a set of male (blue) and female (red) genital ducts. Normal males (b) develop testes, the Müllerian ducts regress and the Wolffian ducts form the male genital ducts. Normal females (c) develop ovaries, the Wolffian ducts regress and the Müllerian ducts develop into the female duct system. If the gonads are removed from ether males or females (d) the Müllerian ducts develop and the Wolffian ducts regress. If only one testis is removed from male embryos (e) Wolffian ducts develop on the side ipsilateral to the remaining testis, whereas Müllerian ducts develop on the opposite side. If a testis is grafted onto one gonad in female embryos (f) the genital ducts masculinize on the side ipsilateral to the tissue graft.

and the Müllerian ducts regress. In females (Figure 4-8c), the Wolffian ducts regress and the Müllerian ducts develop. When either male or female embryos are bilaterally castrated (both gonads are removed), the Wolffian ducts regress and the Müllerian ducts develop (Figure 4-8d). This observation supports the idea that the testes are required for Wolffian development and Müllerian duct regression. If only one gonad is removed from a male embryo (unilateral castration) different duct systems develop on each side (Figure 4-8e). More specifically, Wolffian ducts develop and Müllerian ducts regress on the side ipsilateral to the remaining gonad (testis), whereas Müllerian ducts develop and Wolffian ducts regress on the side contralateral to the gonad. Moreover, if an embryonic testis is grafted onto one female gonad (Figure 4-8f), Wolffian ducts develop and Müllerian ducts regress on the side ipsilateral to the graft; on the side without the grafted tissue, Müllerian ducts and Wolffian ducts regress. These two treatment groups reinforce the notion that the testis is the source of factors that promote Wolffian duct development and Müllerian duct regression. They also lead to the hypothesis that the effects are local. The idea that a humoral mechanism is involved is supported by two observations: 1) Müllerian duct regression does not require direct contact between the testes and Müllerian ducts, and 2) Müllerian duct regression in male embryos is prevented when the testes and Müllerian ducts are separated by dialysis membranes that restrict diffusion of large substances (e.g., peptides) such as AMH. Moreover, implanting testosterone into a female gonad (not shown) promotes development of the Wolffian ducts, but fails to prevent development of the Müllerian ducts. This lead Jost to propose that Wolffian duct development is dependent on testosterone, whereas Müllerian duct regression is caused by some other testicular factor (i.e., a "Müllerian duct-inhibiting substance," now known as AMH). Based on these experiments, Jost proposed a theory for sex determination and differentiation; that is, that chromosomal sex determines gonadal sex which in turn orchestrates the differentiation of the genital ducts (part of the phenotypic sex).

Differentiation of the Gonads

Prior to invasion by the primordial germ cells, the embryonic gonad is sexually indifferent (Figure 4-9). It consists of mesenchymal tissue and epithelial cells that are arranged either in an outer layer covering the surface of the gonad (coelomic epithelium), or in tubules that are branches of the mesonephric duct (mesonephric tubules). Once they enter the genital ridges, the primitive germ cells induce formation of **primitive sex chords;** that is, columns of cells formed by proliferation and inward migration of cells from the mesonephros and coelomic epithelium. Development of the sex cords causes the genital ridges to enlarge and grow into the mesonephros. Once the primordial germ cells have entered the genital ridge, sexual differentiation of the gonad begins.

4

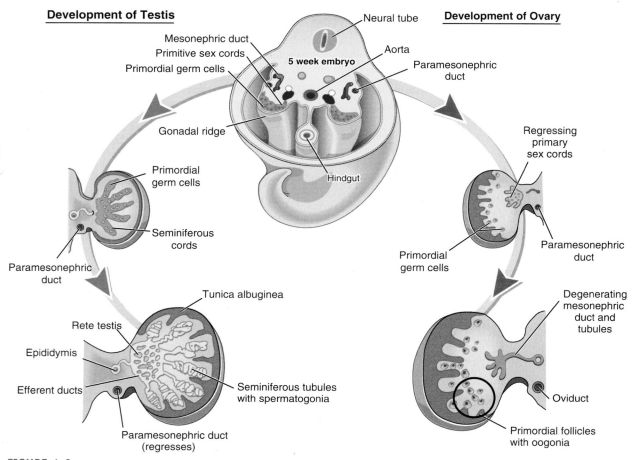

Development of Testis

Development of Ovary

Mesonephric duct
Primitive sex cords
Primordial germ cells
Neural tube
Aorta
Paramesonephric duct
5 week embryo
Gonadal ridge
Hindgut
Primordial germ cells
Seminiferous cords
Paramesonephric duct
Regressing primary sex cords
Primordial germ cells
Paramesonephric duct
Tunica albuginea
Rete testis
Epididymis
Efferent ducts
Seminiferous tubules with spermatogonia
Paramesonephric duct (regresses)
Degenerating mesonephric duct and tubules
Oviduct
Primordial follicles with oogonia

FIGURE 4-9

Sexual differentiation of the gonad. Once the primordial germ cells enter the indifferent gonad, they induce formation of the primitive sex cords. In males (left), the cords become well-organized and form networks of seminiferous tubules, which house spermatogonia, stem cells that give rise to spermatozoa. In females (right), the cords remain disorganized and form primordial follicles, which are located in the cortex of the gonad and house oogonia.

In the presence of the Y chromosome, mesenchymal cells located in the cortical (outer) region of the gonad condense to form the **tunica albuginea,** a thin layer of connective tissue that envelopes the gonad. At this time mesenchymal cells deep in the medullary (center) region of the gonad come in contact with the ingrowing tubules of the mesonephros to form seminiferous cords. The seminiferous cords engulf all of the primordial germ cells and produce a basement membrane which eventually forms a network of **seminiferous tubules.** The seminiferous tubules converge to form the **rete testis,** which

4

connects to the mesonephric duct via the mesonephric tubules, which later become the **efferent ducts.** Within the cords, the primordial germ cells will give rise to **spermatogonia** (precursors for sperm cells), whereas the mesodermal cells from the mesonephros will develop into **Sertoli cells.** Clumps of mesenchymal cells between the sex cords will become vascularized and give rise to **Leydig cells.** It is important to note that the primordial germ cells proliferate via mitosis prior to making contact with the seminiferous cords, but cell divisions cease once these cells become engulfed by the sex cords. Mitosis resumes once spermatogonia develop.

In the absence of the Y chromosome, the primitive sex cords remain disorganized and eventually degrade in the medullary region. The primordial germ cells cluster and continue to divide in the cortical regions of the gonads and become surrounded by remaining clusters of mesenchymal cells. The primordial germ cells, with their surrounding layer of mesenchymal cells, form **primordial follicles.** The mesenchymal cells secrete an outer basement membrane and primordial germ cells stop dividing. Eventually, the primordial germ cells will develop into **oogonia** and ultimately become oocytes, whereas the mesenchymal cells within the membrane will become **granulosa cells.** Remaining clusters of mesenchymal cells located between follicles will become **thecal cells.** Unlike the developing testis, the mesonephric tubules regress away from the ovary. Remnants of these tubules remain in the adult female. Another important difference between the developing ovary and testis is that the paramesonephric duct does not invade the ovary; that is, there is no tubular system connecting the ovary with the paramesonephric duct.

The organization of sex cords and appearance of Sertoli cells in males occurs much earlier than formation of primordial follicles in females. For example, in humans signs of sexual dimorphism in the male appear at 6 weeks of gestation, whereas the female gonad resembles the indifferent gonad until the sixth month of pregnancy. The precise mechanism controlling differentiation of the gonad has not been fully characterized. However, it is clear that something other than the primordial germ cells determine the type of gonad that develops; destruction of primordial germ cells in the developing embryo does not impede development of testes or ovaries.

It is important to note that even though the male and female gonads follow different developmental paths, the organization of mesenchymal and primordial germ cells is similar for both sexes. In each case, gametes are located within a basement membrane and in close association with Sertoli or granulosa cells. Moreover, both gonads contain interstitial cells (Leydig and thecal cells, respectively).

Differentiation of the Genital Ducts

At the time of gonadal differentiation, male and female embryos have indistinguishable urogenital systems consisting of a set of two ducts (Figure 4-10). Sexual dimorphism of the genital ducts involves regression of one or the other of these ducts. Development of the male genital ducts depends on hormones produced by the testes and therefore cannot occur until differentiation of the gonad has been completed. In contrast, development of the female genital ducts is not dependent on production of hormones by the ovary. Thus it appears that there is an inherent tendency for the genital ducts to feminize.

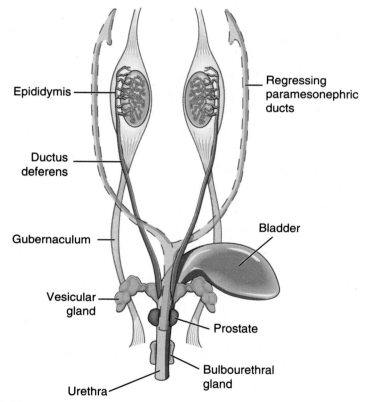

FIGURE 4-10

Sexual differentiation of the urogenital system. Before differentiation, the embryo has two sets of genital ducts. In males, the paramesonephric ducts regress (red) and the mesonephric ducts (blue) develop into the epididymis and ductus deferens. In addition, accessory sex glands develop along the pelvic urethra. Note that the testes are located within the abdominal cavity, supported by the gubernaculum (violet), which is attached to the wall of the scrotum. In females, the mesonephric ducts regress and the paramesonephric ducts form the oviducts, uterus, and cranial vagina.

Differentiation of the genital ducts in males begins shortly after gonadal differentiation (Figure 4-10). In the presence of testes, the Müllerian ducts regress and the Wolffian ducts develop into the epididymis and ductus deferens. In addition, the **vesicular glands, prostate,** and **bulbourethral glands,** accessory sex glands that contribute fluids to the ejaculate, develop at the lower sections of the Wolffian ducts near the urogenital sinus, which in the male contains the urethra.

As noted earlier, development of the male urogenital system is regulated by testicular hormones. Shortly after the organization of spermatic cords, the testes begin producing two hormones. **Testosterone** is produced by the interstitial Leydig cells, whereas the Sertoli cells produce AMH. Testosterone, a steroid, promotes the transformation of the Wolffian ducts into the epididymis, ductus deferens, and seminal vesicles as well as development of the prostate gland along the urogenital sinus. AMH, a peptide, causes regression of the Müllerian ducts by inducing **apoptosis** (programmed cell death). Development of the Wolffian ducts begins after onset of Müllerian duct regression. Testosterone and AMH influence differentiation of the genital ducts via localized actions. In other words, once released by the developing testes, they reach the genital ducts via diffusion, not by entry into the general circulatory system. Therefore, the hormones produced by a testis affect only the ducts ipsilateral to that testis.

In the presence of ovaries, and/or absence of testes, the Wolffian ducts degenerate and the Müllerian ducts differentiate to form the oviducts, uterus, and upper vagina (Figure 4-11). During this process, the rostral Müllerian ducts remain separate and form the oviducts and uterine horns, whereas the caudal ducts fuse to form the uterus and vagina. The fused ducts contact the urogenital sinus to form the uterovaginal plate, which lengthens to increase the distance between the developing uterus and the plate. At a later time the plate canalizes to form the lumen of the vagina. Interestingly, remnants of the regressing Wolffian ducts remain in the female.

Differentiation of the External Genitalia

Differentiation of the external genitalia is depicted in Figure 4-12. Development of male external genitalia begins soon after masculinization of the Wolffian ducts and is completed long before formation of external genitalia in females. As with the genital ducts, there is an inherent tendency for the external genitalia to feminize. Development of male genitalia requires a hormone which is a metabolite of testosterone; 5α-dihydrotestosterone (DHT). Both testosterone and DHT are members of a general class of steroid hormones known as **androgens.**

FIGURE 4-11

Development of the female genital ducts. The left and right paramesonephric ducts (red) fuse and meet with the developing vagina (yellow).

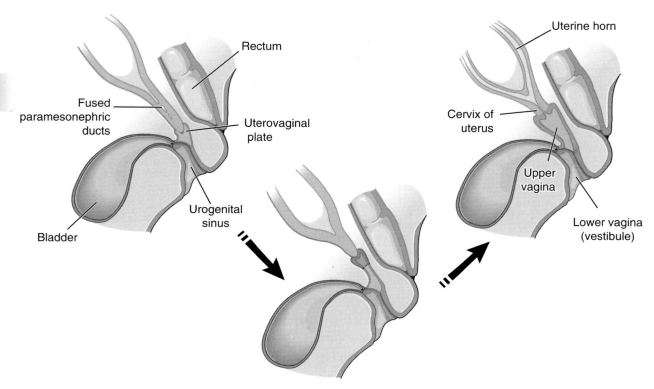

At the indifferent stage, the external genitalia consist of the **genital tubercle, genital fold,** and **genital swellings** (Figure 4-12). The cells of these tissues produce an enzyme (steroid 5α-reductase) that converts testosterone to the more potent DHT. In males testosterone produced by the testes diffuses into these cells and is converted to DHT. The DHT then acts to promote differentiation of these cells to form male external genitalia. Both testosterone and DHT are capable of producing these effects. However, the low concentrations of testosterone produced by the testes at this stage of development are insufficient to induce differentiation of these tissues. The more potent DHT is capable of inducing these changes at low concentrations. Cells of the Wolffian ducts do not express the 5α-reductase enzyme. Therefore it appears that masculinization of the Wolffian ducts is controlled by testosterone.

In the presence of DHT, the genital tubercle elongates and the genital folds fuse around the urethral groove to form the penis and penile urethra. These changes bring the genital swellings closer together to form the scrotum. In the absence of DHT, the genital folds do not fuse, leaving much of

FIGURE 4-12

Development of the male and female external genitalia in a four-legged mammal. Male (left) and female (right) structures develop from the indifferent genital fold (blue), genital swelling (yellow), and genital tubercle (red).

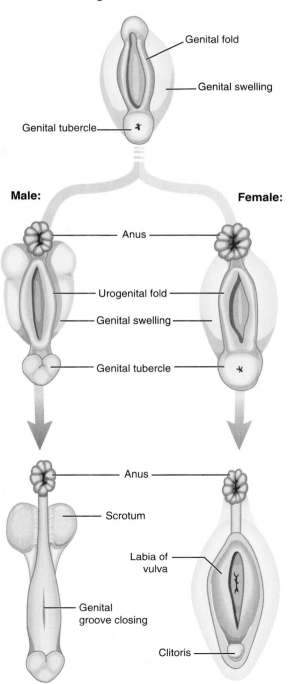

Indifferent Stage

Genital fold

Genital swelling

Genital tubercle

Male: **Female:**

Anus

Urogenital fold

Genital swelling

Genital tubercle

Anus

Scrotum

Labia of vulva

Genital groove closing

Clitoris

4

4

the urogenital sinus exposed. This results in formation of a cleft or **vestibule,** into which the vagina and urethra open. In females, the clitoris develops from the genital tubercle whereas the vulva, consisting of the **labia majora** and **labia minora,** develops from the genital swellings.

Testicular Descent

The testes descend into the scrotum late in development. In some cases (cattle and sheep) this occurs by the middle of pregnancy. In other cases, testicular descent occurs during late pregnancy, or soon after birth (pigs, humans, horses). This process can be divided into three phases; transabdominal movement, formation of the processus vaginalis, and transinguinal descent. Figure 4-13 illustrates the overall process in a schematic manner.

At the beginning of testicular descent, the testes are situated at the level of the ribs, along the mesonephros and are anchored cranially (toward the head) to the abdominal wall by a fold of peritoneum (i.e., a ligament). The testes lie in a retroperitoneal position (outside the peritoneum relative to the viscera). The caudal portion of the testis is attached to the **gubernaculum,** a ligament that connects the testis to the developing scrotum. Movement of the testes from the abdominal region to the inguinal (groin) region is due to three processes:

FIGURE 4-13

Stages of testicular descent: movement of the testis from the abdomen to the inguinal region (a), formation of the inguinal canal (b), and movement of the testis through the inguinal canal (c).

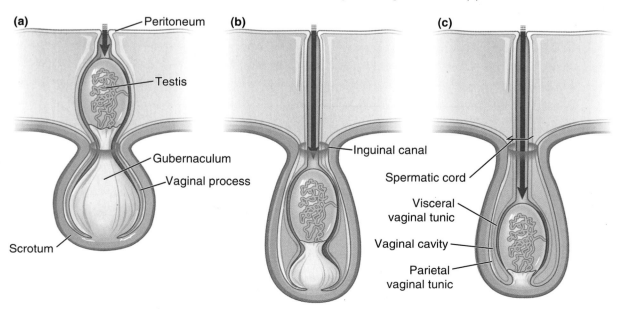

degeneration of the peritoneal fold supporting the cranial part of the gonad, shortening of the gubernaculum, and increased intra-abdominal pressure due to rapid growth of the abdominal-pelvic tissues. These changes bring the testes to rest against the abdominal wall in the inguinal region (Figure 4-13a).

Formation of the **inguinal canals** is necessary for movement of the testes into the developing scrotum. As intra-abdominal pressure increases, there is a **herniation** of the abdominal wall near the gubernaculum. At this point a process of peritoneum pushes outward toward the scrotum (Figure 4-13a). This projection of peritoneum is called the **vaginal process.** Continued pressure causes the vaginal process to enlarge around the gubernaculum and form the inguinal canal. At this point the gubernaculum undergoes rapid expansion which pulls the testis toward the entrance of the inguinal canal (Figure 4-13a).

The final stage of testicular descent involves movement of the testis through the inguinal canal into the scrotum (Figure 4-13b and c). This is largely attributed to progressive degeneration of the proximal gubernaculum (near the testis). As the gubernaculum shortens, it pulls the testis into its final location. As the testis moves into the scrotum, the inguinal canals are constrained by developing **inguinal rings;** that is, openings in the abdominal oblique muscles. These prevent the testis from re-entering the abdominal cavity.

The testis and epididymis reside in the scrotum enveloped by two layers of peritoneum (Figure 4-13c). In the scrotum these tissues are known as the **vaginal tunic.** The layer that lines the interior of the scrotum is the parietal vaginal tunic. This is the abdominal peritoneum through which the testes are pushed during their descent. The layer of peritoneum in direct contact with the testis and epididymis is the visceral vaginal tunic. This is the peritoneum that covered the testis when it resided in the abdominal region. The thin space between these layers is the **vaginal cavity.**

Regulation of testicular descent is poorly understood. The entire process appears to be controlled by the testes, but the factor that regulates this has not been fully characterized. At one time there was consensus that testosterone and AMH regulate growth of the gubernaculum. However, recent work suggests that another hormone may direct this process. A factor that controls shortening of the gubernaculum has not been identified.

There are two common types of anomalies associated with testicular descent; **cryptorchidism** and **inguinal hernias.** Cryptorchidism is the failure of one or both testes to descend into the scrotum. Unilateral cryptorchids have only one undescended testis, whereas neither testis has descended in bilateral cryptorchids. Bilateral cryptorchids are infertile, due to the fact that sperm production by the testes is impeded at normal body temperature. However, these animals retain masculine sex traits since their undescended testes continue to produce testosterone. Inguinal hernias occur when a portion of the intestine penetrates the inguinal canal and enters the scrotum. Swine appear to be particularly prone to this condition compared to other livestock species. Approximately one in 200 male pigs develops an inguinal hernia. In young boars, these are usually repaired at the time of castration.

SECONDARY SEX TRAITS AND SEXUAL BEHAVIOR

At the time of birth, male and females possess all of the anatomic traits necessary to fulfill their reproductive roles. However, they remain incapable of reproducing until they reach puberty. Puberty refers to all the physiologic, morphologic, and behavioral changes that result from the transition of the gonads from the infantile to the adult phase. The adult gonad produces hormones known as sex steroids (e.g., testosterone and estradiol), which bring about the physical and behavioral changes associated with puberty. In individuals of both sexes a growth spurt is associated with puberty onset. This is partly dependent on the sex steroids, but other hormones are involved. Changes in body composition also occur at this time. Although such changes occur in both males and females, there are sex-related differences in these growth characteristics. Moreover, sex steroids bring about other physical changes that are sexually dimorphic; that is, secondary sex traits.

Secondary Sex Traits

You are undoubtedly familiar with development of secondary sex traits in humans. As girls and boys become sexually mature, they experience changes in the breasts, external genitalia, body hair, and voice. Other animals express comparable changes that make it possible to distinguish between males and females. Although the ages at which these changes occur vary among individuals, the sequence of these changes is consistent within a sex. For example, you are likely familiar with the fact that certain boys and girls express mature traits earlier than their peers. Nevertheless, in all children such changes follow a particular developmental sequence; for example, appearance of adult traits can only occur after the testes or ovaries show increases in testosterone or estrogen production.

Sexual Differentiation of the Brain

The brain regulates reproductive activity in two important ways. First, it controls the pituitary gland that produces hormones that govern production of gametes and hormones by the gonads. Second, the brain regulates sexual behavior, which coordinates mating. The neural mechanisms governing gonadal function and sexual behavior are sexually differentiated. More specifically, the secretory patterns of some pituitary hormones as well as behavior patterns associated with mating differ between males and females. Differentiation of the mechanisms controlling hormone secretion and behavior occurs during a so-called critical period of neuronal development. In most mammals this occurs prenatally. However, in rodents, sexual differentiation of the brain occurs during the first 5 days of life. The principles of sexual differentiation of the brain are similar to those governing differentiation of the genital ducts and external genitalia; that is, differentiation is mediated by gonadal hormones.

The brains of mammals are inherently female. If the developing brain is not exposed to testosterone during the critical period, neuronal centers controlling hormone release and behavior will develop in ways that result in female hormone patterns and female sexual behavior. Exposure to testosterone during the critical period masculinizes the brain resulting in neuronal architecture that evokes male hormone patterns and male behaviors later in life (after puberty). It is unlikely that testosterone per se induces these effects. It is generally agreed that testosterone is converted to estradiol in the brain and that it is this metabolite of testosterone that promotes development of sexual dimorphisms responsible for male behavior. This mechanism is analogous to the one controlling development of male external genitalia; that is, a metabolite of testosterone (DHT) induces development of a penis and scrotum.

Special considerations apply to discussions of sexual behavior in primates, especially humans. There is no doubt that the human central nervous system includes regions that are sexually dimorphic, and it is likely that sex steroids play a role in this differentiation. However, we are uncertain about the extent to which gonadal steroids determine the type of sexual behavior expressed by a person. Assessment of a person's sexual behavior involves four categories: 1) gender identity, 2) gender role, 3) gender orientation, and 4) cognitive differences. Gender identity refers to identification of the self as male or female. Gender role deals with differences between male and female behavior as defined by a particular culture in a particular time. Gender orientation refers to one's choice of sexual partners. Cognitive differences refer to differing cognitive abilities between males and females.

Psychosexual differences between male and female humans reflect much more than genetic differences. Environmental and social factors play important roles in determining one's sexual identity and behavior. According to early studies the play behaviors of girls who were exposed to androgens differed from girls who were not exposed to androgens. However, both groups readily identified themselves as female. In other studies, children who were reared as a sex opposite to their chromosomal and/or gonadal sex expressed a gender identity corresponding to their assigned sexes. These types of observations support the theory that gender identity corresponds to the assigned, not the biological sex. However, this notion has been challenged by a recent case involving a 46, XY identical twin whose penis was accidentally ablated during circumcision as an infant. The patient was castrated and assigned a female gender role, which he never accepted. As an adult, the patient underwent sex reassignment and now lives successfully in a male gender role. It should be noted that not all patients who undergo sex reassignment as infants reject their assigned gender roles, especially if reassignment occurs before 30 months of age. A reasonable conclusion from these cases is that androgens exert facultative (taking place under some conditions but not others) rather than deterministic (taking place under all circumstances) roles in establishing gender identity in humans.

ANOMALIES OF SEXUAL DIFFERENTIATION

A large portion of our understanding of sexual differentiation comes from studies of anomalies. Anomalies of sexual differentiation occur in all animal species, but they have been most thoroughly studied in humans. Disorders of sexual differentiation can be divided into four major categories: 1) disorders of gonadal differentiation, 2) female **pseudohermaphroditism,** 3) male **pseudohermaphroditism,** and 4) unclassified forms. We will consider only the first three types of disorders.

Disorders of gonadal differentiation are caused by irregularities in expression of genes regulating development of the gonads. Female pseudohermaphroditism occurs when the genital ducts, external genitalia, and other aspects of phenotypic sex virilize in XX females. Male pseudohermaphroditism occurs in XY males when there is a deficiency of and/or resistance to the testicular hormones that promote development of the male phenotypic sex.

Disorders of Gonadal Differentiation

These types of disorders include **gonadal dysgenesis** (incomplete development of the gonads) and hermaphroditism, presence of gonads that contain both ovarian and testicular tissues. Some of these disorders result in both reproductive and nonreproductive pathologies. However, we will restrict our discussion to the chromosomal, gonadal, and phenotypic sexes associated with these conditions.

Klinefelter's Syndrome

Klinefelter's syndrome is the most common form of gonadal dysgenesis (1 in 800 males). These individuals have a 47, XXY complement of chromosomes. This condition results from nondisjunction of the sex chromosomes during meiosis in parents. Although these individuals develop testes, the presence of an extra X chromosome causes malformation of the seminiferous tubules resulting in extremely low spermatozoa production. However, both the genital ducts and external genitalia are male and appear normal.

4

Turner's Syndrome

Turner's syndrome occurs in one of 5,000 newborn females. Over 90 percent of fetuses with this syndrome die within the first 28 weeks of pregnancy. These individuals have a 45, X complement of chromosomes. The ovaries of 45, X individuals are "streak like" and contain only fibrous stromal tissue that lacks follicles. Genital ducts and external genitalia are female, but remain infantile in appearance due to insufficient production of sex steroids by the ovaries. This also results in short stature and little to no development of secondary sex traits. Thus even adults with this condition appear to be sexually immature.

Other Types of Gonadal Dysgenesis

Gonadal dysgenesis can also occur in 46, XX and 46, XY individuals. These conditions are the result of various mutations of genes located on autosomes and/or sex chromosomes. For example, one type of ovarian dysgenesis is attributed to a mutation of an autosomal gene that regulates hormonal control of follicle growth. In contrast, mutation of the SRY gene on the Y chromosome has been associated with a particular type of testicular dysgenesis. In these cases, development of the genital ducts and external genitalia is consistent with the type of gonad present. However, maturation of the reproductive system is incomplete due to hormone deficiencies.

True Hermaphroditism

True hermaphrodites have both ovarian and testicular tissue. These tissues can be arranged in the following ways: 1) a testis on one side and an ovary on the other (20 percent); 2) two ovotestes (both ovarian and testicular tissues are present in both gonads; 30 percent); 3) testicular and ovarian tissues are present on one side and a testis or ovary is on the other (50 percent). **Hermaphroditism** is rare in humans; slightly more than 400 documented cases. Differentiation of the genital ducts and genitalia is quite variable. External genitalia may resemble those of normal males or females, or may be ambiguous (having characteristics of both males and females). Most hermaphrodites have a large phallus that resembles a penis more than a clitoris. However, in almost all cases there is no penile urethra; that is, the urethra is exposed on the ventral surface of the phallus **(hypospadia).** Labioscrotal folds are prominent on each side of the urethral opening and cryptorchidism is common. A vagina and uterus are present in most hermaphrodites, but the uterus is typically underdeveloped. In patients with ovotestes, the ovarian tissue contains follicles and is functional, whereas the testicular tissue is dysgenic. Thus these individuals express female phenotypes. In cases where the individual has a testis and ovary, genital duct development on each side is consistent with the type of gonad present. This is consistent with the hypothesis that testicular-induced differentiation of the genital ducts involves localized effects of AMH and testosterone. Although the ovaries of these individuals are usually functional, the testes do not usually support spermatogenesis.

Hermaphroditism can arise in several ways: 1) sex chromosome mosaicism, 2) chimerism, 3) Y-to-autosome or Y-to-X chromosome translocation, and 4) mutation of either X-linked or autosomal genes involved with genesis of the testis. Mosaics and chimeras have a mixture of XX and XY cell types. In mosaicism, the different cell types come from different cell lines originating from the same zygote. This is the result of errors in mitosis during early cell divisions. Chimeras also have different cell types, but these originate from different genetic sources. For example, fusion of two zygotes or transfer of cells from one twin to another can result in chimerism. The **freemartin,** commonly seen in cattle, is an example of this condition. This occurs when the placentas of a male and female twin fuse resulting in a conjoined circulation. This permits mixing of primordial germ cells and hemopoietic cells, causing both the male and female twins to become chimeric; that is, each possesses XX and XY cells. Sexual differentiation of the bull calf appears normal, but fertility may be suppressed once it reaches maturity. In contrast, sexual differentiation of the female twin (the freemartin) is clearly abnormal. Freemartins have dysgenic testes or ovotestes, Wolffian-duct derivatives and female external genitalia frequently characterized by an enlarged clitoris. The gonadal and phenotypic sexes of the freemartin can be explained in the following manner. The key to understanding the etiology of this condition is the fact that differentiation of the male gonad occurs before that of the female. By the time the placentas of the twins fuse and permit exchange of blood, the testes have begun to develop. In contrast the ovaries remain largely undifferentiated at this the time. Exposure of the female's presumptive ovaries to

4

XY cells causes partial or complete virilization. Moreover, AMH and testosterone from the male fetus masculinize the genital duct system of the female twin, promoting development of the Wolffian system and degeneration of the Müllerian system.

In humans, most true hermaphrodites have 46, XX karyotypes. A small percentage of these patients are SRY positive, meaning that they have cells to which the SRY gene was translocated to an X chromosome or autosome during parental meiosis. The majority is 46, XX and SRY negative. In these cases hermaphroditism is due to mutations of genes that are activated by SRY to promote genesis of the testes.

Female Pseudohermaphroditism

Pseudohermaphroditism is characterized by discordance between an individual's gonadal and phenotypic sex. Female pseudohermaphrodites have ovaries, female genital ducts, and masculinized external genitalia. The most common cause of this condition is prenatal exposure to androgens resulting from an inherited deficiency in 21-hydroxylase, an enzyme regulating a key step in synthesis of cortisol by the adrenal glands. A deficiency of this enzyme leads to overproduction of androgens by the adrenals. The incidence of this condition is one in 50,000 persons. Virilization of the external genitalia in XX individuals can also result from other biochemical disorders, as well as exposure to exogenous androgens. If exposure by any means occurs early in the sexual differentiation process (before 12 weeks in humans) masculinization is prominent characterized by development of a penis and scrotum. Masculinizing effects are limited to hypertrophy of the clitoris when exposure occurs later in development. It is important to note that the effects of prenatal androgen exposure are limited to the external genitalia. This is due to the absence of testes. Without testes, there is no production of AMH. Thus Müllerian duct development proceeds unimpeded.

Male Pseudohermaphroditism

Pseudohermaphroditism in males is characterized by presence of testes without complete masculinization of the genital ducts and/or external genitalia. This is caused by either a deficiency in testosterone production or insensitivity to testosterone. Various conditions can result in a failure of the testes to produce testosterone. These will become apparent later when we discuss the hormonal control of reproduction. Without testosterone, neither the Wolffian ducts nor the penis and scrotum will develop. In addition, the testes will remain in the abdominal cavity. The same situation occurs in patients whose tissues do not respond to testosterone even though their testes produce the hormone. This latter condition is known as complete androgen resistance. The ability of a hormone to affect a particular cell (target cell) depends on the presence of **receptors;** that is, cellular proteins that specifically bind with a hormone to evoke particular biochemical responses. Resistance to androgens is due to an X-linked disorder that results in a deficiency of androgen receptors or an impaired interaction between androgen and its receptor. This type of male pseudohermaphroditism occurs in one of every 20,000 males.

Genetic males who do not produce testosterone or who are completely resistant to it have bilateral (undescended) testes, but the Wolffian ducts are either absent or underdeveloped. They also lack a uterus, but have a blind vagina. This can be explained by the fact that the testes produce AMH, which causes regression of the Müllerian ducts. Adults with this syndrome develop female secondary sex characteristics, but do not exhibit menstrual cycles.

Resistance to androgens can be incomplete (due to less pronounce disruption of androgen-receptor interactions), resulting in intermediate degrees of feminization, ranging from sexually ambiguous to underdeveloped male external genitalia. In all cases, Müllerian derivatives are absent and Wolffian ducts are underdeveloped.

Deficiencies in AMH can also occur, but this is rare compared to deficiencies in androgens; only 150 documented cases. The condition may be caused by lack of AMH production, and/or resistance to AMH. In either case, males have testes and normal male external genitalia. However, they also express Müllerian-derived genital ducts (oviducts and uterus), which are typically pulled into the inguinal canals and block full testicular descent. The condition is usually discovered when patients are undergoing repair of cryptorchidism or inguinal hernia.

One of the more striking types of male pseudohermaphroditism is 5α-reductase deficiency; that is, the inability to convert testosterone to DHT. Patients with this syndrome are genetic males with functional testes and male genital ducts. However, due to the inability to convert testosterone into the more potent DHT, the external genitalia remain feminized. At birth these individuals express external genitalia that consist of a small phallus with a ventral opening to a blind vaginal pouch. The testes are undescended and remain in the inguinal or labial regions. At puberty, testicular production of testosterone increases producing levels that are sufficient to masculinize the external genitalia. In addition to appearance of secondary sex traits, the phallus enlarges and the testes descend into the labioscrotal folds. Typically surgery is required to repair the hypospadia and to construct a scrotum.

It is not uncommon for students to experience feelings of skepticism or shock when they first learn of these anomalies in sexual differentiation. As noted earlier, many people are accustomed to thinking that there are only two sexes, male and female, and that these are completely separate and opposite conditions. Moreover, the fact that a person's phenotypic sex, the sex with which most of us identify ourselves and others, can be ambiguous or discordant with one's chromosomal and gonadal sex, can be quite disconcerting. Anomalies in sexual differentiation challenge the common, dualistic way of thinking about sex. They force us to view sex in a much more complicated manner. Is a hermaphrodite or pseudohermaphrodite male, female, both or neither? If they are neither, then do we require a new language that accepts more than two sexes?

SUMMARY OF MAJOR CONCEPTS

- The structures and activities of mammalian species are organized to fulfill the basic requirement of sexual reproduction; the production and fusion of two different types of gametes. In other words, mammalian bodies are sexually differentiated.

- The sexual organization of mammalian bodies occurs in a well-defined sequence of developmental events beginning at syngamy and ending at puberty; that is, chromosomal sex determines gonadal sex, which directs sexual differentiation of the genital ducts, external genitalia, and secondary sex traits.

- Initially, mammalian embryos have indifferent reproductive systems, which have the inherent tendency to feminize. In males, the presence of a Y chromosome induces development of testes, which then produces hormones that masculinize internal and external reproductive organs as well as specific regions of the brain.

- Anomalies associated with the sex chromosomes result in gonadal dysgenesis or hermaphroditism (presence of ovarian and testicular tissue), whereas anomalies associated with differentiation of the reproductive system result in pseudohermaphroditism (discordance between gonadal and phenotypic sex).

DISCUSSION

1. Construct a list of characteristics that you typically rely on to distinguish between men and women. Which of these traits arise from biological mechanisms of sexual differentiation? Which ones are social

constructions; that is, characteristics we create in regards to prevailing norms concerning how men and women should look and/or behave? Discuss some of the ethical, social, and political implications of your analysis.

2. Describe the gonadal and phenotypic (genital ducts and external genitalia) sexes that would develop in an XX mouse that received a microinjection of the SRY DNA sequence when it was a single-celled embryo. Explain how these characteristics developed in this case.

3. In 1779, John Hunter provided a written account of "Mr. Wright's freemartin." His publication also included a precise drawing of the gonads and reproductive tract of an animal with this anomaly. The drawing depicts two (apparently undescended) testes, female external genitalia, a vagina, and a genital duct system that does not resemble a normal uterus and oviducts. Twentieth-century reproductive biologists who have examined this rendering agree that it is a representation of an anomaly of sexual differentiation, but assert that is not a freemartin in the sense that we use the term today. On what criteria is their skepticism based? What type of disorder might this actually be? Explain the basis of your answer.

REFERENCES

Diamond, M. and K. Sigmundson. 1997. Sex reassignment at birth. *Archives of Pediatric and Adolescent Medicine.* 151:248–302.

George, F.W. and J.D. Wilson. 1994. Sex Determination and Differentiation. In: E. Knobil and J.D. Neill. *The Physiology of Reproduction Vol. 2. Second Edition.* New York: Raven Press: 3–28.

Grumbach, M.M. and F.A. Conte. 1998. Disorders of Sex Differentiation. In: J.D. Wilson, D.W. Foster, H.M. Kronenberg, P.R. Reed. *Williams Textbook of Endocrinology, 9th Edition.* Philadelphia: W.B. Saunders Company: 1303–1425.

Hunter, R.H.F. 1995. *Sex Determination, Differentiation and Intersexuality in Placental Mammals.* Cambridge: Cambridge University Press.

Marrable, A.W. 1971. *The Embryonic Pig: A Chronological Account.* London: Pitman Medical.

Patten, B.M. *Embryology of the Pig, Third Edition.* New York: McGraw-Hill Book Co., Inc.

CHAPTER 5

Functional Anatomy of Reproductive Systems: Genital Organs

OVERVIEW: REPRODUCTIVE TRACTS

The reproductive system consists of several organs that communicate with one another via neuronal and humoral pathways to regulate reproductive activity. Before we study how these organs interact, it is necessary to learn the structures of these tissues. For convenience, we will divide our analysis into two major components: the genital organs and neuroendocrine systems. This chapter will focus on the anatomy of the genital organs. The following chapter will focus on the anatomy of the nervous system and pituitary gland.

The genital organs include the paired gonads, genital duct system, and external genitalia. Collectively these organs are referred to as the reproductive tract. A tract is a longitudinal arrangement of organs that have related functions. Our discussion begins with general descriptions of the male and female reproductive tracts, followed by more detailed descriptions of the gross and microscopic anatomy of each reproductive organ.

The Male Reproductive Tract

It is convenient to organize the male reproductive organs into three major groups: the testes and their **adnexa** (attached structures), the pelvic reproductive organs, and the external genitalia. The adnexa of the testes include the **excurrent ducts** (efferent ducts, epididymis, and ductus deferens), and scrotum. The pelvic reproductive organs of the male are the urethra and accessory sex glands. The penis and prepuce make up the external genitalia. The gross anatomy of the male reproductive organs is fairly consistent among four-legged mammals (Figure 5-1). Our discussion of the gross anatomy of the male reproductive system will focus on

FIGURE 5-1

Drawings of longitudinal sections of the abdominal-pelvic regions of the bull, ram, dog, boar, and stallion showing major structures of male reproductive tracts.

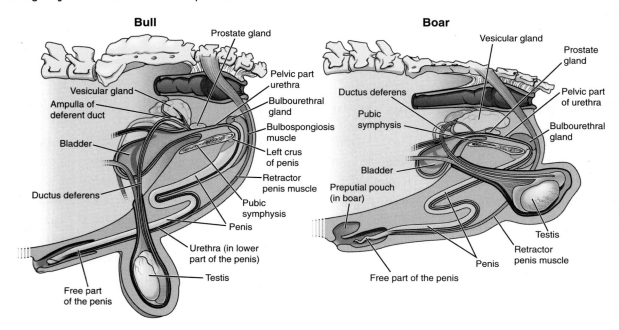

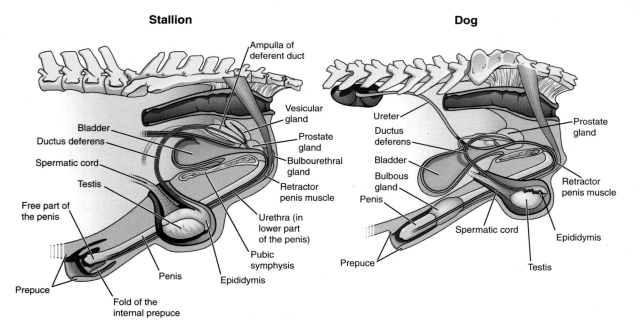

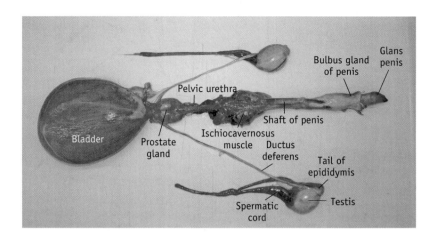

FIGURE 5-2

Photograph of the reproductive tract of a dog. Note that the arrangement of the tract does not correspond to the disposition *in situ*.

5

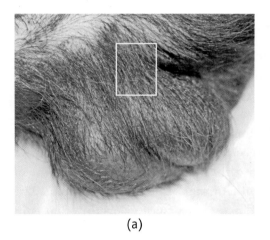

(a)

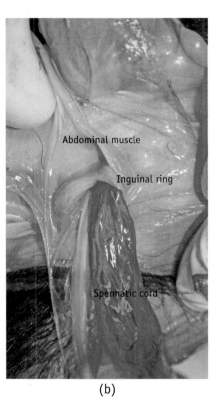

(b)

FIGURE 5-3

Photographs showing the scrotum (a) and spermatic cord descending through the inguinal ring (b) of a dog. The area outlined by the box (a) is the approximate location of the inguinal ring, an opening in the abdominal muscles through which the spermatic cord passes. Removal of the skin and underlying fascia reveals the structure of the inguinal ring and its relationship with the spermatic cord (b).

the dog (Figure 5-2), a species with which most students are familiar. Remembering this anatomy is facilitated by relating structure to function. The major functions of the male genital organs include the production, maturation, packaging, and transportation of spermatozoa. Spermatozoa are produced by the testes, which are suspended between the hind legs in a sac of skin called the **scrotum** (Figure 5-3a). Closely associated with the

FIGURE 5-4

Excised reproductive tract from a dog showing external components of the excurrent duct system.

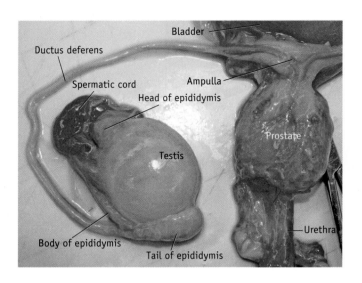

testis is the system of excurrent ducts (Figure 5-4). The **efferent ducts** (not shown) provide a means for spermatozoa to leave the testes. These ducts drain into a larger duct called the **epididymis.** The epididymis concentrates and nourishes the sperm cells and allows them to mature. This duct also serves as a storage depot. During ejaculation, spermatozoa and epididymal fluid enter the **ductus deferens (vas deferens),** which arise from the epididymal ducts and empty into the pelvic **urethra.** At its point of entry into the urethra, the deferent duct widens to form the **ampulla.** The pelvic urethra is continuous with the external or spongy urethra that runs through the penis emptying into an external orifice located at the apex of the penis. The accessory sex glands (prostate gland, vesicular glands, and bulbourethral glands) are aligned along the urethra distal to the point where the deferent ducts empty into the urethra. These exocrine glands produce fluids that are released to the urethra during ejaculation. Accessory gland fluids, together with the spermatozoa and epididymal fluids make up the **semen** (seminal plasma).

The external genitalia of the male includes the penis and scrotum. As noted previously the scrotum encases the testes, which are suspended from the abdominal cavity via the **spermatic cord,** which is connective tissue that supports the ductus deferens, lymphatics, nerves, and blood vessels that carry blood to and from the testes. The spermatic cords extend through the abdominal muscles on each side of the body via the **inguinal rings** (Figure 5-3b). The penis is protected by a layer of skin called the **prepuce.** In primates this is called the foreskin. In most other mammals, the penis retracts into the abdominal cavity during the non-erect state. In these cases, the prepuce is the fold of skin surrounding the opening through which the penis protrudes. Four skeletal muscles are associated with the male reproductive organs. The locations of these muscles are depicted in Figures 5-1, 5-2 and 5-3. The **ischiocavernosus** (see Figure 5-2) and **bulbospongiosus muscles** lie lateral and dorsal to the base of the penis, respectively. These muscles help empty the urethra after urination and play roles in erection. The **cremaster muscles** are fascia-like muscles that are enmeshed in the spermatic cords, and when contracted they lift the testes within the scrotum. The paired **retractor penis muscles** arise from the caudal vertebrae and extend to and connect with each side of the penis. In ruminants, the retractor penis muscles attach at the second bend of the so-called **sigmoid flexure,** an S-shaped curve in the body of the retracted penis. These muscles are less developed in mammals such as the dog and stallion. In this latter case, they originate on the rectum and pass along the ventral surface of the penis, eventually becoming enmeshed with the bulbospongiosus muscle, and ending at the tip (glans) of the penis.

The Female Reproductive Tract

The genital organs of the female include the ovary, uterus (including the cervix), uterine tubes (oviducts), vagina, and external genitalia. Like the male reproductive tract, a major function of this system is the production and transport of gametes. However, unlike the male, this system is involved with transporting both male and female gametes, as well as the developing embryo. Oocytes produced by the ovaries enter the oviducts and move toward the uterus. In addition, spermatozoa deposited into the vagina by the male during copulation are transported to the oviducts, the site of fertilization. The embryo eventually enters the uterus and attaches to the uterine wall, where the placenta develops. This highlights another difference in function between male and female reproductive tracts; the female tract is responsible for maintaining pregnancy and plays an important role in the birthing process.

Although the positions and shapes of the reproductive organs vary among species, the general organization of these tissues is similar among mammals (Figures 5-5, 5-6, 5-7, 5-8, and 5-9). In all cases, the female reproductive tract lies ventral to the rectum. In large species such as cattle and horses, this arrangement permits per rectum examination of the female reproductive tract. Unlike the testes of the male, the female gonads are located internally. Each ovary lies in close proximity to one of the oviducts, which consist of an **infundibulum,** which is closely apposed to the ovary; the **ampulla,** the middle section characterized by a narrowing diameter; and the **isthmus,** the narrow section that connects with the uterus. The shape of the uterus varies among

FIGURE 5-5

Longitudinal sections of the abdominal-pelvic regions of the cow, sow, mare, and bitch showing the dispositions of the female reproductive tracts. The reproductive tracts of the ewe and females of other ruminant species are similar to that of the cow.

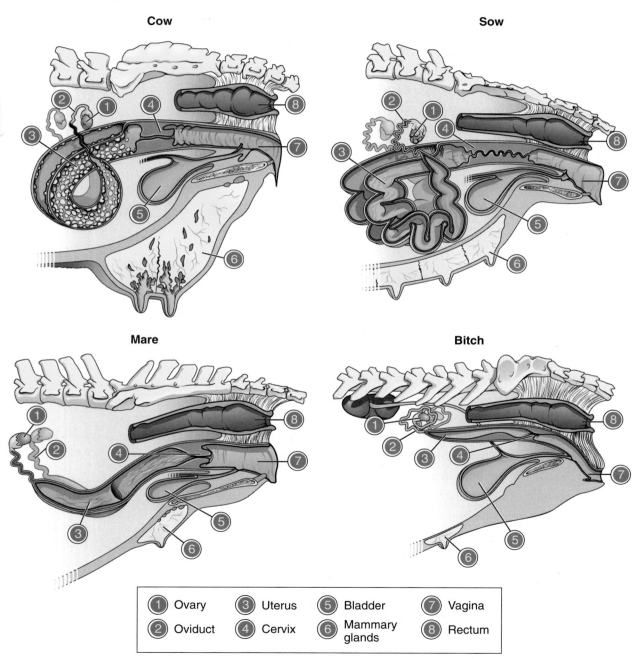

①	Ovary	③	Uterus	⑤	Bladder	⑦	Vagina
②	Oviduct	④	Cervix	⑥	Mammary glands	⑧	Rectum

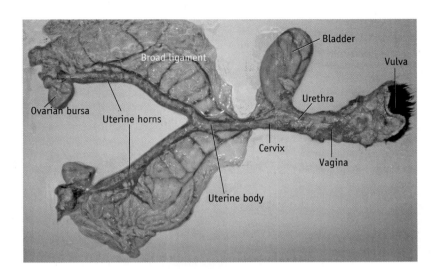

FIGURE 5-6

Dorsal view of an excised reproductive tract from a bitch

5

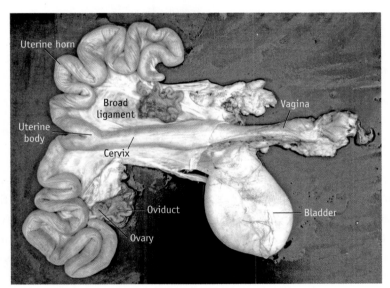

FIGURE 5-7

Dorsal view of an excised reproductive tract from a sow

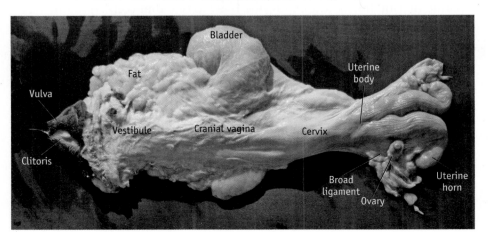

FIGURE 5-8

Dorsal view of an excised reproductive tract from a cow

FIGURE 5-9

Dorsal perspective of a cow's reproductive tract showing areas of demarcation between the vestibule, vagina, cervix, and uterus.

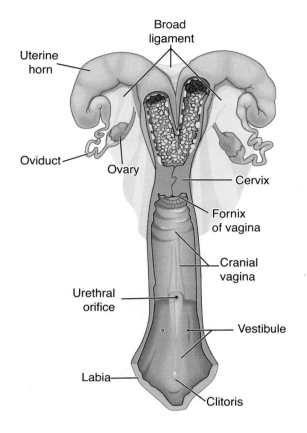

species. In all cases the caudal portion consists of a cervix, a thick-walled segment with a narrow lumen that acts as a sphincter regulating the flow of material to and from the uterus. The cranial portion of the uterus is often referred to as the womb, and is the region in which embryos implant and pregnancy is established and maintained. The shape of cranial uterus varies considerably among mammals, ranging from a structure that consists of two separate uterine horns and no uterine body to one that consists of one large uterine body with no horns. The true vagina extends from the caudal limit of the cervix to the entrance of the urethra (urethral orifice). The vagina is strictly a reproductive passage serving as the female copulatory organ as well as part of the birth canal. The external genitalia or vulva (wrapping or covering) lies caudal to the vagina. This region consists of the labia (lip-shaped folds of skin covering the entry to the female reproductive tract), clitoris, and vestibule. The vestibule extends from the urethral orifice to the inner labia. This portion of the tract serves both urinary and reproductive functions. The clitoris lies in the ventral portion of the external genitalia at the point where the folds of the labia join. In primates the labia is divided into an inner labia minora and an outer labia majora.

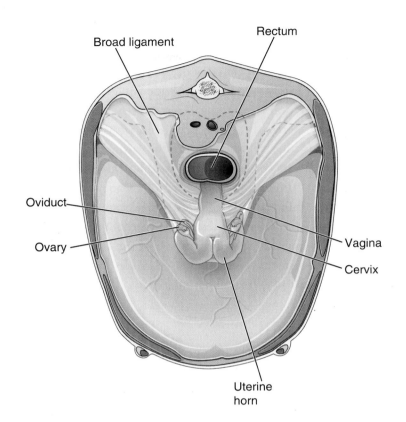

Broad ligament

Rectum

Oviduct

Ovary

Vagina

Cervix

Uterine horn

FIGURE 5-10

Transverse section of the abdominal region of the cow showing a cranial view of the reproductive tract suspended from the body wall by the broad ligament.

5

The female reproductive tract is suspended from the body wall by the **broad ligament,** a peritoneal fold that attaches the ovaries, oviducts, and uterus to the walls of the pelvis (Figure 5-10).

This concludes the general descriptions of the male and female reproductive tracts. The next sections include detailed accounts of the gross and microscopic anatomy of these organs.

For convenience, the reproductive tracts will be divided into the following major subdivisions, each of which is the focus of a separate discussion:

- gonads,
- genital ducts,
- suspension and vasculature,
- and external genitalia.

GONADS

The ovary and testis have homologous anatomic and physiologic features. With respect to function, both of these gonads have gametogenic and endocrine functions. Moreover, the organization of tissues within each of these organs is similar. Each is encapsulated by a white, fibrous tissue known as the tunica albuginea. The parenchyma of both organs can be divided into two major compartments; one concerned with the development of gametes and another consisting of richly vascularized endocrine cells. A major anatomic difference between the ovary and testis lies in the structure of these compartments. In the testis, spermatozoa are produced within a network of small tubules (seminiferous tubules) and exit the testes via an excurrent duct system. The ovary does not contain a duct system. Oogenesis occurs within pouchlike depressions (follicles). The exiting of an oocyte from the ovary involves rupture of the follicular wall. Another major anatomic difference between the testis and ovary is that the testis is located outside of the abdominal wall and is intimately associated with a portion of the external genitalia; that is, the scrotum. In contrast the ovary remains in the abdomino-pelvic cavity and is not located near the external genitalia.

Testis and Scrotum

The testes are paired, ellipsoidal organs located in the scrotum. The position of the testes varies among species, ranging between a pendulous scrotum located in the caudal portion of the abdomen (e.g., ruminants) to more restrained scrotum nestled beneath the anus (e.g., pigs and cats). The orientation of the testes tends to vary with location. In ruminants, the long axis is positioned vertically, whereas the long axis lies more in the horizontal plane in the dog, cat, pig, and horse.

Gross Anatomy of the Testis

The testis is intimately associated with the epididymis (Figures 5-11 and 5-12), which provides reference points for identifying opposing ends of the testis. The end associated with the point of origin (or head) of the epididymis is known as the extremitas capitata, whereas the part of the testis associated with the tail of the epididymis is referred to as the extremitas caudate.

The testis and epididymis are suspended in the scrotum by the spermatic cord (Figures 5-12 and 5-13). The foundation for this structure is a double layer of peritoneum known as the **tunica vaginalis** (see Figures 5-7 and 5-8). As noted earlier, this connective tissue supports the ductus deferens, as well as lymphatics, nerves, and blood vessels. The tunica vaginalis also covers the epididymis and testis. This sheath consists of inner (visceral) and outer (parietal) layers separated by a thin microscopic (vaginal) space (Figure 5-14).

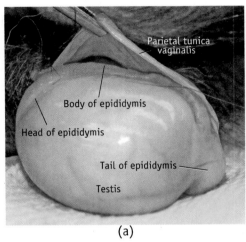 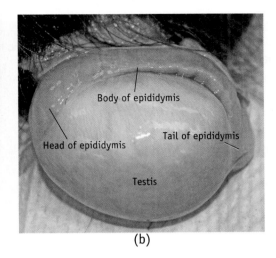

(a)   (b)

FIGURE 5-11

Photographs of a dog's testis excised from the scrotum showing the relationship between the testis and the epididymis. Both structures are encased by two layers of peritoneum known as the tunica vaginalis. In (a) the outer (parietal) layer of this tissue has been retracted, leaving the thin inner (visceral) layer clinging to the organs. A thin space (vaginal space) lies between these layers. (b) shows the testis and epididymis encased by the visceral layer of the tunica vaginalis.

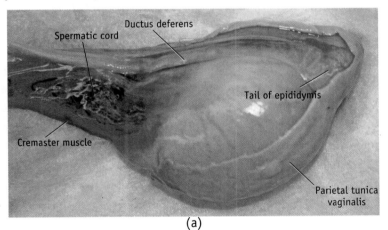

(a)

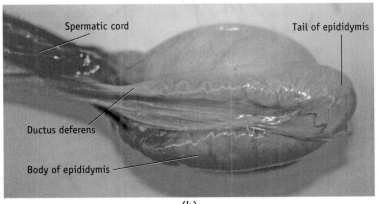

(b)

FIGURE 5-12

Two perspectives of a dog's testis completely excised from the scrotum showing the epididymis, ductus deferens, and spermatic cord. Both the inner and outer layers of the tunica vaginalis can be seen in (a). In both perspectives, the convoluted epididymal duct can be seen through the translucent tunica vaginalis. Note the dense network of blood vessels in the spermatic cord and the cremaster muscle that runs along the length of the cord.

FIGURE 5-13

Schematic drawing of the spermatic cord that houses blood vessels, nerves, and lymphatics as well as the ductus deferens and the cremaster muscle, an extension of the abdominal oblique muscles.

5

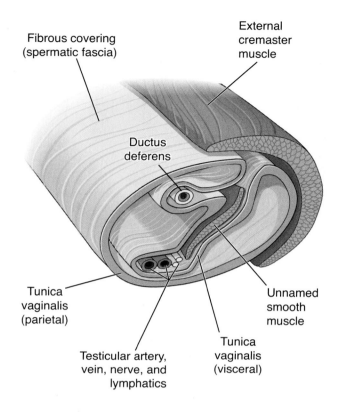

Fibrous covering (spermatic fascia)

External cremaster muscle

Ductus deferens

Tunica vaginalis (parietal)

Unnamed smooth muscle

Testicular artery, vein, nerve, and lymphatics

Tunica vaginalis (visceral)

Beneath the inner layer of the vaginal tunic lies the tunica albuginea, a thick capsule that supports branches of the testicular arteries and veins (Figure 5-14). This testicular capsule is not an elastic tissue, and therefore exerts pressure on the underlying parenchymal tissue. This becomes evident when one cuts into this tissue. A small cut through the tunica albuginea causes the underlying tissue to evert through the opening. Projections of the tunica albuginea extend to the interior of the testis. These septa divide the organ into numerous lobules and provide pathways for blood vessels and nerves to enter and leave the lobules. The septa meet at the center of the testis to form the **mediastinum,** a mass of connective tissue that houses fine tubules. Each lobule contains a small number of seminiferous ("semen-carrying") tubules, which are tortuous **(tubulus contortus)** in the peripheral portion of the lobule. The tubules unite and straighten **(tubulus rectus)** near the mediastinum, and then drain into **rete tubules,** which are embedded in the mediastinum. The rete tubules flow to the extremitas capitata where they penetrate the outer tunica albuginea and connect with the efferent ducts. These small-diameter ducts drain directly into the head of the epididymis. Collectively the system of ducts that transport sperm cells and seminal fluid out of the testes is referred to as the excurrent duct system.

FIGURE 5-14

Schematic drawings of sagittal sections of the testis showing tissue layers (a) and the elaborate system of ducts that transport sperm cells and seminal fluids out of and away from the testis (b).

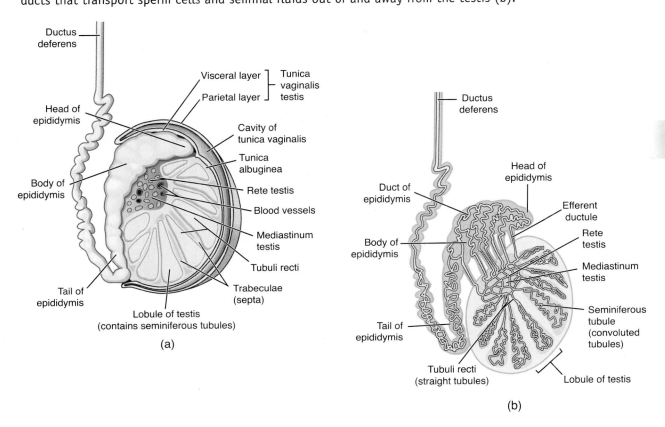

(a)

(b)

Microscopic Anatomy of the Testis

As noted in the previous section, the lobules of the testis are filled with tightly compacted bundles of seminiferous tubules (Figure 5-15a). The tubules are highly convoluted and densely packed within the lobules of the testis. A thin basement membrane forms the wall of the tubule and divides the testicular tissue into tubular and interstitial compartments. The large interstitial cells are called Leydig cells and produce testosterone. Several different types of cells reside in the tubular compartment. Large Sertoli cells extend from the basement membrane to the lumen of the tubules. These cells surround numerous layers of smaller cells, which are the developing germ cells; that is, cells that develop into sperm cells.

Closer examination of a section of seminiferous tubule reveals a more elaborate structure (Figure 5-15b). The walls of these microscopic ducts consists of connective tissue; that is, a thin basement membrane supported by

5

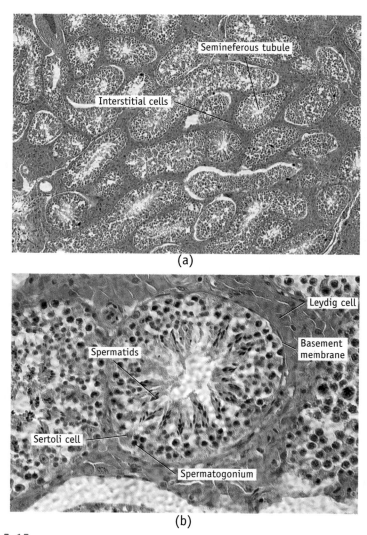

(a)

(b)

FIGURE 5-15

Micrographs of a section of a stallion's testis. The low-power view reveals the numerous seminiferous tubules and interstitial compartments (a). The high-power view (b) shows details of the structure of the tubules including the basement membrane, Leydig cells located in the interstitial compartment, and Sertoli cells located within the tubular compartment. Note the layers of developing germ cells and the elongated spermatids near the lumen of the tubule.

reticular fibers. A thick layer of cells lies along the inner surface of the basement membrane. These cells include spermatogenic germ cells, which are embedded in much larger Sertoli cells. The Sertoli cells are tightly aligned around the tubule, forming a barrier between the outside and inside of the tubule, and creating a narrow lumen. The spermatogenic cells exist in stratified layers engulfed by the Sertoli cells. Each layer contains cells in a particular stage of stage spermatogenesis. Spermatogonia occupy the outermost layers, whereas **spermatids** occupy the innermost layers. **Spermatocytes** reside in intermediate layers. The main function of Sertoli cells is to provide support for the spermatogenic cells. The connective tissue between the seminiferous tubules contains blood vessels, lymphatic vessels, and Leydig cells. The Leydig cells produce androgens that either enter adjacent capillaries or diffuse across the basement membrane into the seminiferous tubules.

Scrotum

As noted earlier, the scrotum is intimately associated with the testes. In addition to housing the male gonads, it plays a role in regulating testicular function. The scrotum consists of several tissue layers. An outer layer of thin skin contains both sweat and sebaceous glands and, depending on the species, may be bare or covered with hair of varying densities. The inner layer of the scrotal skin is adhered to the **tunica dartos,** a fibrous layer of smooth muscle. The tunica dartos envelops each testis and extends between them to form a septum that isolates the testes in separate compartments. Within each testicular compartment, the inner surface of the tunica dartos borders on the tunica vaginalis which, as noted earlier, consists of the outer parietal layer and the inner visceral layer.

The two major functions of the scrotum are protecting and cooling the testes. The temperature in the scrotum is a few degrees cooler than that of the body cavity. This lower temperature appears to be required for adequate production of spermatozoa. In most mammals, production of sperm is markedly reduced at normal body temperatures. The scrotum provides two mechanisms for cooling the testes. First, the presence of sweat glands in the scrotal skin facilitates evaporative cooling. Second, the tunica dartos muscle relaxes during high ambient temperatures, moving the testes away from the warm body cavity. The most important mechanism regulating temperature of the testes involves the vasculature of the spermatic cord.

Ovary

The ovaries are paired organs with an ellipsoid shape and nodular surface (Figures 5-16, 5-17, and 5-18). Unlike the testes, they are located in the sub-lumbar region of the abdominal cavity caudal to the kidneys and their morphology changes depending on the reproductive state of the individual. Each ovary is suspended by the cranial part of the broad ligament. Blood vessels, lymphatics, and nerves run along this tissue and enter the ovary at the **hilus.**

FIGURE 5-16

Photographs of ovaries from a sow (a) and a cow (b) showing follicles that contain oocytes and a corpus luteum, which develops from a follicle after ovulation.

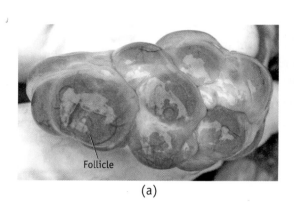

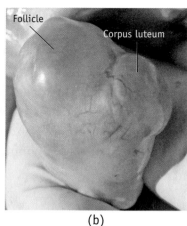

(a)   (b)

FIGURE 5-17

Photographs of several pig ovaries showing structures related to ovulation. Ovulation of a follicle results in formation of a blood clot known as a corpus hemorrhagicum (a), which develops into a corpus luteum (b), which eventually regresses leaving scar tissue known as a corpus albicans (c).

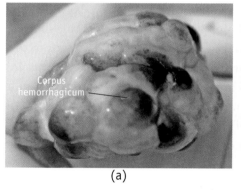

(a)   (b)

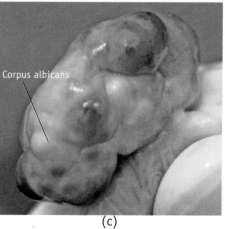

(c)

Cutting a longitudinal section through the ovary reveals the organization of its tissue layers (Figure 5-19). Like the testis, the ovary is encapsulated by the tunica albuginea. However, directly beneath this capsule is the **germinal epithelium,** a single layer of cuboidal cells that is absent from the testis. The underlying tissue of the ovary can be divided into two clearly distinguishable layers. The outer parenchymatous zone, commonly referred to as the **cortex,**

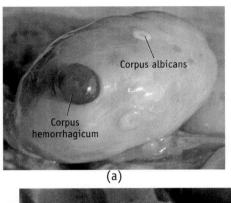

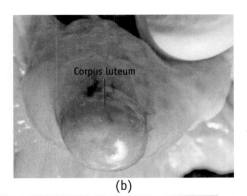

(a)

(b)

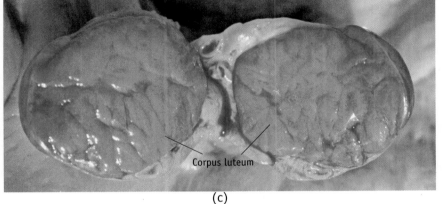

(c)

FIGURE 5-18

Photographs of cow ovaries showing a corpus hemorrhagicum (a), corpus albicans (a), and corpus luteum (b). Transecting the corpus luteum reveals its dense tissue that extends deep into the cortex of the ovary (c).

5

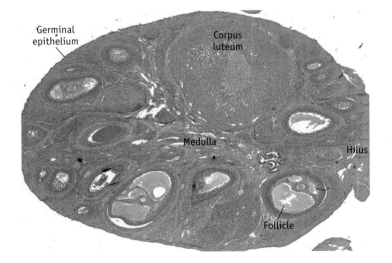

FIGURE 5-19

Micrograph of a longitudinal section of a hamster's ovary showing the hilus, medulla, and cortex, which contains several follicles and a corpus luteum. Note the germinal epithelium, a simple layer of cells that lies beneath the tunica albuginea and covers the surface of the ovary.

is a dense layer of tissue. The central **medulla** is a more loosely organized and vascularized region that is continuous with the hilus.

The ovarian cortex contains important structures that play important roles in the reproductive physiology of females. Ovarian follicles are one type of structure found in this region. At the microscopic level, five types of follicles can be distinguished (Figures 5-20 and 5-21): primordial, primary,

5

FIGURE 5-20

Micrographs of sections of a cow's ovary showing primordial (a), primary (b), secondary (c), and tertiary (d) follicles. Note that the oocyte develops a distinct zona pellucida by the time a follicle reaches the secondary phase. At this stage the oocyte develops numerous lipid droplets, which indicate an increase in its metabolic activity.

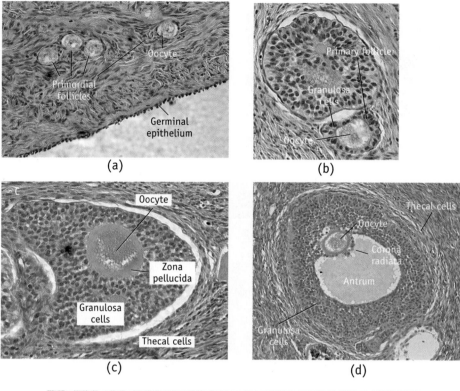

FIGURE 5-21

Micrograph of a follicle undergoing atresia. Note the reduction in number of granulosa cells, dissociation of follicular cells from the oocyte, and disruption of the basement membrane and thecal cells. The small, dense nuclei of the follicular cells indicate that the cells are dying.

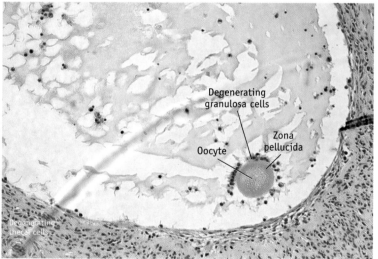

secondary, tertiary, and atretic. Only the tertiary follicles can be seen with the naked eye; the others are microscopic structures. **Primordial follicles** are the least developed type. They consist of an oocyte surrounded by a single layer of squamous cells. A primordial follicle can develop into a **primary follicle,** which can be distinguished by a surrounding layer of cuboidal (follicular) cells. If a primary follicle continues to develop, it forms a **secondary follicle;** that is, an oocyte surrounded by two or more layers of follicular cells. The oocyte of a secondary follicle develops an outer, translucent envelope called the **zona pellucida.** Very few follicles develop into **tertiary follicles.** This type of follicle is also called a "vesicular follicle" due to the presence of a fluid-filled cavity **(antrum).** These follicles appear as fluid-filled blisters and contribute to the nodular appearance of the ovarian cortex (Figure 5-16). Tertiary follicles are surrounded by a capsule of stromal tissue called the **theca.** This capsule consists of two concentric layers of cells: an outer theca externa and an inner theca interna. The inner layer is supplied with capillaries and contains endocrine cells. The theca interna cells rest upon a thin basement membrane. Beneath this membrane is the **granulosa layer;** that is, several layers of follicle cells, also known as granulosa cells. The rapidly expanding population of granulosa cells produces follicular fluid, which accumulates and pushes the granulosa cell to the outer limits of the follicular cavity. The oocyte remains centrally located and is supported by a mound of granulosa cells called the **cumulus oophorus.** The granulosa cells that surround the oocyte are typically referred to as the **corona radiata.** These cells penetrate the space between the oocyte and zona pellucida forming intimate contacts with the germ cells. Like the Sertoli cells of the testis, these cells are thought to provide support to the gametogenic cell.

Tertiary follicles have one of two fates. Most of these follicles undergo degeneration and are resorbed by the ovaries. This process is known as **atresia** (Figure 5-21). Atretic follicles are essentially dying follicles and can be identified microscopically by examining the appearance of follicular cells. A select few tertiary follicles go on to **ovulate;** that is, rupture and release an oocyte.

The other structures located in the ovarian cortex are associated with ovulation (Figures 5-17 and 5-18). Ovulation of a tertiary follicle involves rupture of the follicle wall and release of follicular contents. This leads to the collapse of follicular tissue and hemorrhaging of small blood vessels located within the follicular capsule. The remaining tissue resembles a small blood clot and is called a **corpus hemorrhagicum** ("bloody body"). The remaining granulosa and theca interna cells undergo transformations and re-organize to form a **corpus luteum,** or "yellow body." In most species, this structure is much larger than the tertiary follicle, and protrudes markedly from the surface of the ovary. This structure is highly vascularized and therefore has a red color on its surface. However, the core of the corpus luteum is yellow. As the corpus luteum degenerates, it forms fibrous scar tissue called the **corpus albicans,** or "white body."

It is important to point out that the combination of ovarian structures varies considerably with the physiologic state of the female. At any given time, the ovary contains follicles and/or corpora lutera at different stages of growth and degeneration. It is rare to see all of the aforementioned structures on a single ovary at one time. The pattern of growth of a particular follicle is best described as a **follicular wave;** that is, a period of enlargement followed by a rapid decrease in size due to atresia or ovulation.

GENITAL DUCTS

The basic characteristics of the genital ducts are similar to other tubular organs; that is, the parenchyma consists of concentric layers of tissue: tunica serosa (or tunica adventitia), tunica muscularis, tela submucosa, and mucosa. It is common to refer to these tissue layers simply as the serosa, muscularis, and mucosa. The degree to which these tissue layers are developed varies considerably within the reproductive tracts of males and females.

Genital Ducts of the Male

A system of ducts provides the means for transport of spermatozoa and seminal plasma within the male reproductive system. This duct system consists of the efferent ducts, epididymis, and ductus deferens. The urethra is also involved, but this duct is also part of the urinary system and is therefore considered separately.

Efferent Ducts

The efferent ducts connect the rete testis with the epididymis. There are between six and 20 of these ducts within each testis. The mucosal epithelium consists of columnar cells, some of which are ciliated. These cells exhibit both secretory and absorptive abilities. Beneath these cells is the lamina propria mucosae, which blends in with surrounding connective tissue. Isolated smooth muscle cells are also present.

Ductus Epididymidis

The ductus epididymidis (epididymis) consists of three major regions: head (caput), body (corpus), and tail (cauda). The head and tail of the epididymis are firmly attached to the testis. The attachment to the body may be loose, creating a space between the duct and the testis. The head is attached to the testicular capsule, whereas the tail is fixed to the proper ligament of the testis as well as to the parietal layer of peritoneum that covers it.

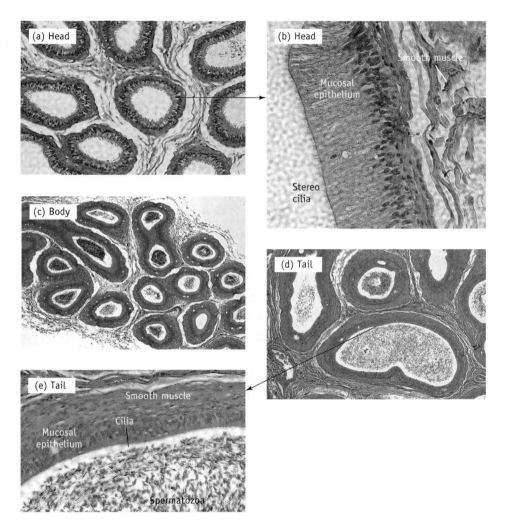

(a) Head

(b) Head

Smooth muscle

Mucosal epithelium

Stereo cilia

(c) Body

(d) Tail

(e) Tail

Smooth muscle

Cilia

Mucosal epithelium

Spermatozoa

FIGURE 5-22

Micrographs of sections through the head (a,b), body (c), and tail (d,e) of the epididymis from an American bison. Note the presence of stereo cilia in the head and cilia in the tail. Also note that the smooth muscle layer is more developed in the tail than in the head.

5

The histology of the epididymis varies from head to tail (Figure 5-22). Throughout the organ, tissue layers include a mucosa consisting of pseudostratified columnar cells, a thin and vascularized lamina propria mucosa, a lamina muscularis mucosa comprised of circular smooth muscle cells, and a submucosa which is continuous with the tunica albuginea. Distinguishing features of the head include a small lumen, very tall mucosal cells with non-motile cilia (stereo cilia), and a thin layer of smooth muscle cells. In contrast, the tail is characterized by a large lumen, more compact epithelial cells with motile cilia, and well-developed layers of smooth muscle cells. The body of the epididymis has characteristics that intermediate between those of the head and tail. The anatomic differences among the various regions of this duct reflect their different functions. The head of the epididymis is involved primarily with absorption of fluid from the testes, whereas the tail of the epididymis serves as a storage depot and is important in the transport of sperm during ejaculation. Spermatozoa emerge from the rete testis suspended in rete fluid. As they make their way through the efferent ducts and head of the epididymis, the fluid is absorbed by the mucosal epithelium, causing the concentration of spermatozoa to increase. Movement of spermatozoa through the head and body of the epididymis is facilitated by rhythmic contractions of the lamina muscularis mucosae. In contrast, the tail remains quiescent unless the male becomes sexually excited. The amount of time required for spermatozoa to move from the proximal head to the distal tail varies among species (4–19 days). Removal of spermatozoa from the tail of the epididymis occurs in two ways. The most familiar (and noticeable) involves ejaculation. During sexual stimulation, smooth muscle cells in the tail contract and push the epididymal contents into the ductus deferens. A second mechanism involves periodic contractions of the epididymal tail and ductus deferens, causing more sustained release.

In addition to transporting and storing spermatozoa, the epididymis plays an important role in the maturation of spermatozoa. When spermatozoa emerge from the testis and enter the head of the epididymis, they are not completely developed; they lack motility and cannot bind to an oocyte. As the cell moves through the body of the epididymis, it takes on a mature form, becomes motile, and gains the ability to bind to the female gamete.

Ductus Deferens

The ductus deferens is coiled where it emerges from the epididymis. It then straightens and runs medial to the epididymis before entering the spermatic cord near the head of the epididymis. After the duct deviates from the epididymis, it runs within the spermatic cord through the inguinal canal and finally joins the urethra distal to the bladder. Throughout most of its length, the ductus deferens has a uniform diameter with a narrow lumen and thick muscular wall (Figure 5-23). In most mammalian species, the terminal end of the duct is enlarged and spindle-shaped. This is due to the presence of mucosal glands, not widening of the lumen. This region is called the ampulla or ampullary gland.

The lamina epithelialis mucosa of the deferent duct is a pseudostratified columnar epithelium. The most striking feature of this duct is the thin submucosal layer and thick, well-organized tunica muscularis varying from two distinct inner and outer layers to intermingled fibers. Unlike the epididymis and efferent ducts, the ductus deferens has a well-defined tunica serosa.

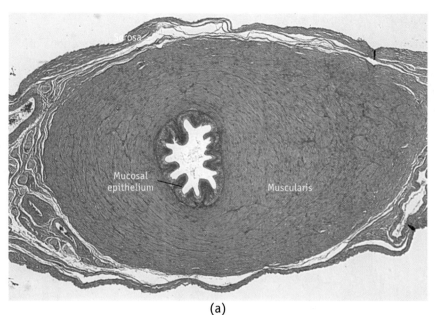

(a)

FIGURE 5-23

Micrographs of a section through the ductus deferens of an American bison at low (a) and high power (b). Note the thick, well-organized muscularis and the presence of mature sperm cells in the lumen.

5

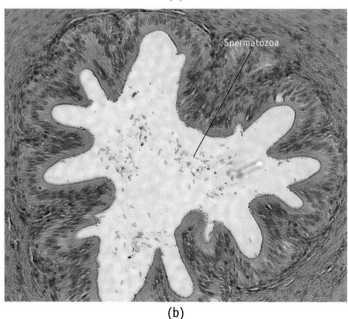

(b)

Urethra and Accessory Sex Glands

In the male, the urethra originates in the neck of the bladder and terminates in an orifice at the end of the penis. Major subdivisions include the internal (pelvic) and external (spongy) urethras. In this section we will consider only the internal urethra and its relationship with the accessory sex glands. It is more convenient to discuss the spongy urethra in relation to the anatomy of the penis.

FIGURE 5-24

Illustrations of the pelvic reproductive organs of the stallion, bull, boar, and dog showing locations of accessory sex glands.

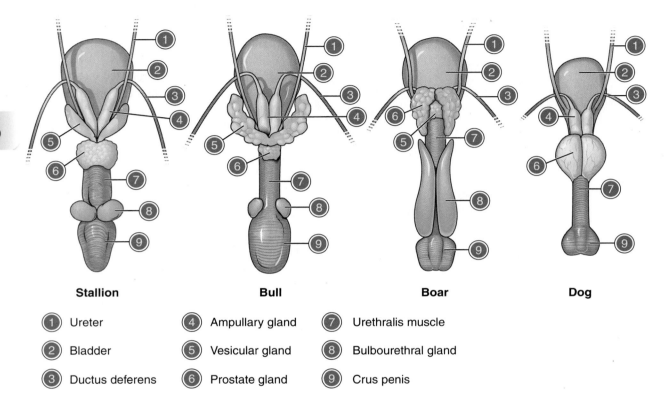

Stallion	**Bull**	**Boar**	**Dog**

① Ureter	④ Ampullary gland	⑦ Urethralis muscle
② Bladder	⑤ Vesicular gland	⑧ Bulbourethral gland
③ Ductus deferens	⑥ Prostate gland	⑨ Crus penis

The pelvic urethra is a tube lined with transitional epithelium that overlays a lamina propria-submucosa that contains glands and some erectile tissue. Erectile tissue in the pelvic urethra is inconspicuous; that is, a minute system of interconnected vascular spaces. This is continuous with the more extensive erectile tissue of the spongy urethra. The structure of the tunica muscularis varies along the length of the duct, ranging between three layers near the bladder and a striated urethral muscle in more distal sections, where it becomes continuous with the bulbospongiosus muscle. The outer covering of the urethra is the tunica adventitia.

The vesicular glands exist in pairs. Each one is associated with the distal portion of its ipsilateral deferent duct. In the horse and human, the glands are smooth and have an appearance similar to the bladder. In contrast, these glands are knobby and have thick walls in most other species. Each seminal vesicle is divided into lobules, each containing numerous pockets that drain into a central duct. The lumens of the pockets are lined with a glandular epithelium arranged in a simple columnar fashion. The central duct is lined

TABLE 5-1 Accessory sex glands present in several species of domestic mammals

Species	Ampullary	Vesicular	Prostate	Bulbourethral
Dogs	Yes	No	Yes	No
Cats	Yes	No	Yes	Yes (vestigial)
Horses	Yes	Yes	Yes	Yes
Cattle	Yes	Yes	Yes	Yes
Sheep	Yes	Yes	Yes	Yes
Swine	Yes	Yes	Yes	Yes

5

with pseudostratified epithelium. In each case a lamina propria-submucosa is present along with a muscularis and serosa.

The **prostate** can exist in several forms that reflect the relative development of the two main parts of the gland: corpus prostate and disseminate prostate. In small ruminants such as sheep, the gland is disseminate; that is, glandular tissue is distributed along the dorsal and lateral surfaces of the pelvic urethra. In carnivores and horses the compact corpus is the dominant portion. The boar has a well-developed disseminate prostate, but the corpus is clearly present. Both portions of the prostate are present in the bull. In all cases, the corpus prostate is located peripheral to the urethralis muscle of the pelvic urethra. Both parts of the prostate consist of many small ducts that drain into the urethra.

A pair of **bulbourethral glands** lies dorsolateral to the urethra near the base of the penis. These glands are of a compound, tubuloalveolar type, meaning that they consist of small sacs (alveoli) each of which is drained by a small duct. The entire gland drains into the urethra via one or more main ducts. Both structures are lined by epithelial cells of varying arrangements.

Genital Ducts of the Female

The genital ducts of the female include the vagina, uterus, and oviducts. As noted earlier, the uterus consists of two main parts: a caudal section that serves as a valve (cervix) and a cranial portion that is the site of pregnancy. In humans, the oviducts are frequently referred to as the fallopian tubes. It is important to emphasize that the vagina is distinct from the vestibule, which is part of the vulva, or external genitalia.

Uterus

The uterus plays a central role in mammalian reproduction by serving several essential functions. First, it is the site of semen deposition in some

FIGURE 5-25

Schematic illustrations of the major types of uteri found in mammals.

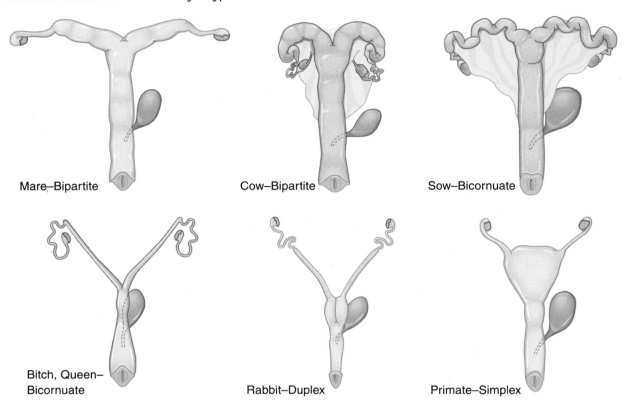

Mare–Bipartite

Cow–Bipartite

Sow–Bicornuate

Bitch, Queen–
Bicornuate

Rabbit–Duplex

Primate–Simplex

mammals (horses, swine). Second, it is important in transporting spermatozoa following insemination. Third, the uterus (and oviducts) capacitate spermatozoa; that is, induce changes that allow spermatozoa to accomplish fertilization of oocytes. Fourth, the uterus is the site of pregnancy; that is, development of the embryo and fetus. Fourth, uterine contractions are essential for the birthing process.

The gross structure of the uterus varies among species (Figure 5-25). Most domestic animals have a **bicornuate** uterus, consisting of two horns, a body, and a single cervix. However, the relative sizes of the horns and body can vary considerably. Some anatomists differentiate between a bicornuate uterus, that has two long horns and no body, and a **bipartite** uterus, that has two horns and a distinct body. In addition, there are differences in the degree of fusion of the paramesonephric ducts during development. Primates have a **simple** uterus, consisting of a large body, two very small horns, and a single cervix. Lagomorphs (e.g., rabbits), monotremes, and marsupials have a **duplex** uterus that is characterized by two separate cervices and uterine compartments.

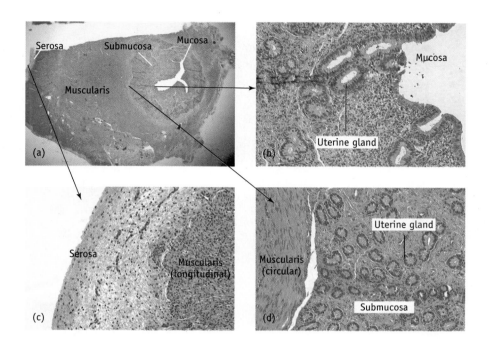

FIGURE 5-26

Micrographs of sections of the uterine horn from a domestic cow showing major tissue layers. Low-power view shows all of the major tissue layers including the mucosa, submucosa, muscularis, and serosa (a). Higher magnification shows the mucosal epithelium that lines glands, which penetrate into the submucosa (b), circular, and longitudinal layers of smooth muscle (c and d).

The uterus is a highly organized organ consisting of all the major tissue layers characteristic of tubular organs (Figures 5-26). In the uterus, these tissues are given special names.

- Tunica mucosa = **endometrium**
- Tunica muscularis = **myometrium**
- Tunica serosa = **perimetrium**

The endometrium contains a simple columnar epithelium, but the height of these cells changes depending on the reproductive state of the individual. Sections of this mucosal epithelium invaginate deep into the lamina propria-submucosa to form **uterine glands,** which secrete mucus, lipids, proteins, and glycogen into the lumen of the uterus. The distal ends of these glands may be highly coiled in some species. In ruminants, certain highly vascularized regions of the submucosa are devoid of uterine glands. These are referred to as **caruncles,** and are points of contact between the maternal and fetal tissue that make up the placenta (Figure 5-27)

The myometrium consists of a thick inner layer of circular smooth muscle, and a thin outer layer of longitudinal smooth muscle. A layer of blood vessels lies between the smooth muscle layers.

The structure of the cervix is considerably different from that of the uterine body and horns. This region serves as a valve between the vagina and body of the uterus. The lumen **(cervical canal)** is narrow and constricted (Figure 5-28).

5

FIGURE 5-27

Photograph of the uterus of a domestic cow. An incision was made in the uterine horn to expose the lumen, which includes caruncles and inter-caruncular tissue. The incised tissue also reveals the rich vas-cularization of the endometrium as well as the myometrium.

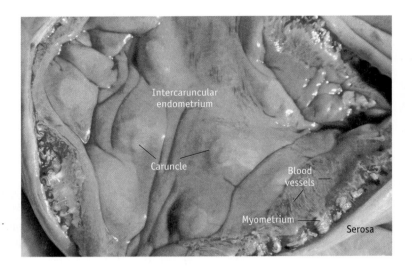

FIGURE 5-28

Photograph of the mouth (os) of the cervix of a domestic cow as viewed from the vagina. Note the small diameter of the opening.

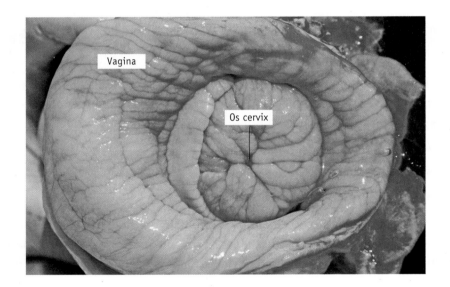

The shape of the cervical canal varies among species. In all cases, the mucosal epithelium is highly convoluted, creating folds that run longitudinally along the length of the cervix. In addition, circumferential folds form several annular rings that extend into the lumen of the cervix, which create a tortuous canal. In the sow the pattern is complementary to the corkscrew shape of the boar's penis, which becomes embedded in the cervix during coitus.

Microscopically, the cervix resembles the other portions of the genital ducts, but with some notable exceptions (Figure 5-29). The mucosal epithe-lium of the cervix ranges from stratified squamous in the bitch to columnar cells in the cow and ewe. In both cases, the mucosa contains goblet cells

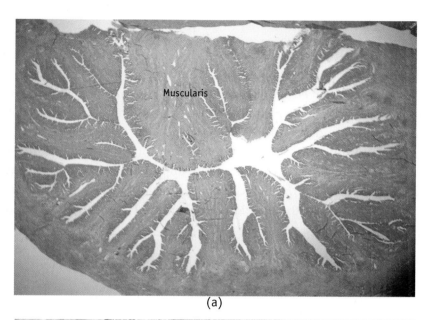

Muscularis

(a)

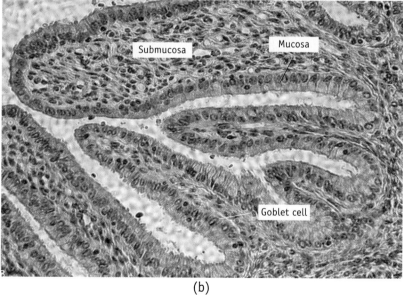

Submucosa

Mucosa

Goblet cell

(b)

FIGURE 5-29

Low- (a) and high- (b) power micrographs of the cervix of a cow. Note the narrow and constricted lumen lined by columnar epithelial cells, which include mucus-producing goblet cells (b).

5

that secrete mucus. The secretory activity of these cells varies with physiologic status. During estrus, the cells secrete thin, clear mucus. In contrast, the cervix produces thick mucus that forms a cervical seal during pregnancy. We will consider the physiologic significance of the cervical mucus in a later chapter. The submucosa contains both loose and dense collagenous tissue and the tunica muscularis is well-developed with extensive fibrous fibers. These characteristics make this region of the uterus feel like a rope or a chicken neck.

Oviducts

The oviducts are narrow extensions of the uterine horns and can be divided into three segments: **infundibulum, ampulla,** and **isthmus** (Figures 5-30, 5-31, and 5-32). The cranial end of the oviduct has a wide opening near the ovary, but then tapers to a smaller diameter distal to the ovary. This region of the oviduct is the infundibulum ("funnel"). The edge of the infundibulum is fringed with irregular processes **(fimbriae),** some of which attach to the ovary. A small orifice, located at the center of the infundibulum, opens to the tubular portion of the oviduct, which can be divided into two segments of similar lengths. In many species these regions have different diameters. The proximal

FIGURE 5-30

Photograph of an excised reproductive tract from a cow showing the relationship between the oviduct and ovary. Note that the diameter of the oviduct decreases as it courses away from the ovary toward the uterus.

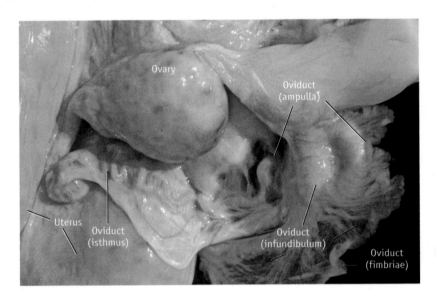

FIGURE 5-31

Low- (a) and high- (b) power micrographs of the infundibular segment of the oviduct of a sow. Note the presence of cilia protruding from the mucosal epithelium. The bleb-like appearance of the cilia is an artifact resulting from the processing of tissue. This segment lacks a prominent muscularis (b).

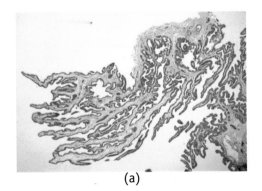

(a)

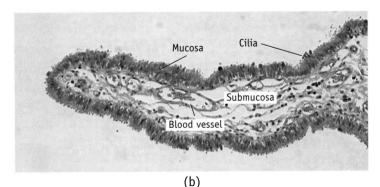

(b)

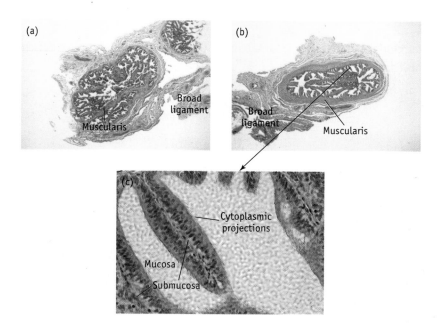

FIGURE 5-32

Micrographs of the ampullary (a) and isthmic (b,c) segments of the oviduct from a domestic cow. Note the presence of a well developed muscularis in both locations. The high-power view (c) of the isthmus reveals cytoplasmic projections emanating from mucosal cells.

5

ampulla is wider than the distal isthmus. The isthmus connects with the apex of the uterine horn forming the **uterotubal** junction. Although this region varies in appearance among species (from gradual to abrupt transitions), in all cases it acts as a physical barrier to impede movement of ascending spermatozoa and descending oocytes or embryos.

The lumen of the oviduct is tortuous due to numerous folds in the mucosa. Major folds run longitudinally along the length of the tube. Smaller secondary and tertiary folds create small clefts that branch from the main canal. The degree of convolution is highest in the ampulla and lowest in the isthmus. The mucosal epithelium of the oviduct consists of columnar cells, some of which have motile cilia. Ciliated cells are most prominent in the cranial region and facilitate movement of the oocyte. Nonciliated cells secrete various substances that nourish the oocyte and capacitate spermatozoa. Unlike the uterus, submucosa does not contain glands. The tunica muscularis is more developed in the isthmus than in the ampulla, and consists of the familiar longitudinal and circular layers. A tunica serosa is present and contains a distinct vascular layer.

Vagina

As noted earlier, the vagina consists of two major parts. The cranial portion (between the uterine cervix and urethral orifice) is primarily a reproductive passage, whereas the caudal portion serves both reproductive and urinary functions. Overall, the vagina is a thin-walled tubular organ that is situated in the medial pelvic cavity ventral to the rectum and dorsal to the bladder and

FIGURE 5-33

Low- (a) and high- (b) power micrographs of the vagina from a sow. Note the stratified epithelium of the mucosa and the well-organized muscularis.

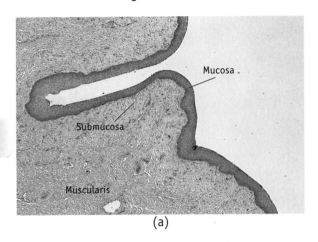
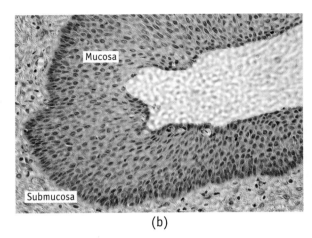

(a)   (b)

urethra. With the exception of the cranial portion, the organ lies in a retroperitoneal position.

The mucosa of the vagina is lined with an epithelial layer that varies in structure. Columnar epithelial cells are characteristic of the cranial vagina, whereas the caudal vagina is lined with stratified squamous epithelial cells (Figure 5-33). The thick layer of epithelial tissue in the mucosa of the caudal vagina helps this portion of the vagina withstand trauma associated with coitus. Mucosal glands are confined to the cranial region. The cervix protrudes into the cranial vagina reducing its lumen and forming a ring-like space; that is, the **fornix.** The muscularis of the vagina is similar to that of the uterus. A transverse fold of mucosal tissue at the junction of the vagina and vestibule **(hymen)** may exist in virgin females (especially in the gilt and filly). This usually breaks down, but is sometimes intact until first coitus.

SUSPENSION AND VASCULATURE OF THE INTERNAL GENITAL ORGANS

The reproductive organs located in the abdomino-pelvic cavity are suspended by the **genital fold,** a ligament formed by the peritoneum. In females the genital fold is called the **broad ligament** (Figures 5-34). The cranial portion supports the ovary and is called the **mesovarium.** The oviducts are supported by the middle section known as the **mesosalpinx** (referring to the "trumpet-shaped" oviduct). The **mesometrium** is the caudal end of the broad ligament and it supports the uterus. The mesovarium and mesosalpinx form a pouch around the ovary; that is, the **ovarian bursa** (Figure 5-35). The relationship

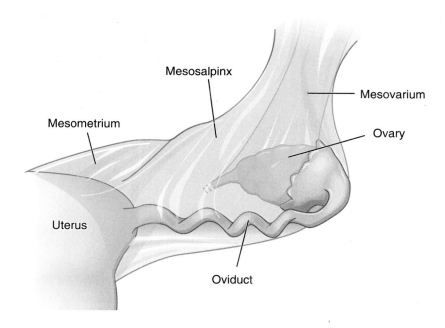

FIGURE 5-34

Lateral view of the broad ligament suspending the uterus, oviduct, and ovary.

Mesosalpinx

Mesovarium

Mesometrium

Ovary

Uterus

Oviduct

5

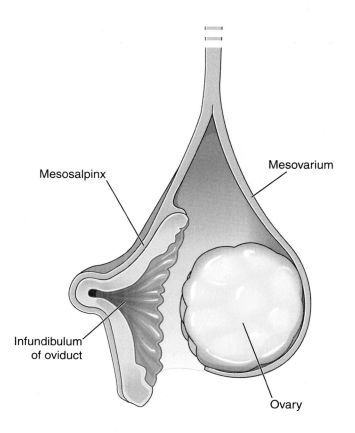

FIGURE 5-35

Caudal view of the broad ligament forming the ovarian bursa in a cow.

Mesosalpinx

Mesovarium

Infundibulum of oviduct

Ovary

FIGURE 5-36

Photograph of the well-developed ovarian bursa in the bitch. The broad ligament forms a purse-like structure that completely envelopes the ovary (a). Retraction of the bursa reveals the ovary (b).

5

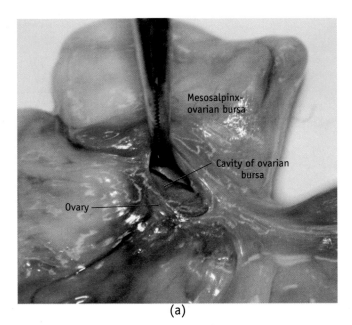

(a)

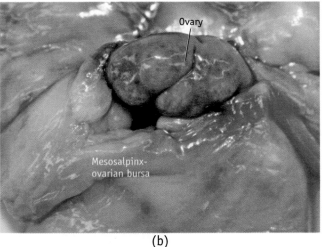

(b)

between the ovary and its bursa ranges from a shallow sac that does not contain the ovary (e.g., the mare) to a deep enclosure that completely surrounds the ovary (e.g., the bitch; Figure 5-36). In rodents, the ovary is completely trapped by the bursa leaving no contact between the cavity and the surrounding peritoneum.

The blood vessels that supply and drain the female reproductive tract (Figure 5-37) are supported by the broad ligament. A branch of the abdominal aorta known as the **ovarian artery** supplies the ovary, oviducts, and the cranial uterine horn. This blood vessel is highly convoluted and intertwines

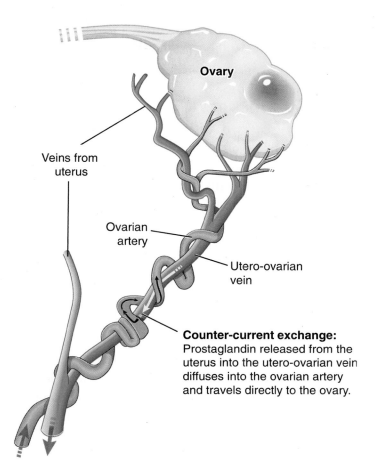

Ovary

Veins from
uterus

Ovarian
artery

Utero-ovarian
vein

Counter-current exchange:
Prostaglandin released from the
uterus into the utero-ovarian vein
diffuses into the ovarian artery
and travels directly to the ovary.

FIGURE 5-37

Schematic illustration
of the vasculature of
the ovary. Note the
close relationship
between the ovarian
artery, supplying blood
to the ovary, and the
utero-ovarian vein,
draining the uterus and
ovary. Contact between
the vessels allows
a counter-current
exchange mechanism
whereby $PGF_{2\alpha}$, released
into the uterine vein by
the uterus, can enter
the ovarian artery and
travel directly to the
ovary.

5

with the ovarian vein. Blood flows to the body and horns of the uterus via the
uterine artery, which branches from the internal iliac artery. Blood is supplied to caudal portion of the tract via the vaginal artery. Branches of the
uterine artery run cranially and caudally and form anastomoses with the
uterine branch of the ovarian artery as well as with the vaginal artery. Smaller
arteries branch from these main branches forming an arterial arcade that
supplies blood along the entire length of the reproductive tract. Blood drains
from the ovary, uterine horn, and uterine body via the **ovarian vein** (also
called the **utero-ovarian vein). The accessory vaginal vein** and the **vaginal
vein** drain the caudal uterine body and vagina. The ovarian vein is plexiform
(made up of a web of interconnected vessels) near the ovary, oviducts, and
uterine horn. These smaller vessels converge on a larger vessel that runs laterally to the uterine body.

The ovarian artery is intimately associated with the utero-ovarian **vein**
(Figure 5-37). The convoluted artery wraps around and makes numerous

contacts with the utero-ovarian vein. This anatomic relationship has tremendous physiologic significance. The close proximity of the uterine artery to the utero-ovarian vein permits the transfer of an important uterine hormone from the utero-ovarian vein to the ovary artery via a **counter-current exchange mechanism. Prostaglandin F$_{2\alpha}$ (PGF$_{2\alpha}$),** present in high concentrations in the blood of the utero-ovarian vein, is transported via diffusion to the ovarian artery, which has little or no prostaglandin. In this way, high concentrations of this hormone are shunted directly from the uterus to the ovary.

As with the female, ligaments support the reproductive tract in males. The structure that is analogous to the broad ligament in females is much smaller in males and supports only the ampullae of the ductus deferens and the vesicular glands. The testis, epididymis, and most of the ductus deferens are enveloped by the vaginal tunic, an evagination of abdominal peritoneum that extends through the inguinal canal into the scrotum. The proximal portion is narrow and surrounds blood vessels, lymphatics, nerves, and the ductus deferens. As noted earlier, these tissues are known collectively as the spermatic cord.

The dominant feature of vasculature of the testis is the spermatic cord, which supports the **testicular artery** surrounded by a network of veins (Figure 5-38). As noted earlier, these structures are encased in an inward fold of the vaginal tunic. The outer layer is the parietal layer of the vaginal tunic,

FIGURE 5-38

Schematic illustration of a small section of the spermatic cord showing the pampiniform plexus, a network of testicular veins within which the testicular artery is enmeshed. This provides the basis for a counter-current mechanism that cools the testis.

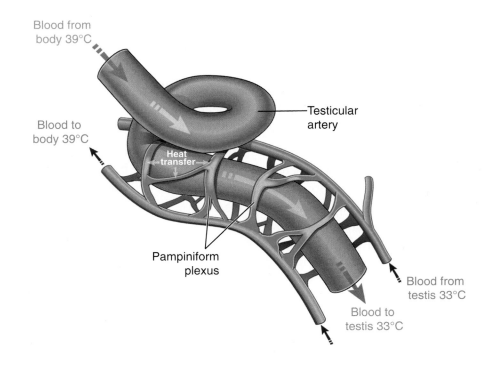

whereas the inner layer is the visceral layer. The vaginal cavity lies between these layers. This arrangement extends from the **vaginal ring** to the tail of the epididymis.

The arrangement of blood vessels in the spermatic cord is remarkable and provides the primary means for thermoregulation of the testis. Each testicular artery branches from the abdominal aorta, courses directly to the vaginal ring, and then follows the spermatic cord to the testis. In the spermatic cord, the artery is highly convoluted (7 m of artery compacted into 10 cm of spermatic cord) in a way that is similar to that of the ovarian artery within the broad ligament in the female. The contortions of the artery are embedded in a mesh of veins called the **pampiniform plexus** (Figures 5-38), the structure of which is homologous to the ovarian artery/utero-ovarian vein complex in the female. The testicular veins eventually converge and flow into the vena cava. Arteriovenous anastomoses exist between the testicular artery and surrounding veins permitting the direct shunting of blood between vessels. This intimate association facilitates heat exchange between the blood vessels. Heat from the artery is lost to the veins, which are cooled in the suspended testis. Thus blood flowing to the testis falls below body temperature creating a temperature gradient along the testis; that is, the temperature of the dorsal portion (near the spermatic cord) can be as much as 6°C warmer than the temperature of the ventral portion. In addition to cooling the testis, the pampiniform plexus allows testosterone, secreted into the arterial circulation by the testes, to be shunted back to the testes via the testicular veins. This contributes to the high concentrations of testosterone normally found in the testes. The physiologic significance of this mechanism lies in the fact that testosterone is necessary for spermatogenesis.

EXTERNAL GENITALIA

The external genitalia of males and females develop from the same embryonic tissues. Therefore, it should come as no surprise that the external sex organs of adult males and females consist of homologous structures. The penis and clitoris arise from the genital tubercle, whereas the scrotum and vulva arise from the genital folds (see Chapter 4). Although the external genitalia of males and females are anatomically related, their functions are less comparable. For example, the penis is the copulatory organ in males, whereas the clitoris is not necessary for the female to copulate. Likewise, the scrotum and vulva have different functions. The scrotum is intimately associated with the testis and influences its activity. In contrast, the vulva has nothing to do with ovarian functions.

Vestibule and Vulva

The external genitalia of the female include the **vestibule** and **vulva.** The vestibule lies caudal to the ischial arch (the ventral and caudal portion of the hip bone) and slopes downward to the vulva. The urethra opens at the floor of the vestibule in most species. However, in some cases (e.g., the cow) the urethra opens into a small sac called the **suburethral diverticulum.** Familiarity with this anatomy is particularly useful when performing procedures such as artificial insemination. For example, it is common for novice inseminators to place the tip of an inseminating gun into the suburethral diverticulum, instead of the vagina. A pair of **vestibular glands** empties into the vestibule lateral and ventral to the urethral opening. These glands produce mucus that helps lubricate the passage during coitus and parturition. In addition, these secretions contain aromatic compounds **(pheromones)** that arouse the sexual interest of males during mating. The vestibule is a highly vascularized tissue. In particular, a lateral **vestibular bulb,** consisting of a network of veins, forms a patch of erectile tissue that is homologous to the bulb of the penis (see subsequent discussion).

The walls of the vestibule are continuous with the vulva, a vertical opening below the anus (Figure 5-39). The labia, lateral folds of tissue on either side of the vulva, meet at the dorsal and ventral commissures. In humans, the labia can be divided into the inner labia minora and the outer labia majora. The labia minora is not distinct in domestic mammals. The clitoris lies within

FIGURE 5-39

Photograph of the vulva of a domestic cow. The labia was lightly retracted to reveal the clitoris and vestibule.

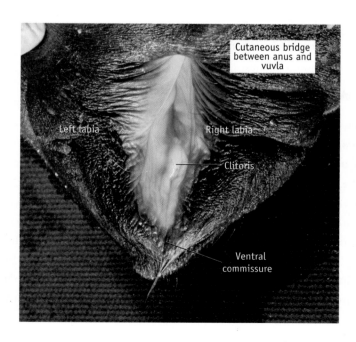

the ventral commissure and consists of a body and glans. It is homologous to the male's penis and contains spongy erectile tissue analogous to that found in the male phallus.

Penis and Prepuce

The scrotum, penis, and prepuce make up the external genitalia of the male. The anatomy of the scrotum was considered in an earlier section dealing with the gross anatomy of the testis. The penis is located between the thighs, and is suspended below the trunk of the body. In the larger species (bull, boar, ram, and stallion) the penis is anchored to the floor of the pelvis by a ligament. In mammals other than primates, the penis is concealed by an invagination of abdominal skin called the **prepuce.** The penis is made up of three independent columns of erectile tissue (Figure 5-40). The cura of the penis **(crus penis)** consists of a pair of columns situated dorsally. At their site of origin, near the ischial arch, they are separated. A short distance distal to this location, the columns converge, bend cranially, and then run along the

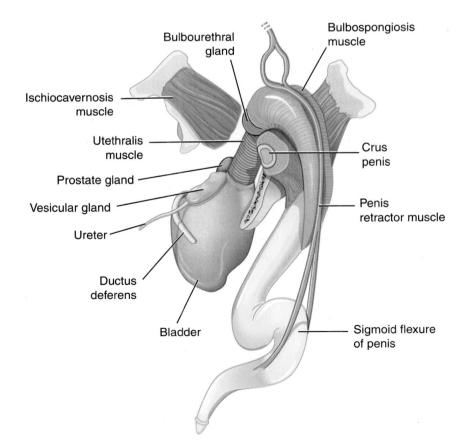

Bulbourethral gland

Bulbospongiosis muscle

Ischiocavernosis muscle

Utethralis muscle

Prostate gland

Vesicular gland

Ureter

Ductus deferens

Bladder

Crus penis

Penis retractor muscle

Sigmoid flexure of penis

FIGURE 5-40

Schematic illustration of the major structural components of the bull's penis.

FIGURE 5-41

Transverse section through the penises of a bull and stallion showing erectile tissue.

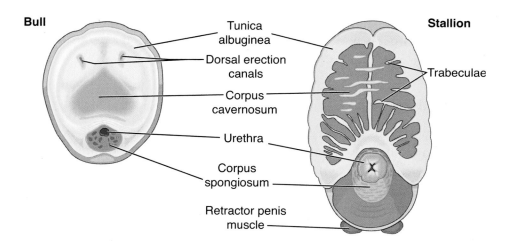

Bull   Tunica albuginea   Stallion

Dorsal erection canals

Trabeculae

Corpus cavernosum

Urethra

Corpus spongiosum

Retractor penis muscle

5

floor of the pelvis, where they unite. Each of these columns contains a core of cavernous tissue (**corpus cavernosum**) covered by the tunica albuginea (Figure 5-41). A separate erectile compartment (**corpus spongiosum**) lies beneath the corpus cavernosum and surrounds the urethra. A groove of connective tissue separates these two vascular compartments. The origin of the corpus spongiosum lies at the base of the penis and is characterized by a bi-lobed enlargement called the **bulb of the penis.** It continues along the shaft of the penis and expands over the distal end, where it is called the **glans penis.** The corpus cavernosum does not extend to the tip of the penis. During sexual arousal, each of these cavernous structures fills with blood to cause erection of the penis.

There are several variations in penis structure among mammals. In the dog and cat (and some other species of mammals), the distal corpus cavernosum becomes ossified and exists as a boney tissue called the **os penis** (bacculum). The shape of the glans also varies considerably. Perhaps the most extraordinary structure is the filiform appendage of the ram's and buck's penis. Another major anatomic difference deals with the type of tissue found in the corpus cavernosum. In swine and ruminants, this tissue contains strands of fibroelastic tissues that divide the space into numerous small compartments (**sinusoids**). This so-called **fibroelastic penis** exists in a semi-erect state and requires little blood to achieve full erection. In contrast, the corpus cavernosum of the stallion, dog, and primate contains much larger sinusoidal compartments separated by septa that are more muscular and less elastic than those of the fibroelastic penis; that is, a **musculocavernous penis.** The shapes of these two types of penis differ considerably. The fibroelastic penis has a **sigmoid flexure;** that is, the shape is bent in the shape of an S. This is maintained by a pair of

retractor penis muscles that run bilaterally from the coccygeal vertebrae to the ventrolateral sides of the penis. These are made up of smooth muscle fibers that keep the penis retracted most of the time. However, upon sexual arousal, they relax and allow the penis to protrude from the prepuce.

In both types of penis, erection results from the blood flowing into the sinusoids of the corpora cavernosum and spongiosum (see Figure 5-41). The artery of the penis supplies all of the blood flowing into these tissues. Before entering the penis, the artery splits into three branches: one enters the bulb of the penis to supply the sinusoids of the corpus spongiosum; another other passes through the tunica albuginea to supply the corpus cavernosum; a third runs dorsally along the shaft to supply the apex of the penis.

SUMMARY OF MAJOR CONCEPTS

- The genital organs of the male and female include the gonad, genital ducts, and external genitalia.
- The testis and ovary are similar in that they both consist of gametogenic and endocrine tissues. They differ with respect to location, structure, and how the developing gametes are housed and released from them.
- The genital ducts of the male and female are tubular organs consisting of concentric layers of tissue. The number of tissue layers and the organization of cells within each layer vary among the organs of the reproductive tracts.
- The genital ducts of the male are primarily concerned with the transport of spermatozoa, whereas the female genital ducts are involved with the transport of male and female gametes, as well as transport and nourishment of the conceptus.
- The reproductive tracts of the male and female are suspended by evaginations of peritoneum, which also support afferent and efferent blood vessels and nerves. In each sex, the main artery providing blood to the gonad is intimately associated with a venous plexus, which permits a counter-current exchange of heat and/or hormones.
- The external genitalia of males are homologous to that of females, but there are marked anatomic and physiologic differences between the external genitalia of the two sexes.

DISCUSSION

1. Discuss the major similarities and differences between the reproductive tracts of male and female mammals. Restrict your discussion to the major components and the arrangement of these components.

2. Compare and contrast the histologies of a section of seminiferous tubule and a tertiary ovarian follicle. Pay particular attention to the basement membrane and cell types.

3. Compare and contrast the arrangement of tissue layers in the ductus deferens and the oviduct.

4. Describe the counter-current vascular systems found in the male and female reproductive tracts. Explain the significance of each one.

REFERENCES

Banks, W.J. 1993. *Applied Veterinary Histology (Third Edition)*. St. Louis, MO: Mosby Year Book.

Dyce, K.M., W.O. Sack, and C.J.G. Wensing. 2002. *Textbook of Veterinary Anatomy (3rd Edition)*. Philadelphia: W.B. Saunders Company.

Evans, H.E. and G.C. Christensen. 1979. *Miller's Anatomy of the Dog (2nd Edition)*. Philadelphia: W.B. Saunders Company.

Mullins, K. J., and R.G. Saacke. 2003. *Illustrated Anatomy of the Bovine Male and Female Reproductive Tracts, From Gross to Microscopic*. Blacksburg, VA: Germinal Dimensions, Inc.

6

Functional Anatomy of Reproductive Systems: Neuroendocrine Systems

CHAPTER OBJECTIVES

- Review the general organization of the nervous system.

- Describe the gross anatomy of the brain and pituitary gland.

- Describe the anatomic basis for how the brain coordinates reproductive activity.

OVERVIEW

The reproductive success of an individual is largely determined by its ability to coordinate its reproductive activity with conditions that are favorable for the production and rearing of offspring. For example, it is in the best interest of an animal to avoid reproducing during times of illness, poor nutrition, and extremes in environmental conditions. The nervous system plays a central role in mediating the effects of environment on reproduction. In this chapter we will study the anatomy of the nervous system in relation to its major physiologic functions, which include:

- Detection of stimuli that provide information about the animal's internal and external environments.

- Transmission of information to the brain via neural pathways.

- Integration of neuronal inputs.

- Generation of neural and humoral responses that affect reproductive activity.

Organization of the Nervous System

In order to understand how the nervous system coordinates an animal's reproductive activity with changes in its environment you must have a rudimentary understanding of how the nervous system is structured (Figure 6-1). Anatomically, the nervous system can be divided into the central nervous system, which includes the brain and spinal cord, and the peripheral nervous system, which includes the cranial and spinal nerves (nerves leaving and entering the central nervous system). Functionally, the nervous system can be subdivided based on the direction in which nerve impulses travel and the type of information conveyed by the impulses. With respect to direction of transmission, nerve systems can be classified as afferent (toward the spinal cord

6

Schematic illustration of the nervous system showing its major anatomic subdivisions and how nerves are classified based on function, direction of transmission, and type of information transmitted.

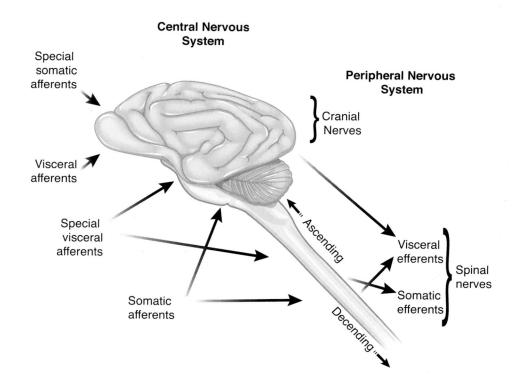

and/or brain) or efferent (away from the spinal cord and/or brain). With respect to type of information transmitted, the nervous system can be divided into the somatic system, which deals with the organism's relationship with the external environment (e.g., movement and cognition), and the visceral system, which conveys information about the organism's internal environment (e.g., blood pressure, heart rate, digestion, and so on). The latter is also referred to as the autonomic nervous system. These subdivisions lead to a more elaborate classification scheme—one that includes four types of afferent nerves and two types of efferent nerves. This classification scheme reveals a great deal about the function of the nervous system. Somatic afferent pathways arise from pressure, temperature, and pain receptors in the skin and deeper tissues of the body and send information to the brain and spinal cord via cranial and spinal nerves, respectively. Special somatic pathways originate in the sense organs of the eye and ear and enter the brain via two cranial nerves. Information from receptors in blood vessels, glands, and viscera of the head and body trunk enters the central nervous system via visceral afferent pathways, which include five cranial nerves and all spinal nerves. There are only two major efferent neural pathways. The somatic efferent pathways leave the central nervous system via spinal nerves and innervate various skeletal muscles. The visceral efferent pathways originate in the brain and spinal cord and innervate smooth muscles of viscera and blood vessels, muscles of the heart, and various glands.

Afferent nerves that enter the brain or spinal cord may synapse with efferent nerves that leave the central nervous system and terminate in various effectors. In addition, afferent neurons may synapse with other afferent nerve cells within the central nervous system. In the spinal cord, these ascending nerve tracts enter various regions of the brain, allowing it to monitor activity of the spinal afferents. Afferent neurons that enter the brain via the cranial nerves may synapse with nerves that innervate other regions of the brain. The brain also has the ability to intervene in the activity of spinal afferents. This is due to descending efferent nerve tracts that synapse with spinal efferent neurons.

Sensory receptors also exist within the brain. The **circumventricular organs** are chemosensitive cells that are highly innervated with fenestrated capillaries. It is likely that these tissues monitor the chemical composition of blood and respond to changes via neural mechanisms as well as by releasing hormones into the blood and/or **cerebrospinal fluid.**

In light of the previous discussion, it is appropriate to view the central nervous system as a collection of stimulus-response systems that are orchestrated to regulate the activity of peripheral tissues. The particular characteristics of such systems are described in the next section.

Stimulus-Response Systems

The neural reflex arc is the simplest and most familiar type of stimulus-response system (Figures 6-2a and 6-3). Its components include: 1) a sensory receptor that detects a stimulus (e.g., heat or pressure) and generates a neuronal impulse; 2) a sensory (afferent) neuron that carries an impulse to the spinal cord or brain; 3) a motor (efferent) neuron conveying an impulse away from the spinal cord or brain; 4) an effector (a muscle, gland, and so on), which responds to the efferent impulse. In many cases, an interneuron exists between the afferent and efferent neurons (Figure 6-3).

A second type of stimulus-response system is the neuroendocrine reflex arc (Figure 6-2b). In this case, the efferent neuron is replaced by an endocrine cell, which produces a hormone that induces a response in an effector or target cell. The release of milk by the mammary gland in response to a suckling stimulus is an example of such a reflex. In this case suckling is detected by pressure receptors in the teat and this stimulates afferent pathways that ultimately impinge upon hypothalamic neurosecretory cells that release the hormone oxytocin into the blood. Oxytocin then acts on the alveoli of the mammary gland (the effector) to stimulate milk ejection.

Reproductive activity is also regulated by endocrine reflexes (Figure 6-2c). In this type of reflex, an endocrine cell responds to a stimulus by releasing

FIGURE 6-2

Three types of reflexes involved with regulation of reproductive activity. A neural reflex (a) involves only nerve cells, whereas a neuroendocrine reflex (b) involves both neurons and endocrine cells. An endocrine reflex (c) does not involve neurons and consists of an endocrine cell that produces a hormone that regulates a target cell.

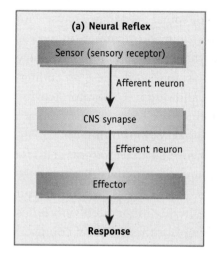
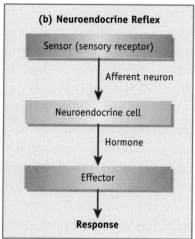
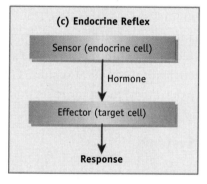

FIGURE 6-3

Schematic illustration of the neural reflex arc governing the movement of skeletal muscles in response to a stimulus, which in this case is stimulation of a stretch receptor in the leg muscle of a dog. This type of reflex would be important in regulation of a reproductive behavior such as mounting.

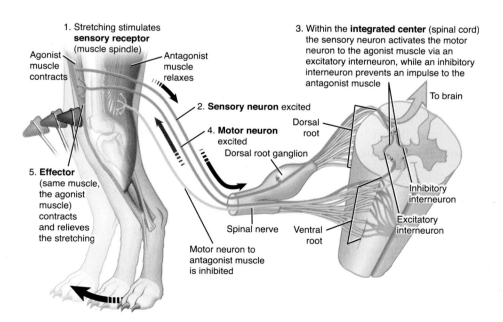

a hormone into the blood. The hormone then interacts with a target cell to elicit a response. An example of this type of regulatory system is the relationship between blood glucose and the hormone insulin. An increase in circulating levels of glucose stimulates endocrine cells in the pancreas to release insulin, which then acts on several types of target cells to enhance clearance of glucose from the blood. We will discuss such systems in greater detail when we study reproductive endocrinology.

It is important to keep in mind three important concepts regarding stimulus-response systems. First, it is not uncommon for several of these systems to work together to regulate a particular effector. Second, a particular stimulus-response system may influence several different effectors. Third, two types of efferent responses are possible—stimulatory and inhibitory. These characteristics permit a tight and graded regulation of physiologic processes.

ANATOMY OF THE BRAIN

The stimulus-response systems of the brain regulate a wide array of physiologic functions. Each of the systems described in the previous section plays a role in some aspect of reproduction. An appreciation of how the brain orchestrates these various mechanisms can be gained through studying the structure of this complex organ.

Major Subdivisions of the Brain

Figure 6-4 summarizes the major subdivisions of the brain, the embryonic vesicles that give rise to them, and the adult structures that make up these regions. Each of the major subdivisions in the adult brain consists of neural tissue organized around a central cavity known as a **ventricle.** Most of our attention is focused on the diencephalon, especially the hypothalamus, a small area that surrounds the third ventricle in the ventral portion of the brain. We are also concerned with the pituitary gland, a bilobed organ that protrudes below the hypothalamus.

Gross Anatomy of the Brain

In order to appreciate the structure and function of the hypothalamus and pituitary gland, it is necessary to review some basic neuroanatomy. Figure 6-5 shows a sagittal view of the sheep's brain. This is a particularly useful perspective because it portrays the anatomic relationship of the hypothalamic-pituitary system to the central nervous system. Note the location of the pituitary gland,

6

FIGURE 6-4

Major anatomic structures of the brain. Terminology refers to the vesicles that make up the embryonic brain, the major subdivisions that arise from these vesicles, and the specific areas that develop within these subdivisions.

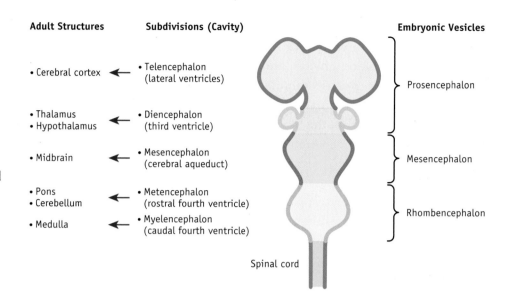

Adult Structures	Subdivisions (Cavity)	Embryonic Vesicles
• Cerebral cortex	• Telencephalon (lateral ventricles)	Prosencephalon
• Thalamus • Hypothalamus	• Diencephalon (third ventricle)	
• Midbrain	• Mesencephalon (cerebral aqueduct)	Mesencephalon
• Pons • Cerebellum	• Metencephalon (rostral fourth ventricle)	Rhombencephalon
• Medulla	• Myelencephalon (caudal fourth ventricle)	

Spinal cord

FIGURE 6-5

Sagittal section through the sheep's brain showing location of the hypothalamus and pituitary gland relative to major brain structures. The ventricles are highlighted in purple.

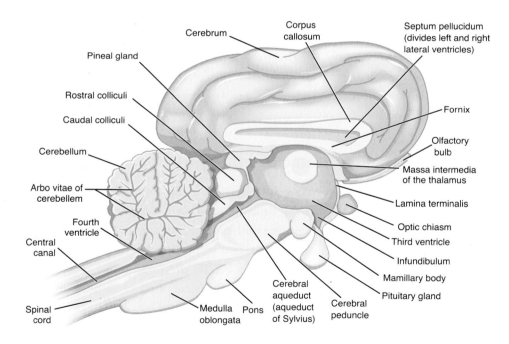

which is suspended from the ventral surface of the brain. The hypothalamus is located dorsal to the pituitary gland in the ventral and medial region of the diencephalon, directly beneath the massa intermedia of thalamus. Its rostral limit lies just in front of the optic chiasm. From there it extends bilaterally around the **infundibulum** (the funnel-shaped stalk from which the pituitary is suspended) and ends caudally at the mammillary body. The third ventricle runs along the median plane of the hypothalamus dividing into left and right halves. It extends from its rostral boundary, the lamina terminalis to the mamillary bodies.

Circulation of the Brain and Spinal Cord

As with all organs, normal function of the central nervous system requires the delivery of nutrients as well as the transport of wastes away from tissues. The central nervous system is unique in the sense that it relies on two circulatory systems to accomplish these functions: 1) blood flowing through arteries, capillaries, and veins and 2) cerebrospinal fluid flowing through the ventricles and the subarachnoid space (a narrow space lying between the thin layers of connective tissue that encase the central nervous system). Much of the central nervous system is encased by a **blood-brain barrier,** which prevents blood from circulating through these tissues. The blood-brain barrier is attributed to the predominance of nonfenestrated capillaries that restrict diffusion of solutes into neural tissues. The bulk of tissues that make up the brain and spinal cord are served by **cerebrospinal fluid,** a cell-free fluid that is produced via ultrafiltration (excluding proteins and other large molecules) of blood plasma. The principle functions of cerebrospinal fluid are to provide chemical buffering capacity to stabilize concentrations of various constituents, transport nutrients and wastes, and provide a medium for neurohormones and neurotransmitters to circulate within the central nervous system.

Figure 6-6 summarizes the circulation of fluid through the central nervous system. Briefly, blood flows to the brain via several arteries. Smaller arteries penetrate the brain and merge with specialized networks of capillaries (chorioid plexus), which line the ventricles. These structures filter blood, resulting in the production of cerebrospinal fluid. The cerebrospinal fluid flows through the ventricular system and then into the subarachnoid space via specialized vents located at the brain stem beneath the caudal portion of the cerebellum. Here, the cerebrospinal fluid becomes the extracellular fluid supplying the bulk of tissue in the central nervous system. This fluid is returned to the blood via specialized organs (arachnoid granulations), which protrude into the venous sinuses of the brain. These sinuses comprise the venous system that carries blood away from the brain and into the jugular veins, which then return blood to the heart.

Several areas adjacent to the cerebral ventricles are not protected by the blood-brain barrier. In these **circumventricular organs** (Figure 6-7), blood from the arteries supplying the brain enters capillary networks consisting of fenestrated capillaries and then returns to the venous circulation via the venous sinuses. These organs arise from the lining of the ventricles and are comprised

6

FIGURE 6-6

Schematic illustration depicting circulation of fluids through the brain. A few regions known as circumventricular organs are served by blood flowing directly to capillary plexes with fenestrated capillaries. Most of the brain is served by cerebrospinal fluid which is a filtrate of blood.

6

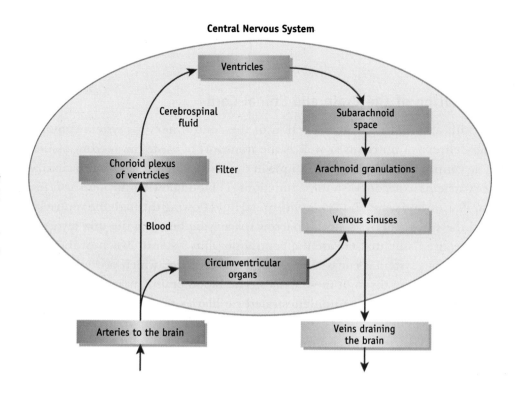

FIGURE 6-7

Location of the circumventricular organs (shaded in red) in the brain. These areas are believed to play roles in monitoring circulating concentrations of various substances that influence brain activity because they contain chemoreceptors and lie outside the blood-brain barrier.

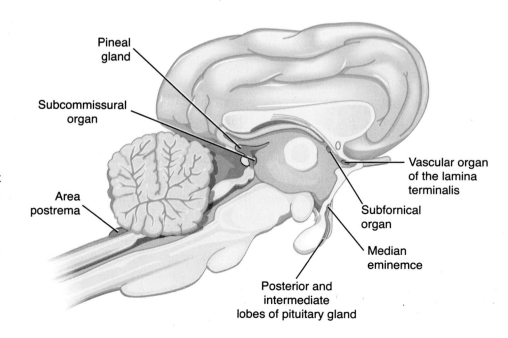

of chemosensitive nerve cells (chemoreceptors), which secrete various substances into the surrounding vasculature and/or ventricles. Circumventricular organs surrounding the third ventricle include the subfornical organ, subcommissural organ, organum vasculosum of the lamina terminalis, pineal gland, posterior pituitary gland, and part of the median eminence. The area postrema is located in the roof of the fourth ventricle. Several of these structures play important roles in regulating reproductive activity in mammals.

ANATOMY OF THE HYPOTHALAMIC-PITUITARY AXIS

The hypothalamus and pituitary gland are intimately associated in both anatomic and functional ways. Figure 6-8 shows an expanded view of the hypothalamic-pituitary interface. The ventral portion of the hypothalamus forms a mound (**median eminence**) that is continuous with the infundibulum (stalk), which connects with and supports the pituitary gland.

Anatomy of the Pituitary Gland

The hypothalamic-pituitary unit is located deep within the skull in the ventral cranial cavity (Figure 6-9). The pituitary gland is embedded in the **sella turcica** (Turkish saddle), a prominence in the dorsal surface of the sphenoid bone

6

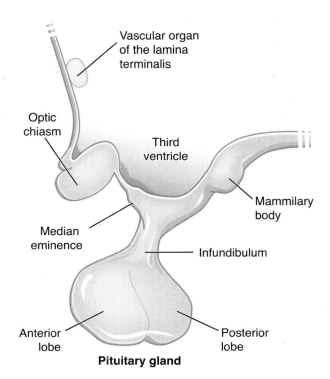

Pituitary gland

FIGURE 6-8

Enlarged view of the hypothalamus and pituitary gland showing major anatomic landmarks.

FIGURE 6-9

Schematic illustration of the pig's skull showing the cranial cavity and pituitary fossa. Note that the pituitary resides in a boney depression known as the sella turcica.

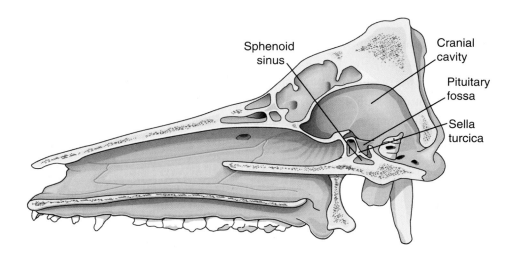

forming portions of the sides and base of the skull near the orbits (eye sockets). In four-legged animals, the infundibular stalk angles caudally, so that the pituitary gland lies in a ventral and caudal position relative to the hypothalamus. In primates, the pituitary lies directly beneath the hypothalamus.

The pituitary gland (also known as the **hypophysis**) consists of two major parts (lobes), which are derived from different embryonic tissues (Figure 6-10). The posterior lobe (**neurohypophysis** or **pars nervosa**) is derived from the brain, whereas the **adenohypophysis (or pars distalis)** develops from the oral ectoderm (epithelium forming the roof of the mouth) of the embryo. In many species the adenohypophysis can be further subdivided into a large anterior portion and a smaller **intermediate lobe** (or **pars intermedia**). Some species lack a distinct intermediate lobe.

Functional and Structural Relationships between Hypothalamus and Pituitary Gland

Blood is supplied to the pituitary gland by branches of the arteries that surround the base of the infundibulum (Figure 6-11). The superior hypophysial artery enters at the interface between the ventral hypothalamus and infundibulum and supplies the anterior pituitary gland. The inferior hypophysial artery enters the ventral region of the posterior pituitary gland. Each of the lobes is drained by hypophysial veins.

The vasculatures of the anterior and posterior lobes of the pituitary gland are remarkably different and reflect differences in the anatomic and physiologic relationships with the hypothalamus (Figure 6-12). In the posterior lobe, blood from the inferior hypophysial artery flows into a capillary plexus, which supplies blood to surrounding tissues. The posterior pituitary gland is made up predominantly of neural tissue; in particular the axons of neurons, which originate in the hypothalamus and terminate near the

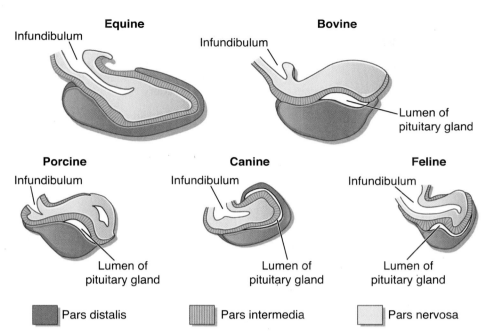

FIGURE 6-10

Structures of the pituitary glands of several species of domestic mammals. In all cases the gland can be divided into three lobes: pars distalis (dark grey), pars intermedia (purple), and pars nervosa (light grey).

6

FIGURE 6-11

Ventral view of the sheep's brain showing major arteries (shaded in purple) supplying the organ (a). A more detailed view (b) reveals that the pituitary gland receives blood from the superior and inferior hypophysial arteries (b).

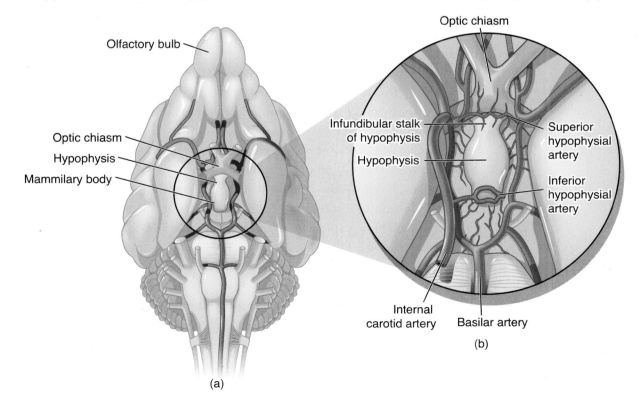

FIGURE 6-12

Schematic illustration of the major blood vessels supplying the pituitary gland. Note that the posterior lobe receives blood directly from the inferior hypophysial artery, whereas the anterior lobe is supplied primarily by long portal vessels that receive blood from the superior hypophysial artery. The venous drainage of these areas is not shown in the diagram.

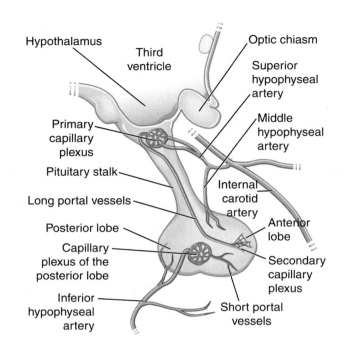

capillaries. Blood draining from these capillaries can flow in two directions. Much of the blood flows into the hypophysial vein, which carries blood to the jugular vein. Some blood enters so-called short portal vessels, which are sinusoidal vessels that converge on a capillary bed in the anterior pituitary gland.

In contrast to the posterior pituitary gland, the anterior pituitary gland receives only a little blood directly from arteries. The bulk of blood flowing into the anterior pituitary gland arrives via a portal vascular system. The superior hypophysial artery feeds into a capillary plexus that is located in the median eminence of the hypothalamus. Neurons from the hypothalamus terminate on the median eminence capillaries. This capillary plexus is sometimes referred to as the **primary capillary plexus.** These capillaries drain into "long portal blood vessels" that run longitudinally along the infundibular stalk and drain into a **secondary capillary plexus** in the anterior lobe of the pituitary gland. The capillaries of this region drain into a hypophysial vein.

Having described the circulation of the pituitary gland, it is now possible to explain how the hypothalamus communicates with the anterior and posterior lobes of this important organ (Figure 6-13). Communication between the hypothalamus and posterior lobe is primarily neuronal. Two peptides (oxytocin and vasopressin) are produced in the cell bodies of neurons located in the hypothalamus, and are transported to axon terminals that lie adjacent to capillaries. Stimulation of these neurons causes the peptides to be released into the surrounding extracellular space. Because the posterior

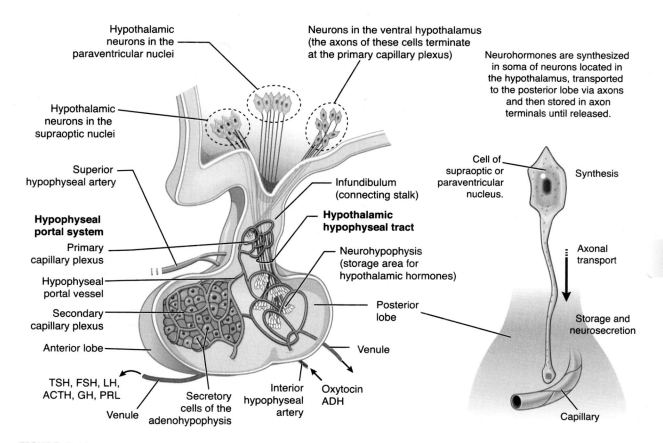

FIGURE 6-13

Schematic illustration of the hypothalamic-pituitary system showing how the hypothalamus communicates with each lobe of the pituitary gland. In the case of the anterior pituitary, neurohormones produced in the hypothalamus are released into portal vessels and then carried to the gland. In contrast, neurohormones from hypothalamic neurons are transported along axons directly to the posterior pituitary gland.

lobe lies outside the blood-brain barrier (contains fenestrated capillaries), the peptides can diffuse into the capillaries and enter the general circulation. Alternatively, these neurohormones can enter the anterior pituitary gland via the short portal vessels.

Communication between the hypothalamus and anterior lobe occurs through neurovasculature connections. The capillaries of the median eminence are fenestrated. Therefore, the secretory products of hypothalamic neurons that impinge upon the primary capillary plexus enter the blood and are carried via the portal vessels to the secondary capillary plexus. These chemicals diffuse out of the capillaries and stimulate or inhibit release of hormones by endocrine cells located in the anterior lobe. These pituitary hormones diffuse into the secondary capillary plexus and enter the general circulation through the hypophysial vein.

6

BOX 6-1   Focus on Fertility: Discovery of the Hypothalamic-Hypophysial Portal System

One of the major problems with textbooks is that they often take too much for granted. For example, the idea that the release of hormones from the anterior pituitary gland is regulated by various hypothalamic hormones is one of the fundamental assumptions underlying reproductive physiology. How did we come to accept this hypothesis? The elaborate arrangement of the portal blood vessels found along the pituitary stalk was described in detail during the 1930s. However, a full appreciation of its functional importance was not attained until the 1970s.

Although the arrangement of blood vessels connecting the hypothalamus and anterior pituitary gland were recognized as portal in nature (i.e., running between two capillary plexes), early investigators erroneously concluded that the direction of blood flow was upward from the pituitary gland toward the hypothalamus. However, several subsequent avenues of research soon challenged this conclusion. First, a study in the 1930s showed that surgical transection of the infundibular stalk in toads caused necrosis of the anterior pituitary. It was therefore suggested that blood flowed in a downward direction from the hypothalamus to the pituitary gland. Studies in mammals followed and confirmed this conclusion. Second, in the 1940s and 1950s several investigators documented this downward flow of blood by directly observing blood flow in portal vessels of rats, mice, dogs, cats and monkeys. Finally, direct evidence supporting this view came from an elegant experiment done with rats in the early 1970s. When dyes were injected into individual portal vessels they eventually infiltrated the tissue of the anterior pituitary gland.

From the moment investigators embraced the idea of a downward movement of blood within the portal vessels, there was growing support for the hypothesis that the hypothalamus regulated the anterior pituitary gland via humoral (blood-borne) factors. A series of classic endocrine experiments generated empirical support for this hypothesis. As noted earlier, disruption of the portal vascular system causes deterioration of anterior pituitary tissue. The idea that the hypothalamus is the source of humoral factors that support the functions of anterior pituitary cells came from the fact that grafts of anterior pituitary tissues remained functional when transplanted to the median eminence, but not when transplanted to other regions of the brain (e.g., temporal lobe of cerebrum) or in the highly vascularized renal capsule. At the time of these studies it was becoming clear that the anterior pituitary gland regulated gonadal function via two gonadotropins. Work concerning the vascular relationship between the hypothalamus and pituitary gland gave rise to the idea that a hypothalamic gonadotropin–releasing factor regulates secretion of LH and FSH. Support for this hypothesis came from experiments that showed that surgical disconnection of the portal vascular system, or placement of an impermeable barrier between the hypothalamus and pituitary gland resulted in decreased production of LH and FSH by the pituitary gland as well as infertility. By the 1970s the structure of this factor was determined and became known as gonadotropin-releasing hormone (GnRH). Today, this and other hypothalamic hormones are produced synthetically and are used clinically to treat a variety of endocrine disorders including infertility.

Organization of the Hypothalamus

The hypothalamus is an organ of integration. It receives diverse neural inputs from the peripheral and central nervous systems as well as blood-borne signals. All of these signals converge on the hypothalamus and are blended to generate humoral signals that regulate the pituitary gland. The tissue of the hypothalamus consists primarily of nuclei and nerve tracts (see Figure 6-13). The hypothalamic nuclei appear as bundles of neuron cell bodies arranged in bilateral pairs along the walls of the third ventricle. Only a few of these play a role in reproduction. The others control other autonomic functions as well as release of pituitary hormones that regulate systems other than the reproductive system. Large cell bodies (magnocellular) localized in the paraventricular nuclei and supraoptic nuclei produce the hormone **oxytocin,** which plays a role in milk ejection, parturition, and regulation of the ovary. The axons of these cells course ventrally and caudally into the infundibulum and terminate in the posterior pituitary gland. This nerve tract is called the **tuberohypophysial** tract. The medial preoptic nuclei, located in the anterior hypothalamus, play a role in sexual behavior. A hypothalamic hormone called **gonadotropin-releasing hormone (GnRH)** controls the release of two pituitary hormones (luteinizing hormone [LH] and follicle-stimulating hormone [FSH]) that regulate gonadal function. Neurons that produce GnRH are not localized in particular nuclei or discrete regions of the brain (Figure 6-14). They are scattered sparsely throughout the forebrain from the olfactory bulbs, rostrally, to the hypothalamus, caudally. Within the hypothalamus, cell bodies of GnRH neurons are found in the medial preoptic area, anterior hypothalamic area, and medial basal hypothalamus. The majority of this type of cell body is found in the medial preoptic area. Axons from these cells course laterally and caudally and finally terminate in the median eminence, forming the **tuberoinfundibular tract.** Neurosecretory cells from other locations in the hypothalamus (e.g., ventral region) also send axons toward the median eminence and contribute to this tract. The products of these neurons regulate release of adenohypophysial hormones other than those regulating reproduction.

The neurons that produce oxytocin and GnRH, as well as those producing other hypothalamic hormones, are regulated by neurons that originate in regions of the brain outside the hypothalamus. It is beyond the scope of our discussion to describe these neuro-pathways in any detail. For our purposes it is sufficient to understand only that neurons from these regions synapse with hypothalamic neurons and communicate with them via various neurotransmitters.

6

6

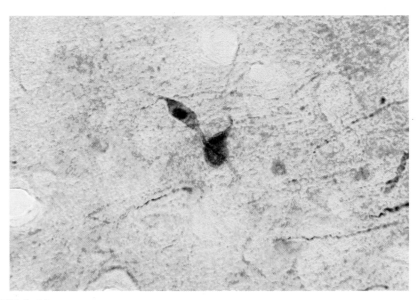

FIGURE 6-14

Microscopic view of a section of rat hypothalamus showing two GnRH-producing neurons (stained brown). Note the prominent cell nucleus, dendrite, and axon of one neuron. Compared to other types of hypothalamic nerve cells, GnRH neurons exist in relatively small numbers and are widely dispersed. (Photograph provided by Dr. Sandra J. Legan, Department of Physiology, University of Kentucky).

SUMMARY OF MAJOR CONCEPTS

- With respect to reproduction, the brain plays a central role in timing reproductive activity with favorable environmental conditions.

- The brain regulates bodily functions, including reproduction, via various stimulus-response systems that involve neuronal and humoral mechanisms.

- A small portion of the diencephalon, the hypothalamus, integrates neuronal and humoral inputs and regulates reproductive activity via regulation of the pituitary gland.

- Most of the central nervous system is protected by a blood-brain barrier, meaning that only some blood constituents enter the extracellular fluid of the brain (cerebrospinal fluid).

- The pituitary gland consists of two major lobes, each of which plays a role in reproduction. The anterior lobe interacts with the hypothalamus through a neurovascular connection whereas the posterior lobe is connected to the hypothalamus by a neuronal system.

DISCUSSION

1. When male mammals encounter sexually receptive females, they express particular sexual behaviors (e.g., investigation, mounting, erection of the penis, ejaculation). Outline a stimulus response system that explains how the male behavior is generated.

2. Suppose you take simultaneous blood samples from a carotid artery and a jugular vein and measure the concentration of LH in each sample. Which sample would you expect to have the greater concentration of LH? Why?

3. Scientists have discovered several peptide hormones that are produced in various peripheral tissues, but affect appetite, which is regulated by the central nervous system (specifically, the hypothalamus). It is unlikely that these large molecules can pass the blood-brain barrier. If so, then how can these blood-borne signals affect activity of the central nervous system? Where might these peptides act?

4. In rare occasions, human infants as young as one year of age undergo precocious puberty; that is, they display adult secondary sex traits. One of the major causes of this anomaly is a tumor in the medial-basal hypo-thalamus. Explain how such a tumor might cause this. (Hint: Recall that secondary sex traits develop in response to production of high levels of gonadal hormones).

REFERENCES

Carpenter, M.B. 1976. *Human Neuroanatomy (Seventh Edition)*. Baltimore: Williams and Wilkins.

Dyce, K.M., W.O. Sack and C.J.G. Wensing. 2002. *Textbook of Veterinary Anatomy*, Third Edition. Philadelphia: Saunders.

Evans, H.E. and G.C. Christensen. 1979. *Miller's Anatomy of the Dog*, Second Edition. Philadelphia: W.B. Saunders Company.

Everett, J.W. 1994. Pituitary and Hypothalamus: Perspectives and Overview. In: E. Knobil and J.D. Neill, *The Physiology of Reproduction Vol. 2.*, Second Edition. New York: Raven Press, pp. 1509–1526.

CHAPTER

7

Principles of Reproductive Endocrinology

HOMEOSTASIS

All living things have the ability to metabolize various nutrients to support vital life functions. The numerous biochemical reactions that regulate an organism's metabolism occur in an internal environment that is maintained in a constant state relative to the external environment in which an organism lives. The processes through which living bodies maintain such conditions are collectively referred to as **homeostasis.** In animals, homeostasis requires complex interactions among various organs. Communication among various organ systems involves the nervous and the endocrine systems. Various neuronal reflexes between sense and effector organs provide a means of communication within the nervous system. In contrast, communication within the endocrine system involves **hormones,** which can be thought of as extracellular chemical messengers. These systems do not operate independently; rather, they interact with each other to maintain homeostasis. Like other vital functions, the mechanisms controlling the reproductive activity of mammals are subject to homeostatic regulation involving complex interactions between nervous and endocrine tissues. We have already examined the anatomic basis for neuronal regulation of reproduction. In this chapter we will consider how hormones regulate reproductive activity. The field of study that deals with hormones is known as endocrinology.

WHAT IS ENDOCRINOLOGY?

A traditional approach to studying the regulation of reproductive tissues is to distinguish between the nervous and endocrine systems. This distinction is largely based on the way biologists originally defined a hormone. Based on the work of Starling in the early 1900s, a hormone is a chemical messenger that is released into the blood and exerts actions on cells that are distant from its site of origin (Figure 7-1).

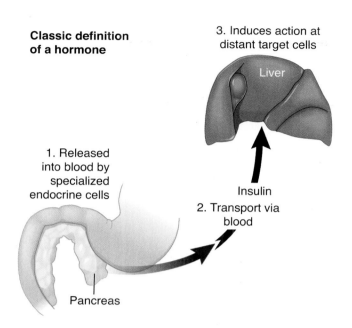

Classic definition of a hormone

3. Induces action at distant target cells

Liver

1. Released into blood by specialized endocrine cells

Insulin

2. Transport via blood

Pancreas

FIGURE 7-1

The effects of the hormone insulin on the liver illustrates the so-called classic definition of a hormone.

7

Neurotransmitters–chemical messengers that allow communication between adjacent nerve cells–are not considered to be hormones in the traditional sense of the word because these chemicals are released into synapses and do not enter the blood.

The distinction between the endocrine and nervous systems has blurred during the past 30 years primarily because of two major discoveries. First, some nerve cells were found to release neurohormones, chemicals that enter the circulation and affect distant cells (e.g., neurosecretory cells that release oxytocin). Second, scientists learned that substances traditionally classified as hormones do not have to enter the bloodstream to affect other cells. A hormone can also diffuse within the extracellular fluid to effect changes in nearby cells and, in some cases, the very cell that produces the hormone. Based on these insights, the definition of hormone has been broadened to include any chemical that acts as an extracellular messenger.

Based on the current understanding, it may seem that almost any biochemical (e.g., vitamins, energy substrates, and minerals) can be classified as a hormone. However, this is not the case. One important characteristic that distinguishes hormones from other chemical messengers is that hormones circulate in extracellular fluids in very low concentrations; that is, at concentrations ranging from picograms (10^{-12} g) to nanograms (10^{-9} g) per mL of blood. This is much lower than concentrations of vitamins, minerals, and metabolites, which are present in concentrations of micrograms (10^{-6} g) and milligrams (10^{-3} g) per mL of blood.

FIGURE 7-2

Classification of hormone action based on the relationship between the hormone-producing cell and the target cell.

Ways hormones can act:

a. Endocrine

b. Paracrine

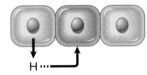

c. Autocrine

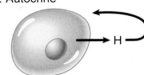

Hormones coordinate the activities of cells in several ways (Figure 7-2). When hormones act via the so-called classic or **endocrine** mode of action, they are released into the circulation and exert effects on distant target cells. If the hormone is released into the blood by a neurosecretory cell, the mode of action is called **neuroendocrine.** A hormone that is released into the extracellular fluid and effects changes in nearby cells is said to act via a **paracrine** mechanism. A specialized case of this type of action is neurocrine; that is, involving chemical messengers that are released by nerves and exert local effects on nearby cells. Finally, a cell can release a chemical that regulates its own activity; that is, **autocrine** regulation.

Contemporary endocrinology is concerned with all types of hormones and all modes of action, as well as the tissues that produce and respond to hormones. Reproductive endocrinology is a branch of endocrinology that is concerned specifically with the hormones produced by reproductive tissues.

Hormones and Reproduction

Before the mid-nineteenth century, studies of the reproductive system were confined to anatomic descriptions of reproductive organs. This changed in 1849 when Berthold completed a series of experiments that attributed development of secondary sex traits in roosters to a humoral factor that is produced by the testes and carried to target tissues via the blood (see Box 7-1).

BOX 7-1   Focus on Fertility: The First Reported Endocrine Experiment

In 1849, the French scientist A.A. Berthold reported what appears to be the first endocrine experiment. The focus of his work was the development of secondary sex traits in cockerels (Figure 7-3). Male chicks that were bilaterally castrated did not display male sexual and aggressive behaviors, failed to develop a comb and wattles, and had a weak crow (masculine vocalizations). When Berthold implanted one or two testes from the same or a different animal into the abdominal cavities (the location of testes in birds) of castrated cockerels, the birds developed as normal males. The loss of masculine traits following removal of the testes together with the fact that implanted testes prevented these effects demonstrated that these organs were the source of some factor that regulates development of male secondary sex traits in chickens. The fact that the transplanted testes were effective suggests that the relationships between the testes and target tissues are humoral rather than neuronal. Years later, researchers demonstrated that testicular extracts alone replaced the activity of the testes. In the 1930s, testosterone, the chemical responsible for these actions, was purified from testicular extracts.

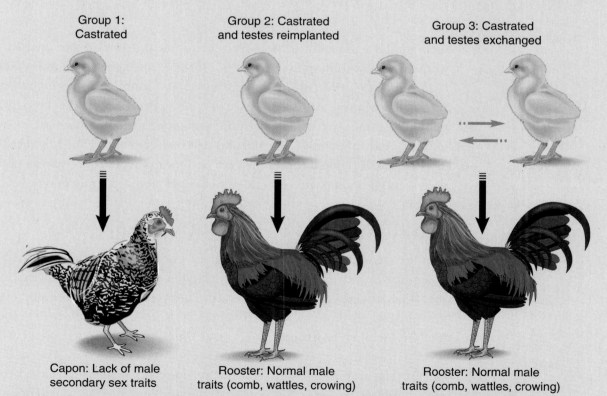

Group 1: Castrated

Group 2: Castrated and testes reimplanted

Group 3: Castrated and testes exchanged

Capon: Lack of male secondary sex traits

Rooster: Normal male traits (comb, wattles, crowing)

Rooster: Normal male traits (comb, wattles, crowing)

Figure 7-3 Summary of what most physiologists believe to be the first endocrine experiment. In 1849, Berthold reported the results of an experiment that examined the effects of the removal and re-implantation of the testis on secondary sex traits of cockerels.

The methods employed by Berthold represent a standard experimental approach to demonstrating that a particular physiologic event is controlled by a particular hormone, and for studying the physiologic effects of hormones. This approach is commonly called the ablation-replacement paradigm. For example, experiments designed to characterize the role of estradiol on LH secretion in females typically involve measuring LH concentrations before and after injection of estradiol in ovariectomized (surgical removal of the ovaries) animals. In this case the source of estradiol is *ablated* by surgery, and the hormone is then *replaced* by injecting it. If ablation of a particular hormone produces some physiologic effect *and* the effect is reversed by treatment with the hormone, then the effect can be attributed specifically to the action of the hormone. Neither ablation nor hormone injection alone can link a particular response with a particular hormone. Today, methods other than surgery are used to ablate endocrine tissues. Some of the more common approaches include chemical ablation of the endocrine tissue, chemical disruption of hormone synthesis and secretion, immunization against specific hormones, use of receptor antagonists, and use of transgenic animals that do not express genes for certain hormones ("genetic knockouts").

Such studies shifted focus away from anatomy and toward the analysis of mechanisms controlling the activity of reproductive tissues. The importance of hormones in the control of reproduction cannot be overemphasized. Reproductive hormones control virtually all aspects of reproductive activity including: gametogenesis, sexual differentiation, sexual maturation, sexual behavior, and function of internal and external genitalia. Thus it is virtually impossible to develop a comprehensive understanding of reproductive physiology without understanding basic principles of endocrine physiology.

CLASSIFICATION OF REPRODUCTIVE HORMONES

An understanding of how a particular hormone regulates reproductive activity requires knowledge of where the hormone is produced, where it acts, and what it does as well as insight into *how* the hormone is synthesized, secreted, transported, metabolized, and acts, on target tissues. As we explore the reproductive physiology of mammals, we will encounter 15 hormones that play significant roles in regulating reproductive activity (Table 7-1). The task of learning the aforementioned information for each of these hormones may seem daunting. However, this process can be facilitated by first learning some general characteristics of reproductive hormones and then learning specific details about them as we study the physiologic processes in which they play particularly important roles.

TABLE 7-1 Classification of Major Reproductive Hormones

Hormone	Major Source(s)	Class of Molecule	Major Female Target Tissue(s)	Major Male Target Tissue(s)
Gonadotropin-Releasing Hormone (GnRH)	Hypothalamus	Decapeptide	Anterior lobe of pituitary gland	Anterior lobe of pituitary gland
Luteinzing Hormone (LH)	Anterior lobe of pituitary gland	Glycoprotein	Ovary (theca cells of follicle and cells of corpus luteum)	Testes (Leydig cells)
Follicle-Stimulating Hormone (FSH)	Anterior lobe of pituitary gland	Glycoprotein	Ovary (granulosa cells of follicle)	Testes (Sertoli cells)
Prolactin (PRL)	Anterior lobe of pituitary gland	Protein	Mammary gland (glandular epithelium); ovary (corpus luteum) in some species	Testes
Oxytocin (OT)	Hypothalamus	Octapeptide	Uterus (myometrium and endometrium) and mammary gland (myoepithelial cells)	Cauda epididymis, ductus deferens, and ampulla (smooth muscle cells)
Estradiol (E_2)	Ovaries (granulosa cells), placenta, and testes (Sertoli cells)	Steroid	Internal and external genitalia, brain, pituitary gland, and mammay gland	Brain and pituitary gland
Progesterone (P_4)	Ovaries (corpus luteum) and placenta	Steroid	Brain, pituitary gland, uterus (myometrium and endometrium), and mammary gland	No major target tissue
Testosterone (T)	Testes (Leydig cells) and ovaries (theca interna cells)	Steroid	Brain and ovaries (granulose cells)	Accessory sex glands, external genitalia, and testes (seminiferous epithelium)
Inhibin	Ovary (granulosa cells) and testes (Sertoli cells)	Glycoprotein	Anterior lobe of pituitary gland	Anterior lobe of pituitary gland
Activin	Ovary (granulosa cells) and testes (Sertoli cells)	Glycoprotein	Anterior lobe of pituitary gland	Anterior lobe of pituitary gland
Prostaglandin $F_{2\alpha}$ ($PGF_{2\alpha}$)	Uterus (endometrium) and vesicular glands	Fatty acid	Ovaries (corpus luteum and follicles) and uterus (myometrium)	Epididymis

7

continued

TABLE 7-1 Classification of Major Reproductive Hormones (*continued*)

Hormone	Major Source(s)	Class of Molecule	Major Female Target Tissue (s)	Major Male Target Tissue(s)
Prostaglandin E$_2$ (PGE$_2$)	Ovaries, uterus and extra-embryonic membranes	Fatty acid	Ovaries (corpus luteum) and oviduct	No known targets
Human chorionic gonadotropin (hGC)	Trophectoderm of blastocyst	Glycoprotein	Ovaries (corpus luteum)	
Equine chorionic gonadotropin (eCG)	Chorion (girdle cells)	Glycoprotein	Ovary	
Placental lactogen	Placenta	Protein	Mammary gland	

Making general inferences about the reproductive hormones is facilitated by various schemes for classifying hormones. Table 7-1 illustrates a few of these methods. Classifying hormones based on their chemical characteristics is the most useful for our purposes. The hormones listed in Table 7-1 fall into one of several chemical classes: polypeptides (including proteins and peptides), steroids, and fatty acids. These classes can be further reduced to two major categories; that is, hydrophilic (water-soluble) and hydrophobic (water-insoluble or fat-soluble). In general, the polypeptide and prostaglandin hormones are hydrophilic, whereas the steroid hormones are hydrophobic. We will use this classification scheme to explore some general features of hormone synthesis, secretion, transport, and action. However, before we can do this we must establish some basic principles of endocrine physiology.

OVERVIEW OF ENDOCRINE PHYSIOLOGY

Whether a hormone is acting in an endocrine, paracrine, or autocrine manner several concepts of fundamental importance apply to its regulatory activity (Figure 7-4):

- Cells that produce hormones can also serve as targets for hormones.
- A particular hormone can induce the same or different biological effects in several different target cells.
- Different hormones can exert the same or different actions on the same target cell.

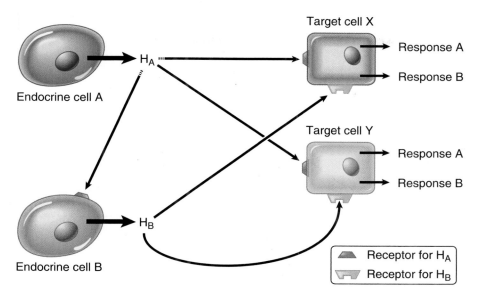

FIGURE 7-4

Schematic illustration depicting how two hormones can interact to regulate two types of target cells. Note that a hormone can act on endocrine and nonendocrine cells and interact with another hormone to regulate a particular cell. A hormone can produce the same effects or different effects in different target cells. In addition, two hormones can produce the same or different effects in the same target cells.

Two hormones can interact by affecting the cells that produce them and/or the cells upon which they act.

- Whether or not a particular target cell responds to a particular hormone depends on whether or not the target cell has the ability to respond to the hormone.

The fourth concept may seem obvious, but it is of central importance for understanding endocrinology. Specifically, this idea raises the following question: What determines whether or not a cell can respond to a particular hormone?

Hormone Receptors

The biological activity (i.e., the ability to evoke a biological response) of a particular hormone depends on it interacting with its **receptor.** Thus, a particular hormone can affect only those target cells that express the receptor for the hormone. A hormone receptor is a protein that interacts chemically with a hormone. They are located on the cell membrane or within the cytosolic or nuclear compartments, depending on the receptor type. Receptors mediate the effects of hormones by triggering intracellular processes that regulate activities of enzymes and/or expression of genes. All receptors have the following characteristics:

- A receptor has *high affinity* for the hormone.
- A receptor is *specific* for the hormone.
- Hormone-receptor binding is a *saturable* phenomenon.

FIGURE 7-5

The relationship between hormone concentration and amount of hormone bound to a fixed amount of receptor for three different hormones.

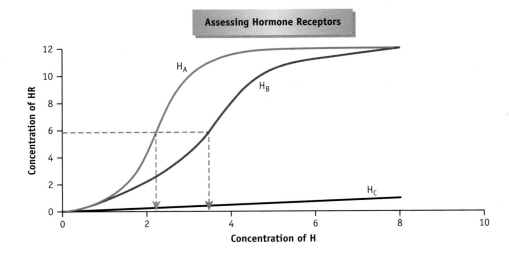

One approach to conceptualizing these characteristics is to consider the following experiment. Imagine that different amounts of a hormone (H_A) are allowed to react with a constant amount of receptor (e.g., a fixed volume of cellular extract from target tissue). After a period of time the reaction is stopped and the amount of hormone bound to the receptor is measured for each concentration of hormone. Figure 7-5 shows data derived from this type of experiment. Note that the relationship between hormone concentration and amount of binding is not linear. Initially the amount of hormone bound to the receptor is directly proportional to the amount of hormone added. Eventually, the curve flattens; that is, the amount of bound hormone is about the same no matter how much hormone is added. This demonstrates the characteristic of saturability. In other words, at a particular concentration, the receptors cannot bind additional hormone. This is also referred to as the binding capacity of the sample, which is an indirect measure of the number of receptors present.

Affinity is the chemical attraction between the hormone and its receptor. In our example, affinity is defined as the concentration of hormone that results in 50 percent of the binding capacity. To further illustrate this concept, imagine that we run a second experiment using a second hormone (H_B), one that isn't as biologically potent as the first. The binding curve for H_B is similar to the one for H with one important exception; that is, the curve for H_B has shifted to the right. Thus the concentration of H_B required to produce 50 percent of the binding capacity is greater than that for H_A. In other words, the receptor has less affinity for H_B than for H_A. A particular receptor might not have any affinity for a hormone (e.g., H_C in Figure 7-5). In this case, the hormone will never saturate the receptor. When endocrinologists speak of the potency of a hormone, they are referring its affinity with the receptor relative to other hormones. You have already encountered an example of this phenomenon. Recall that a potent metabolite of testosterone (DHT) induces development of the male external genitalia. DHT is approximately 100 times more potent than testosterone, meaning that 100 times more testosterone is required to produce the same effects as DHT; that is, the affinity of the receptor for DHT is 100 times that for testosterone.

Specificity is measured by determining what hormones will bind to a particular receptor. High specificity means that a particular receptor binds only one type of hormone with high affinity. For example, receptors for androgens will bind hormones such as testosterone and DHT with high affinity, but not other hormones such as estradiol or progesterone. In our example H_A and H_B could be DHT and testosterone, whereas H_C could be progesterone, a hormone for which the receptor has no affinity.

7

Kinetics of Hormone-Receptor Interactions

The binding of a hormone to its receptor can be understood as a chemical reaction that is governed by the law of mass action, where the hormone (H) and receptor (R) are the reactants and the hormone-receptor complex (HR) is the product:

$$H + R \leftrightarrows HR$$

At equilibrium,

$$k_f[H][R] = k_r[HR]$$

or

$$[H][R]/[HR] = k_r/k_f = K_D$$

where k_f is the rate constant for HR formation, k_r is the rate constant for HR dissociation, K_D is the equilibrium dissociation constant (a measure of affinity) and [H], [R], and [HR] are concentrations of hormone, receptor, and the hormone-receptor complex, respectively. It is important that you take a minute and think about these equations because they provide the basis of all endocrine experiments. At equilibrium, the rate of HR formation is equal to the rate of HR dissociation, and the ratio of [H][R] to [HR] is K_D, a measure of the affinity of between the hormone and its receptor. In the case of a hormone with high affinity for a receptor, the ratio of [H][R] to [HR] at equilibrium is low. In the case of a hormone with lower affinity for a receptor, the ratio of [H][R] to [HR] at equilibrium is higher.

ASSESSING A HORMONE'S EFFECTS

As noted earlier, receptors mediate the effects of hormones on target cells. The degree of the response induced by a hormone is directly related to the amount of HR complex generated by the interaction between H and R. The law of mass action implies that the amount of HR is a function of the concentration of hormone, concentration of receptor, and the affinity of the receptor. This provides a theoretical basis for evaluating endocrine systems. Assessing the biological effects of a hormone on its target cell requires the following information (Figure 7-6):

- Concentration of the hormone in extracellular fluids (e.g., blood).
- Concentration of receptors in target tissues.
- Affinity of the receptor.

Any change in activity of an endocrine system is the result of changes in one or more of these parameters.

One of the most common approaches to assessing hormonal control of reproduction is to measure concentrations of hormones in the blood. Much of what we know about the physiologic mechanisms regulating the reproductive activity in mammals is based on experiments that have characterized patterns of reproductive hormones in blood during various physiologic states as well as ones that have assessed the concentrations and affinities of receptors in target tissues. Our study of the reproductive physiology of

FIGURE 7-6

Schematic illustration summarizing the various factors that influence the amount of hormone that binds to its receptor, which in turn determines the extent of the biological response induced by the hormone.

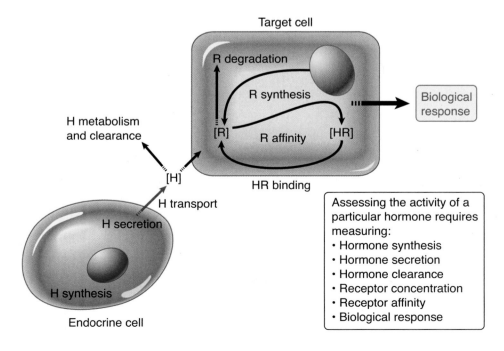

mammals relies heavily on understanding circulating patterns of reproductive hormones. Therefore, some cautions regarding interpreting such data are warranted. First, it is important to emphasize that a change in circulating concentrations of a hormone can reflect changes in synthesis and secretion of the hormone by endocrine cells, transport of the hormone in blood and clearance of the hormone from the circulation. Second, a change in concentration of a hormone will not induce changes in target tissues unless the target tissue is responsive to the hormone. Changes in tissue response to hormones can be attributed to changes in number of receptors (which is a function of synthesis and degradation) and affinity of receptors (which is largely attributed to its physical-chemical properties) as well as the ability of receptors to induce changes within the target cell (so-called postreceptor mechanisms). The major implication of these caveats is that assessment of the activity of a particular hormone may require more than simply measuring its concentration in the blood. A more comprehensive analysis of a hormone's activity may require assessing its synthesis, secretion, clearance, the number and affinity of its receptor, and finally the mechanisms linking hormone-receptor binding and induction of a biological response.

ACTIVITY OF HYDROPHILIC HORMONES

Having established a theoretical framework for assessing the activity of endocrine systems, we can now turn our attention to analyzing how the two major classes of hormones (hydrophilic and hydrophobic) are synthesized, secreted, transported, cleared, and exert their effects on target cells. We will first deal with hydrophilic hormones (Figure 7-7).

Synthesis

Polypeptide hormones are synthesized and packaged in membrane vesicles which act as storage depots until the hormones are released. We will not concern ourselves with the details of polypeptide synthesis. Only the highlights of this process will be addressed. Briefly, a particular gene that codes for precursor of a hormone is transcribed to yield messenger RNA (mRNA) that is translated along the rough endoplasmic reticulum (RER) to produce a hormone precursor known as a pre–pro-hormone. As the emerging peptide enters the RER, a small (pre-) fragment is cleaved off, leaving the pro-hormone. As the pro-hormone travels along the RER and is packaged into vesicles by the Golgi apparatus, a second (pro-) fragment is removed, leaving the actual hormone. The pro-peptide and the hormone are packaged and stored in

FIGURE 7-7

Overview of the synthesis, secretion, transport, and action of hydrophilic hormones.

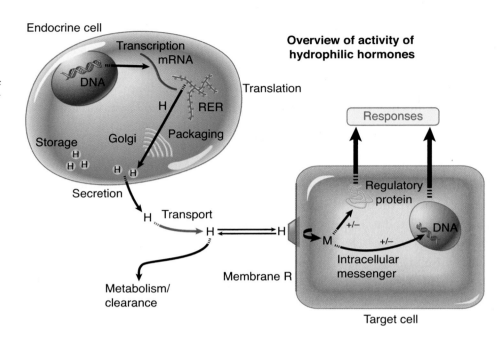

7

secretory vesicles, which accumulate in the cytosol in peripheral regions of the cell.

The synthesis of prostaglandins (not shown) is quite different from that of polypeptide hormones. Prostaglandins are derived from fatty acid derivatives found in the cell membrane. These hormones are not packaged in vesicles and leave the cell by facilitated diffusion.

Secretion

Secretion of polypeptide hormones involves exocytosis. Stimulation of endocrine cells causes secretory vesicles to move towards and fuse with the cell membrane thereby releasing their contents into the surrounding extracellular space. The packaging and storage of hormones in vesicles allows the processes of synthesis and secretion to be uncoupled. In other words, the synthesis and secretion of a hormone can occur independently. On the one hand, a cell may be secreting, but not producing a hormone. On the other hand, synthesis can occur while secretion has ceased. The former situation results in a decrease in intracellular concentrations of the hormone, whereas the latter situation results in an increase in cellular stores of the hormone.

Due to a lack of storage, prostaglandin synthesis is tightly coupled to secretion. In other words, a change in synthesis results in a corresponding change in secretion of the hormone.

Transport and Clearance

Hydrophilic hormones are readily soluble in the aqueous blood. Therefore, most of these types of hormones circulate in the free form; that is, not associated with hydrophilic serum proteins. Polypeptide hormones do not last long in the circulation. The half-life (time required for the concentration to decrease by half) of polypeptide hormones is typically no longer than several hours and in some cases as short as several minutes. Clearance of polypeptide hormones from the blood involves two processes. Small amounts of these hormones are degraded by blood proteases. The major route for removal of polypeptide hormones is via uptake and degradation by target cells.

The half-lives of prostaglandins are also very short. Unlike the polypeptide hormones, these hormones are rapidly degraded by various enzymes that are widely distributed in tissues such as the lungs.

Action

Water-soluble hormones do not diffuse freely across the lipid bilayer of cell membranes. These hormones exert their effects on target cells by interacting with receptors that are located on the membranes of target cells (Figure 7-8). Membrane receptors are very complex proteins consisting of several major domains. The binding domain extends into the extracellular region and consists of hydrophilic amino acids that bind to the hormone. An intracellular domain is also hydrophilic, but is linked to membrane-associated enzymes. These receptors also contain a transmembrane domain. The transmembrane domain of the receptor consists of alternating sequences of hydrophobic amino acids (within the membrane) linked by hydrophilic sections that protrude outward and inward into the extracellular and intracellular compartments, respectively.

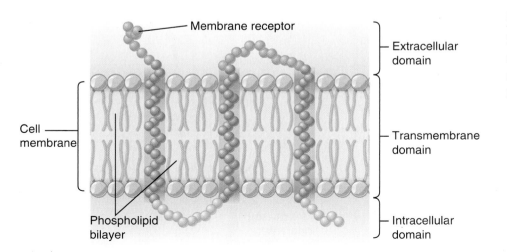

FIGURE 7-8

Schematic depiction of one class of membrane receptors for certain hydrophilic hormones.

7

Binding of the hormone to the binding domain of the receptor induces conformational changes in the receptor, allowing it to activate intracellular enzymes that are associated with the inner surface of the cell membrane. This leads to a chain reaction that ultimately results in generation of **intracellular (second) messengers.** These compounds induce changes in the target cell by activating and/or inhibiting various regulatory proteins. For example, the interaction between LH and its receptor in Leydig cells of the testes generates a second messenger that enhances activity of the rate-limiting enzyme in the synthesis of testosterone.

As noted in the previous section, uptake of polypeptide hormones by target cells accounts for most of the clearance of hormone from the circulation. Target cells internalize hormone-receptor complexes via endocytosis; that is, sections of cell membrane containing hormone-receptor complexes invaginate to form vesicles. The hormone is typically metabolized within the endocytotic vesicle and the unoccupied receptors are recycled to the cell membrane.

ACTIVITY OF HYDROPHOBIC HORMONES

The synthesis, secretion, transport, and mode of action of hydrophobic hormones (Figure 7-9) is markedly different from that of hydrophilic hormones. Many of these differences reflect the different solubilities of these two classes of hormones.

FIGURE 7-9

Overview of the synthesis, secretion, transport, and action of hydrophobic hormones. These hormones move freely across the membranes of endocrine and target cells, but have low solubility in blood. Therefore, most of these hormones are bound to serum binding proteins (B) in the circulation.

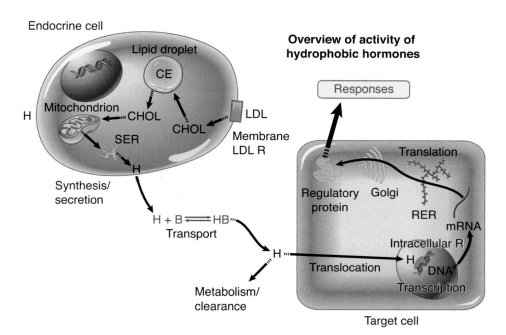

Synthesis

All of the hydrophobic hormones regulating reproductive activity are steroids. Steroids are members of a chemical class of compounds that are characterized by a four-ring (tetracyclic cyclopental[a]phenanthrene) skeleton. Reproductive tissues produce three major classes of steroid hormones: progestins, estrogens, and androgens. Corticosteroid hormones, another general type of steroid, are produced by the adrenal gland and play a role in parturition. We will be concerned with only one to three specific steroids within each class. Figure 7-10 shows the biosynthetic pathway for these steroid hormones. Each

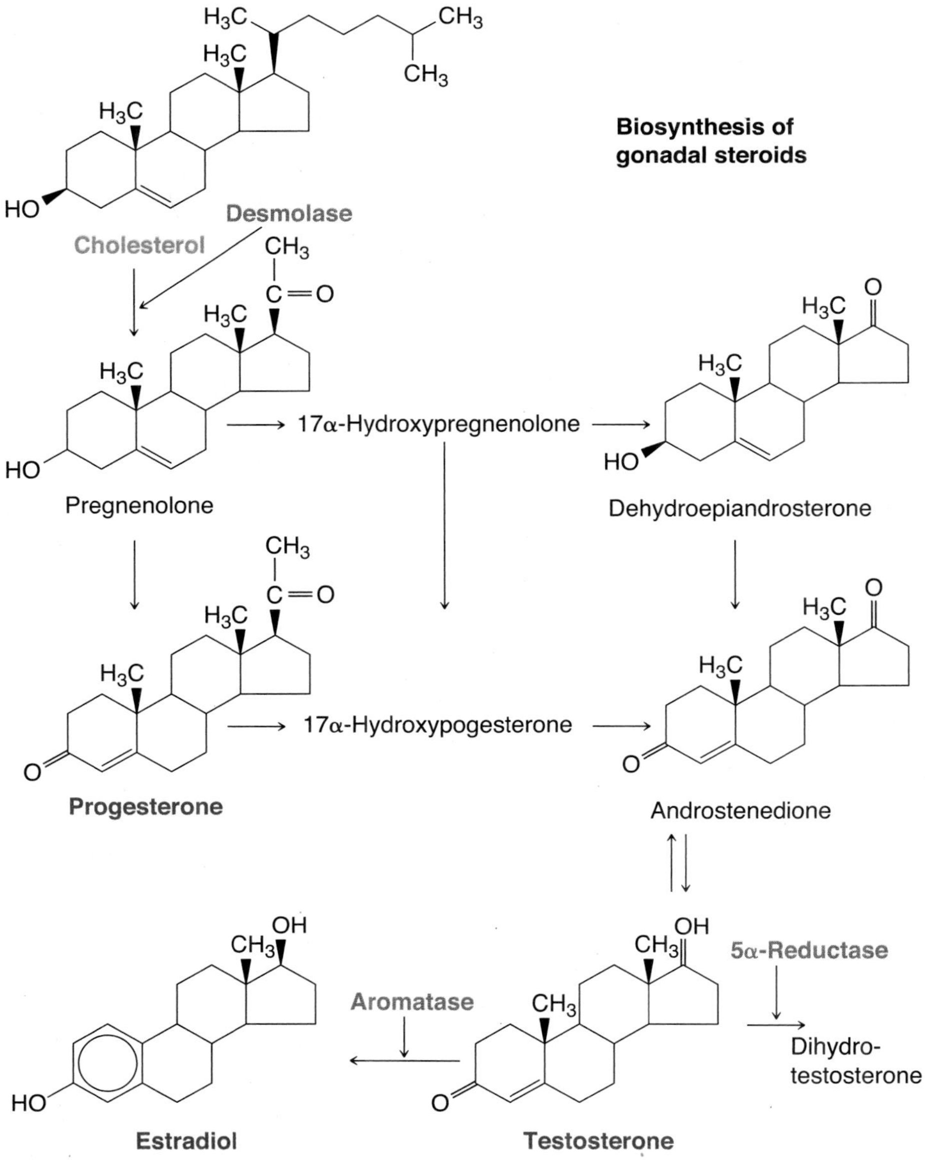

Biosynthesis of gonadal steroids

FIGURE 7-10

Biosynthetic pathway for the most common steroid hormones controlling reproductive activity.

7

of these hormones is derived from cholesterol. It is important to note that the complement of steroid hormones produced varies among steroidogenic tissues. For example, the corpus luteum produces only progesterone whereas the ovarian follicle produces large amounts of estradiol and some androgens. The type of steroid hormones produced by a particular cell type is a function of which steroidogenic enzymes are expressed by the cell. For example, the corpus luteum produces only progesterone because it lacks the enzymes necessary to convert progesterone to estradiol.

The major source of cholesterol is low-density lipoproteins (LDL), which consist of a lipid shell containing phospholipids, cholesteryl esters, and a protein. The protein binds to membrane receptors on many cell types including cells that produce steroid hormones. The LDL-receptor complex is internalized by the cell via endocytosis. Vesicles containing this complex fuse with lysosomes which contain enzymes that liberate free cholesterol. The cholesterol is then re-esterified and stored in lipid droplets. Although cells are capable of *de novo* cholesterol synthesis, most of the cholesterol used for steroid hormone synthesis is derived from pools stored in lipid droplets.

Synthesis of steroid hormones from cholesterol involves enzymes located in the mitochondria and smooth endoplasmic reticulum. The rate-limiting step in steroid hormone production is the transfer of cholesterol to the mitochondria, the site of the first step in steroid hormone synthesis. Because steroids are not very soluble in water, only a minute amount exists in the free form in cytosol. Most of the cholesterol and steroid hormones are bound to low-affinity carrier proteins.

Secretion

Unbound steroid hormones readily diffuse across the cell membrane into extracellular fluids. Unlike cells that produce polypeptide hormones, steroid-producing cells do not have a large capacity for storing their hormones. Aside from a small storage pool provided by carrier proteins, most steroids leave the cell soon after they are synthesized. Thus, secretion of steroid hormones closely reflects the synthetic activity of these cells.

Transport and Clearance

Once steroid hormones enter the blood they interact with several **serum-binding proteins.** The affinity of these proteins for steroid hormones is much lower than that of receptors. However, because binding proteins are present in large concentrations most (>90 percent) of the circulating steroid

**General structure of
intracellular receptor**

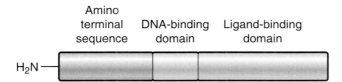

FIGURE 7-11

Schematic depiction
of the molecular struc-
ture of an intracellular
receptor.

7

hormones exist in the bound form. Serum-binding proteins protect steroid hormones from metabolic degradation, thereby providing a substantial storage pool for these chemical messengers. Unbound steroid hormones are readily metabolized to water-soluble forms by peripheral tissues. These metabolites are then eliminated in the urine and feces.

Action

Unbound molecules of steroid hormones that are not metabolized diffuse across the hydrophobic cell membrane of target cells, where they interact with their receptors. There has been considerable controversy regarding the precise intracellular location of steroid hormone receptors. Some steroid hormones interact with a receptor located in the cytosol and then the hormone-receptor complex is translocated to the nuclear compartment where it binds to DNA. The receptors for other steroid hormones reside in the nuclear compartment associated with specific regions of DNA. In either case, interaction of the hormone-receptor complex with the DNA enhances transcription of certain genes that give rise to proteins that regulate activity of the target cell. The intracellular receptors for hydrophobic hormones are complex proteins that express conformational changes upon binding with a hormone. One notable change is that the receptors form dimers and then activate specific genes. These receptors consist of three major domains (Figure 7-11): 1) an amino terminal sequence, 2) a DNA binding domain, and 3) a ligand-binding domain. The ligand-binding domain interacts with the hormone, whereas the DNA-binding domain interacts with the DNA to regulate gene transcription.

REGULATION OF ENDOCRINE SYSTEMS

Our analysis of interactions between endocrine and target cells raises an important question: How are the relationships between these two cell types regulated? We will consider two conditions that require communication between endocrine and target cells. The first case deals with homeostasis: that is, what prevents a hormone from overstimulating its target cell? The second case deals with situations where a rapid response is required; that is, how can a hormone induce a rapid and robust stimulation of its target cell?

Negative Feedback

Homeostasis in endocrine systems is typically maintained by a **negative feedback** relationship between target and endocrine cells (Figure 7-12a). In its simplest form, a negative feedback system consists of a hormone inducing a response in a target cell, and the response in turn inhibiting further release of the hormone. The response that inhibits hormone secretion can be another hormone or some other physiologic change. The negative feedback signal can inhibit hormone release by suppressing synthesis and/or secretion of the hormone.

A negative feedback system with which you are undoubtedly familiar involves home heating systems (Figure 7-12b). In this example, a thermostat, sending an electronic signal to a furnace, is analogous to an endocrine cell producing a hormone that evokes some physiologic effect in a target cell. In the case of the heating system, a drop in temperature is detected by the thermostat. This opens an electronic circuit that causes the furnace to generate

FIGURE 7-12

Schematic illustrations showing the negative feedback relationships between an endocrine and target cell (a) and between a thermostat and furnace (b). In each case the feedback systems serves the purpose of minimizing variations in the controlled variable (biological response or temperature).

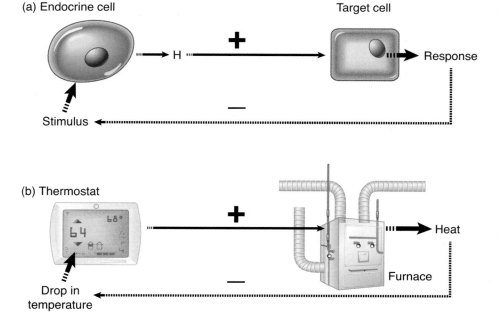

(a) Endocrine cell   Target cell

Stimulus

(b) Thermostat

Drop in temperature

heat. Once a rise in temperature is detected by the thermostat the signal to the furnace is shut off and the furnace turns off.

Positive Feedback

Imagine that instead of shutting off in response to a temperature increase, a furnace responds by accelerating its production of heat (Figure 7-13b). This would be an example of a **positive feedback** system. Such a system is potentially dangerous; that is, the furnace could overheat and self-destruct. Even though such systems seem to defy homeostasis, they exist in some biological systems and play important regulatory roles. A simple example of such a system is shown in Figure 7-13a. In this case an endocrine cell releases a hormone that evokes a response in a target cell which in turn further stimulates the endocrine cell. We will encounter two positive feedback systems in reproductive physiology. In each case these systems serve the purpose of causing rapid increases in hormone levels. Positive feedback systems shut down because the sharp increase in hormone concentrations eliminates the source of the response. For example, during the birthing process, stimulation of the cervix by the emerging fetus stimulates release of oxytocin by the posterior pituitary gland, which induces uterine contractions, which stimulate a greater release of oxytocin. The system shuts down once the fetus leaves the birth canal (i.e., no stimulation of the cervix).

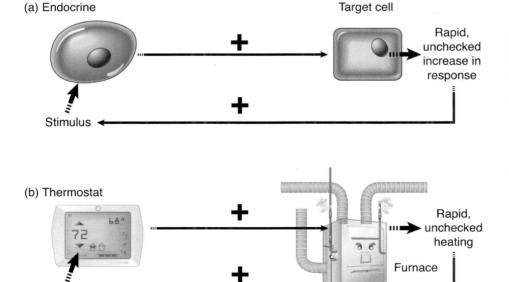

(a) Endocrine — Target cell — Rapid, unchecked increase in response — Stimulus

(b) Thermostat — Rapid, unchecked heating — Furnace — Increase in temperature

FIGURE 7-13

Schematic illustrations showing positive feedback relationships between an endocrine and target cell (a) and between a thermostat and furnace (b). Positive feedback systems produce rapid changes which can be catastrophic.

OVERVIEW OF THE ENDOCRINE SYSTEMS CONTROLLING REPRODUCTION

Having described the major reproductive hormones as well as general principles of endocrine physiology it is now possible to provide an overview of the endocrine system that regulates reproductive activity in mammals. Figure 7-14 shows major endocrine tissues, reproductive hormones, target tissues upon which these hormones act, and the feedback systems that sustain homeostasis within this systems. For the time being, it is only necessary for you to become familiar with the general flow of information between the various organs controlling reproduction. Information from higher brain centers converge upon the hypothalamus which integrates numerous neuronal inputs and transduces this type of information into two major hormone signals: oxytocin and gonadotropin-releasing hormone (GnRH). Oxytocin affects the reproductive tract and the mammary gland (not shown). GnRH stimulates the release of luteinizing hormone (LH) and follicle-stimulating hormone (FSH), which in turn regulate gonadal activity. In some species (rodents) another pituitary hormone (prolactin) also influences gonadal activity. Release of prolactin by the anterior pituitary gland is controlled by dopamine, a biogenic amine that suppresses release of prolactin. One of the major effects of these gonadotrophic hormones is to stimulate release of various hormones from the ovaries and testes. These gonadal hormones regulate the reproductive ducts and

FIGURE 7-14

Schematic illustration summarizing the endocrine communications among the major reproductive organs.

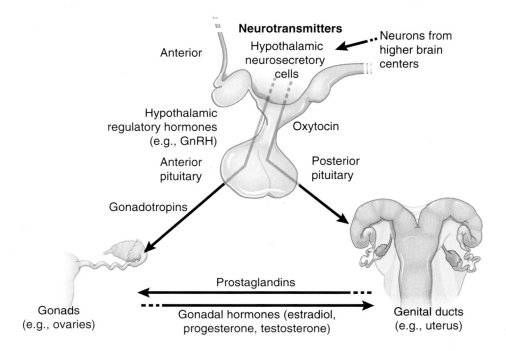

exert positive and negative feedback effects on the brain and pituitary gland to regulate gonadotropin secretion.

ASSESSING ENDOCRINE PHYSIOLOGY

As indicated earlier, most of our understanding of reproductive endocrinology is based on interpretation of hormone patterns in blood. Characterizing patterns of hormones is one of the most commonly-used methods for characterizing activity of endocrine cells. The experimental approach for doing so involves measuring concentrations of a hormone in blood samples taken sequentially over a period of time. Because so much of our knowledge of reproductive physiology is based on analysis of hormone patterns it is necessary to understand how hormone concentrations are determined as well as how they should be interpreted.

Measuring Hormones

A standard approach to evaluating the reproductive status of an animal is to collect blood samples and measure concentrations of particular reproductive hormones in serum or plasma. Before the 1960s it was impossible to characterize patterns of hormones in blood. None of the traditional chemical methods for measuring various compounds were sensitive enough to detect the low concentrations of hormones found in biological fluids. This all changed in 1969 when Berson and Yalow discovered that it was possible to generate highly-specific antibodies to hormones. This breakthrough led to development of the **radioimmunoassay (RIA),** a highly sensitive and specific method for measuring physiologic levels of hormones in biological fluids.

Unlike chemical assays, the RIA indirectly measures the amount of hormone present in a sample. Reagents for the assay include a highly purified form of hormone that is labeled with a radioactive isotope (^{125}I or ^{3}H), purified (nonradioactive) hormone in known concentrations (standard), and an antibody that specifically binds radioactive and nonradioactive forms of the hormone with the same affinity. The interaction among these reagents can be expressed as a chemical reaction, as follows:

$$H + H^* + Ab \leftrightarrows HAb + H^*Ab$$

Where H is the nonradioactive hormone, H* is the radioactively labeled hormone (label) and Ab is the antibody. The reaction is allowed to proceed to equilibrium. Due to the law of mass action, the amount of product formed depends on the affinity of the antibody and the initial concentration of reactants. Suppose we conduct this reaction in a number of different test tubes

FIGURE 7-15

Schematic illustration depicting the fundamental principle of radioimmunoassays. An antibody specific for a particular hormone can bind a labeled and an unlabeled hormone with the same affinity. Therefore, the proportion of labeled hormone bound to the antibody is inversely proportional to the amount of unlabeled hormone present in a sample.

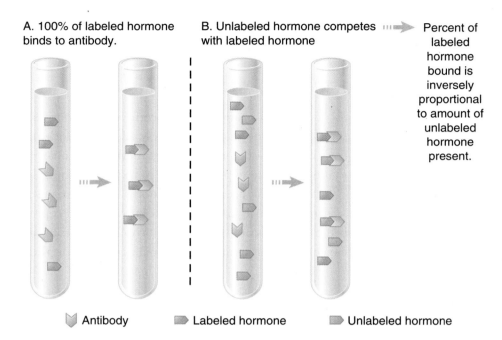

A. 100% of labeled hormone binds to antibody.

B. Unlabeled hormone competes with labeled hormone

Percent of labeled hormone bound is inversely proportional to amount of unlabeled hormone present.

Antibody   Labeled hormone   Unlabeled hormone

where the amount of H is varied among tubes, but the amount of antibody and the amount of H* are held constant. Under these circumstances, the amount of H is the only variable that determines the amount of H*Ab formed. Thus the amount of H*Ab formed is inversely proportional to the initial amount of H (Figure 7-15). Now suppose we measure the amount of H*Ab in each of the test tubes and create a graph of the concentration of H added verses the concentration of H*Ab formed (Figure 7-16). From this graph it is clear that there is a mathematical relationship between the amount of H present in the assay tube and the amount of H*Ab formed by the reaction. Thus we can represent the concentration of a hormone in a blood sample as a function of the amount of radioactive label bound to the antibody. This means we can subject a sample containing an unknown amount of H to the assay, and measure the resulting H*Ab. This corresponds to a particular concentration of standard H, which provides an accurate estimate of hormone concentration in our sample.

Radioimmunoassay is still the most common way to assess hormone concentrations. However, a derivative of this technique called the enzyme-linked immunosorbent assay (ELISA) has become more popular in recent years. The only difference between the RIA and ELISA is the nature of the labeled hormone. Instead of a radioactive label, the hormone is labeled with an enzyme that catalyzes a chemical reaction that produces a color change. After the labeled hormone, unlabeled hormone, and antibody are allowed to react, and antibody-bound hormone is separated from free hormone, the enzyme

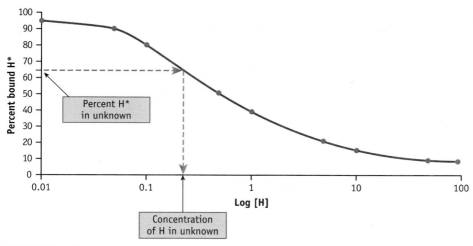

FIGURE 7-16

A radioimmunoassay standard curve showing the relationship between the concentration of unlabeled hormone (H) and the percent of labeled hormone (H*) bound to the antibody. The concentration of unlabeled hormone in biological sample (the unknown) is a function of the percent binding of H*.

substrate is added, and the amount of color generated is directly proportional to the amount of labeled hormone bound to the antibody. This of course is inversely proportional to the amount of unlabeled hormone in the sample.

Interpretation of Hormone Profiles

The ability to monitor circulating concentrations of reproductive hormones made it possible for researchers to describe patterns of hormones associated with a variety of physiologic states including sexual development, puberty, reproductive cycles, and pregnancy. As noted earlier, changes in blood concentrations of a hormone can be affected by various mechanisms. Therefore, it is important that you understand how to interpret circulating hormone profiles.

Hormones usually fluctuate at basal levels within a particular homeostatic range (Figure 7-17). These tonic or basal concentrations reflect negative feedback control of hormone secretion as well as the rate of hormone clearance from the circulation. In some cases hormone concentrations rise rapidly to amounts that greatly exceed basal levels. If these elevations are prolonged, lasting several hours, they are referred to as surges. Some surges can be induced by positive feedback (Figure 7-17). The ascending portion of the

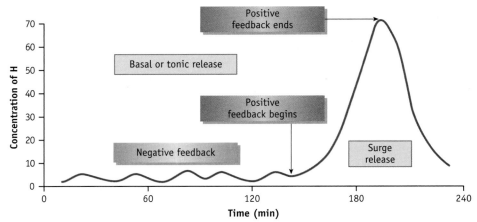

FIGURE 7-17

Two patterns of hormone concentrations observed in the blood reflecting different types of release. Negative feedback systems typically sustain a basal or tonic release where concentrations exhibit small fluctuations within a narrow range. In contrast, positive feedback systems usually evoke a rapid increase in hormone concentrations.

response is attributed to enhanced release of the hormone. Concentrations begin to decrease once release of the hormone stops. In a true positive feedback system a stimulus enhances hormone release and the resultant increase in hormone concentration further enhances generation of the stimulus. Not all hormone surges are generated by positive feedback. In some cases a stimulus induces massive release of the hormone, but the hormone does not influence the stimulus.

It is important to understand that the pattern of hormone observed depends on how often and how long blood samples are collected. Measuring hormone concentrations in samples taken infrequently (e.g., once per day or once per week) permit characterization of long-term changes (over days, weeks, or months), but does not reveal short-term (over hours or minutes) fluctuations. Assessing hormone concentrations in blood samples taken at frequent intervals for several hours can reveal a great deal about regulation of hormone secretion. Figure 7-18 depicts a typical pattern of LH in blood. The most striking feature of this type of profile is that concentrations of the hormone fluctuate in a pulsatile or episodic manner. This type of pattern is characterized by a rapid increase (pulse) in hormone concentration, followed by a more gradual decrease. The highest concentration achieved within a particular pulse is called the peak, whereas the lowest concentration between pulses is called the nadir. Such patterns are usually described in terms

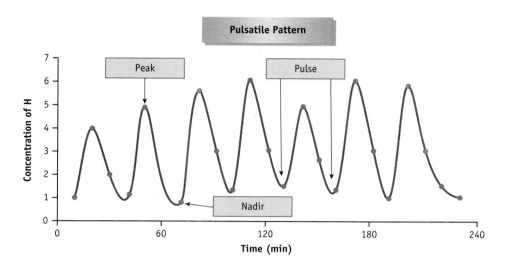

FIGURE 7-18

The pulsatile pattern of hormone concentrations typical for many hormones.

of the frequency and average amplitude of the pulses. Frequency refers to the number of pulses per unit of time. Amplitude is the difference between the peak and nadir. In some cases, the average time between pulses is quantified. This is known as the period, and it is the inverse of the pulse frequency.

Other terms refer to other types of hormone patterns. In some cases, hormone concentrations fluctuate in accordance with time of day or time of year. When a hormone rhythm has a period of about 24 hours, it is called a **circadian rhythm.** A rhythm with a period of approximately 1 year is called a **circannual rhythm.**

Circulating patterns of hormones are of little use unless we understand their physiologic bases; that is, understanding how they are generated and what effects they perpetuate. It is generally assumed that hormone patterns in blood reflect secretion of the hormone. This is not a reasonable assumption because mechanisms such as metabolism and clearance also play insignificant roles in determining hormone patterns in blood. With respect to the pulsatile pattern of LH, it appears that each pulse is the result of a short-term (several minutes) release of LH by the pituitary gland, followed by clearance from the blood. Another concern associated with interpreting hormone patterns deals with the specificity of the assay used to quantify hormone concentrations. Some assays detect only the free form of the hormone, whereas others measure bound and free forms. Other considerations include the extent to which assays detect metabolites or isoforms of hormones. These considerations are important because only some forms of the hormone have biological activity. Thus if one is unaware of what the assay is measuring, it is possible to generate hormone profiles that have little to do with the biological status of the animals studied.

7

SUMMARY OF MAJOR CONCEPTS

- Homeostatic relationships among various tissues are maintained by the endocrine and nervous systems.
- Hormones are intercellular messengers present in low concentrations in extracellular fluids.
- Mechanisms of synthesis, secretion, transport, metabolism, and action differ between hydrophilic and hydrophobic hormones.
- The ability of a hormone to induce a biological effect depends on its concentration, the concentration of its receptors in target cells, and the affinity of its receptor for the hormone.
- Endocrine systems are typically regulated by negative feedback loops; positive feedback loops exist, but are rare.
- Circulating patterns of hormones can be easily monitored, but their physiologic significances may be unclear in the absence of other information.

DISCUSSION

1. Glucose can be released from the liver and transported to many other tissues where it is taken up and induces a variety of biological effects. Glucose concentrations in animals range between 40 and 150 mg/100 mL. Based on this information, would you consider glucose to be a hormone? Why or why not?

2. GnRH, a peptide hormone produced by hypothalamic neurosecretory cells, and estradiol, a steroid hormone produced by the ovary, each exert effects on LH-producing cells of the anterior pituitary gland. In general terms, compare and contrast how these hormones act to affect LH release by pituitary cells.

3. Plasma concentrations of LH increase in males following removal of both testes. If the castrated male is then injected with testosterone, concentrations of LH fall to levels observed before removal of the testes. Explain these results based on your understanding of how endocrine systems are regulated.

4. Suppose you want to assess the potency of a synthetic estrogen (relative to the naturally occurring estradiol) with respect to effects on epithelial cells of uterine endometrium. Describe how you would do this. Show a graph of your expected results.

5. Suppose you measure plasma concentrations of testosterone in bull elk throughout the year and discover that concentrations increase during the rutting season. You conclude that this is due to increased secretion by the testes. Is this a valid conclusion? Why or why not?

REFERENCES

Berthold, A.A. 1849. Transplantation der Hoden. *Archives of Anatomy and Physiology. Wissenschaftliche Medicin.* 16:42–46.

Cannon, W.B. 1960. *The Wisdom of the Body.* New York: W.W. Norton.

Hadley, M.E. and J.E. Levine. 2007. *Endocrinology,* sixth edition. Upper Saddle River, NJ: Pearson Education.

Starling, E.H. 1905. The chemical correlation of the functions of the body. *Lancet* 1:340–341.

Wilson, J.D., D.W. Foster, H.M. Kronenberg and P.R. Larsen. 1998. Principles of endocrinology. In *Williams Textbook of Endocrinology,* ninth edition. Philadelphia: W.B. Saunders, pp. 1–10.

7

CHAPTER 8 | Puberty

CHAPTER OBJECTIVES

- Define and characterize puberty in mammals.

- Describe the biological significance of puberty.

- Provide a theoretical framework for understanding puberty.

- Describe the physiologic events leading to onset of puberty.

- Describe the physiologic mechanisms regulating timing of puberty onset.

CHARACTERISTICS AND SIGNIFICANCE OF PUBERTY

This chapter marks the beginning of detailed discussions concerning physiologic mechanisms controlling reproduction in mammals. These discussions rely heavily on the background information provided in earlier chapters. Therefore, you should make sure you have a firm understanding of terminology, anatomy, sexual differentiation, and basic endocrinology. We begin with an analysis of sexual maturation, which can be viewed as a continuation of the previous discussion of sexual differentiation.

What is Puberty?

It is likely that humans became cognizant of the biological changes associated with sexual maturity long before recorded history. The transition between childhood and adulthood is recognized as a significant event in virtually all human societies, and in many cases is celebrated by rites of passage. The physical and emotional changes that children express as they become adults are attributed to development of various secondary sex characteristics. One of the more noticeable changes is the development and distribution of body hair. In fact, the term puberty is derived from the Latin word *pubescere,* which means "becoming covered with hair." This literal translation is quite anthropocentric because it refers to a change that occurs only in humans. In other mammals a change in pelage does not accompany sexual maturation.

In the modern sense of the word, puberty refers to all of the physiologic, morphologic, and behavioral changes that occur in association with developing the ability to reproduce. This involves *both* the ability to produce viable gametes (gonadal maturation) and the behavioral capacity to engage in sexual activity (behavioral maturation).

Specifically, these abilities require maturation of the genital organs and development of secondary sex traits. It is important to understand that puberty is a process involving temporal changes in the reproductive system. Thus, there is no simple definition of puberty. For most female mammals (those that express estrous cycles as adults), puberty is typically assumed to be completed when an individual first expresses sexual receptivity (estrus). In the so-called higher primates, puberty is usually thought of as the time at which first menstruation occurs. However, neither of these definitions is adequate because the first estrus and first menstruation aren't necessarily correlated with fertility. A more useful definition of puberty in females is the time at which an individual ovulates and experiences a fertile estrous or menstrual cycle. In males, puberty occurs when the individual expresses copulatory behavior and produces sufficient viable spermatozoa to impregnate a female.

Although development of the gonads and reproductive behavior are intimately related, they are distinct processes involving different neurobiological mechanisms. This is illustrated by the fact that in some species such as cattle, the first expression of estrus is not normally accompanied by ovulation. In this chapter we will view puberty in terms of maturation of the gonads and focus our attention on mechanisms controlling development of the gonads.

Biological Significance of Puberty

The reproductive success of an individual depends on the number of offspring it produces in a lifetime. According to this Darwinian perspective, one might conclude that an animal that begins reproducing at an early age will have a high reproductive rate in its lifetime. However, there are significant biological risks associated with early sexual maturity. Recall from our earlier discussions that reproduction, like all other biological activities, depends on availability of metabolic energy. Although animals that express reproductive activity at early ages have the opportunity to produce more offspring in a lifetime than later maturing animals, they might also have their reproductive abilities compromised by not having sufficient energy to find and defend territories and mates as well as care for their young. Moreover, they may experience higher mortality rates because they divert less energy to life-sustaining processes. In light of these trade-offs it would seem that the timing of sexual maturation is a critical determinant of lifetime reproductive success and therefore is subject to natural selection. Indeed, selection pressure seems to have promoted considerable plasticity in age at puberty among mammals. Age at puberty varies considerably within a particular species, and depends largely on environmental conditions. As a rule, mammals become pubertal

8

only when the opportunity for successful reproduction arises; for example, when the individual has access to adequate dietary energy to support vital processes and lives in social and physical conditions that promote high reproductive success. This means that mechanisms governing sexual maturation are responsive to cues that provide information about the individual's internal and external environments.

Management of domestic livestock relies heavily on the principles discussed in the previous paragraph. For example, the greatest source of variation in lifetime production of beef cows is the age at which the cow first produces a calf. The same thing can be said for sheep, swine, or any animal used for food production. This realization is responsible for the fact that most modern production systems emphasize management of the developing female in ways that minimize age at puberty. However, as noted previously, there are significant costs associated with this type of management. In order to reach puberty at an early age, an animal must be fed a high plane of nutrition to ensure that it maintains adequate growth as well as other vital processes. Feed costs represent one of the major expenses of livestock producers. Unless there is abundant energy available at low costs, it may not be profitable to manage animals to reach puberty at an early age. In some climates (e.g., parts of North America and Europe) it may be economically feasible to feed livestock high planes of nutrition to minimize age at puberty. Such practices may not be possible under different circumstances.

THEORETICAL FRAMEWORK FOR PUBERTY

The previous discussion on the biological importance of puberty illuminates an extremely important concept; that is, the timing of puberty is a major determinant of the lifetime reproductive success of an individual. From a theoretical perspective, this means that a comprehensive understanding of the mechanism of puberty onset requires insight into how sexual development proceeds, as well as how the time of puberty onset is determined.

Permissive Signals and a Developmental Clock

Decades of research involving laboratory animals, domestic animals, and humans has lead to development of a general theory for puberty onset (Figure 8-1). The theory consists of two parts. First, it is generally accepted that multiple "permissive signals" determine when onset of puberty begins. These signals provide information about an individual's metabolic status, its social relationship with other individuals, as well as the physical environment in

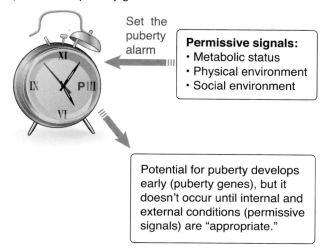

Developmental clock:
Rate of ticking is set internally and represents sequential expression of "puberty genes."

Set the puberty alarm

Permissive signals:
• Metabolic status
• Physical environment
• Social environment

Potential for puberty develops early (puberty genes), but it doesn't occur until internal and external conditions (permissive signals) are "appropriate."

FIGURE 8-1

A well-accepted model for understanding the regulation of puberty onset in mammals. The rate of sexual development is determined by a developmental clock which times the expression of genes (puberty genes) that regulate the physiologic mechanisms governing maturation of the reproductive system. Activation of these physiological mechanisms is dependent on several "permissive signals." In this way, an animal is capable of attaining puberty at any time after a particular age, but does not become reproductively active until various internal and external signals are present.

which it lives. These signals work in consort with each other and collectively determine when the individual becomes pubertal. The second part of the theory involves a "developmental clock." According to this idea, the development and coordinated activities of the reproductive organs unfold in an ordered fashion and results from expression of particular "puberty genes" (Figure 8-2). Some of the more important puberty genes govern:

• Hypothalamic release of gonadotropin-releasing hormone (GnRH).

• Release of luteinizing hormone (LH) and follicle-stimulating hormone (FSH) by the pituitary gland.

• Release of gonadal hormones.

• Positive and negative feedback systems.

Of these regulatory genes, those mediating responses to permissive signals may be most important. For example, the extent to which an increase in

FIGURE 8-2

Onset of puberty is dependent on expression of various puberty genes that regulate key components of the hypothalamic-pituitary-gonadal axis. The order in which these genes are expressed may differ among species.

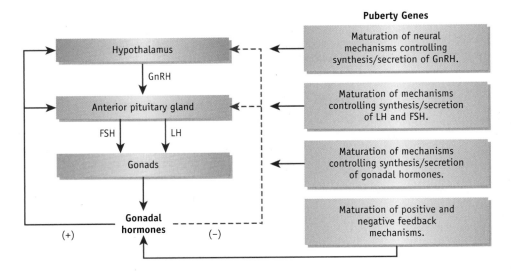

FIGURE 8-3

Application of the prevailing model of puberty onset to explain how dietary energy influences timing of puberty onset in heifers. According to the model, the hypothalamic-pituitary-gonadal axis is developed by 6 months of age. Animals on a high plane of nutrition generate permissive signals earlier than those on a low plane of nutrition causing them to attain puberty at an earlier age.

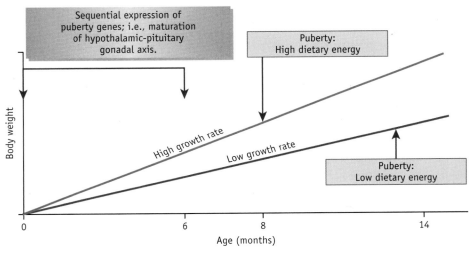

dietary energy intake enhances puberty depends on the extent to which feedback mechanisms controlling LH secretion are developed. Using the clock metaphor, timing of puberty is regulated by permissive signals that influence when the puberty alarm sounds, rather than the rate at which the clock ticks.

Theories are considered true only if they have high explanatory power. How well can the aforementioned theory of puberty explain well-documented claims concerning the effects of various environmental factors on sexual development? For example, how is it that animals fed high-energy diets reach puberty at earlier ages than those fed low-energy diets? Consider beef heifers for instance (Figure 8-3). These animals typically attain puberty at an average of 12 months of age. However, if they are raised on a low plane of nutrition (e.g., grazing poor-quality pastures or rangeland), or if they are depleting energy reserves to cope with internal parasites or stress, they may not become sexually mature until much later. In contrast, feeding an extremely high plane of nutrition induces puberty as early as 6 months of age. These observations support the previously mentioned idea that age at puberty is plastic; that is, it can occur at any time after a certain minimum age. According to the aforementioned theory, the developmental clock appears to be fully developed (i.e., puberty genes are expressed) by 6 months of age, but the puberty alarm doesn't sound until nutritional (permissive) signals allow it to (i.e., until certain nutritional signals appear). Other signals also influence timing of puberty. For example, demographics (e.g., sex ratio and presence of mature individuals) and seasonal changes in environment (e.g., temperature and day-length) play roles in determining when the puberty alarm sounds.

Importance of GnRH in Puberty

The precise nature of the developmental clock controlling puberty onset remains elusive. This notion is complicated by the fact that each reproductive organ undergoes a series of developmental changes leading up to puberty. In addition, successful reproduction depends on the coordinated activity of all components of the reproductive system. Thus, activity of one reproductive tissue affects and is affected by other reproductive tissues. There is variation among mammalian species regarding the nature of the developmental clock. In later sections, we will examine three major developmental patterns: rodents, domestic ungulates, and primates. Although the sequence of puberty gene activation appears to differ among these groups, mechanisms regulating secretion of GnRH and LH are of critical importance in all cases.

In spite of differences in developmental steps leading to puberty among mammals, there are some important similarities. A voluminous amount of puberty research supports the hypothesis that development of the hypothalamic-pituitary system is of pivotal importance in sexual maturation. In particular, a high-frequency pattern of LH release appears to be a necessary condition for puberty onset in all species of mammals. This depends on the ability of the hypothalamus to release GnRH in an episodic manner. Thus research that seeks to illuminate the mechanisms controlling sexual maturation focuses heavily on regulation of GnRH release.

Regulation of GnRH Release

GnRH is a neurohormone produced by neurosecretory cells that are sparsely distributed in the anterior hypothalamus. These cells release GnRH into the hypothalamic-hypophysial portal vessels in pulses. The hormone then travels to the anterior pituitary gland and induces the release of LH. The pattern of LH release from the pituitary gland is also pulsatile and corresponds closely

8

FIGURE 8-4

Depictions of hypothetical patterns of GnRH (red) and LH (blue) in the portal and peripheral circulations. These types of patterns have been documented in several species of mammals including laboratory rodents, monkeys, sheep, and cattle.

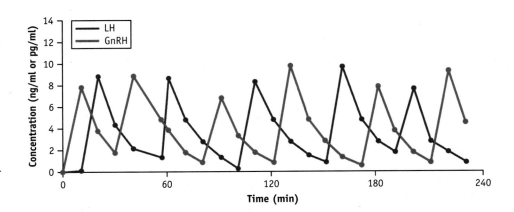

to the pattern of GnRH release. This relationship can be seen when concentrations of GnRH and LH are measured in sequential blood samples collected simultaneously from hypophysial portal and peripheral blood vessels (Figure 8-4). The close association between patterns of GnRH and LH means that changes in LH concentrations in peripheral blood provide an accurate characterization of GnRH release from the hypothalamus. Because of the difficulty in measuring GnRH concentrations in portal blood, most studies rely on LH patterns to assess GnRH release.

GnRH Pulse Generator

The neural mechanisms regulating the pulsatile pattern of GnRH release are thought to involve a "GnRH pulse generator." Although this is an extremely useful concept, the anatomic nature of such a system has not been adequately described. Precise characterization of the GnRH pulse generator is difficult because 1) the number of GnRH neurons (1,000–3,000 cells) is very small compared to other types of nerve cells, 2) GnRH neurons are scattered and not clustered in particular hypothalamic nuclei, and 3) the innervation of GnRH neurons is extremely sparse compared to neighboring nerve cells. Although we know very little about the physical nature of the GnRH pulse generator, we do know that the system is subject to regulation by various neuronal and humoral mechanisms. As you will soon learn, maturation of the GnRH pulse generator appears to be an important rate-limiting step leading to onset of puberty. In other words, animals do not attain puberty until they can exhibit adult patterns of GnRH release. Using language consistent with the prevailing theory of puberty onset, one can say that expression of genes that comprise the pulse generator is a prerequisite for puberty onset, but the precise timing of the onset of adult patterns of GnRH depends on the presence of appropriate permissive signals.

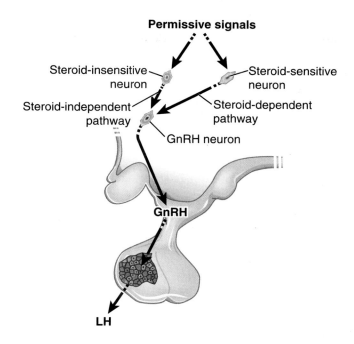

Permissive signals

Steroid-insensitive neuron

Steroid-sensitive neuron

Steroid-independent pathway

Steroid-dependent pathway

GnRH neuron

GnRH

LH

FIGURE 8-5

Schematic illustration of neuro-pathways whereby various permissive signals might regulate the pulsatile secretion of GnRH. Signals reflecting environmental and social status can influence the GnRH pulse-generating system via steroid-independent and steroid-dependent neuronal pathways.

8

Steroid-Independent and Steroid-Dependent Control

Two major types of neural mechanisms regulate the pulsatile secretion of GnRH: steroid-independent and steroid-dependent (Figure 8-5). The former type regulates GnRH release in the absence of gonadal steroids, whereas the latter type requires the presence of gonadal steroids. Steroid-independent changes in patterns of LH secretion typically involve changes in patterns of GnRH release. In contrast, changes in sensitivity of the hypothalamic-pituitary axis to the feedback actions of gonadal steroids mediate steroid-dependent control of LH release. A change in sensitivity of a target cell to a particular hormone means that the response to a particular amount of hormone changes. This concept can be expressed mathematically as $T_s = \Delta R / \Delta[H]$, where T_s is the sensitivity of the target cell, ΔR is the change in response of the target cell, and $\Delta[H]$ is the change in concentration of the hormone. For example, the amount of estradiol required to suppress LH concentrations in ovariectomized females increases during the late prepubertal period. In terms of the sensitivity formula, $\Delta R / \Delta[H]$ becomes smaller as females approach sexual maturity. Changes in sensitivity to hormones are undoubtedly attributed to changes in hormone receptors. In some cases, this involves changes in receptor number. However, changes in receptor type and postreceptor events might also be involved.

The sites at which gonadal steroids act to influence GnRH release have not been precisely identified in most species. However, it is clear that these hormones do not act directly on GnRH neurons: few if any receptors for steroid hormones have been found on GnRH neurons. Therefore, steroid-dependent effects appear to be mediated by other neurons that innervate GnRH-producing cells.

Two steroid-dependent mechanisms control GnRH release; that is, positive and negative. Gonadal steroids feed back on the hypothalamus to influence GnRH release in both positive and negative ways. You should understand that positive and negative feedback mechanisms are separate and most likely involve different populations of GnRH neurons. With respect to puberty, we will be concerned with the negative and positive feedback actions of estradiol in females and the negative feedback actions of testosterone in males. The negative feedback effects of estradiol and testosterone result in a low-frequency pattern of GnRH release. The positive feedback action of estradiol produces an LH surge.

MECHANISMS CONTROLLING PUBERTY

The vast majority of research on puberty has focused on the female. Although much of this information is applicable to the male, there are important differences. In the next several sections, we will examine mechanisms controlling puberty in several types of females, and then consider some general characteristics that are unique to males.

Regulation of Puberty in Females

As noted earlier, sexual maturity involves both behavioral and physiologic development. Our current discussion will emphasize the physiologic changes that result in the ability to reproduce. With respect to females, the ability to ovulate is the physiologic endpoint of greatest concern. Therefore, our analysis of puberty in mammalian females will be restricted to events leading to first ovulation. We know virtually nothing about the physiology of sexual maturation in most mammalian species. Almost all of the research in this area has been confined to laboratory rodents, the so-called higher primates (Old World monkeys, apes, and humans), and domesticated ungulates. Because there are important differences among these groups, we will consider each group separately.

Rodents

Virtually all of our knowledge of rodent reproduction is based on research done with rats. Rats were first used for research in the mid-nineteenth century, and were domesticated during the early twentieth century. Since that time, laboratory rats have been used extensively for all types of biological and medical research, including reproductive physiology. Much of our understanding of mammalian reproduction stems from experiments involving this species.

As with all species, the sexual development of rats encompasses both the pre- and postnatal periods. The gestation period of rats is 22 to 23 days in length. Studies of postnatal development are facilitated by dividing the period between birth and puberty into neonatal (0 to 7 days), infantile (8 to 21 days), juvenile (21 to 35 days) and peripubertal (35 to 60 days) periods. Rat pups are born at a highly immature stage of development (comparable to a 150-day-old human fetus) and reach puberty at an early age (35 to 45 days).

In spite of the fact that the neonatal rat is underdeveloped compared to other mammals, components of its reproductive system are intact and functional at the time of birth or shortly thereafter. Neurosecretory cells in the hypothalamus are producing GnRH by day 17 and 18 of gestation, and LH and FSH can be detected in the anterior pituitary gland by 21 days of gestation. However, circulating concentrations of these gonadotropins remain low until birth. Production of steroid hormones and the appearance of primordial follicles occur within the first 2 days after birth.

The sequence of major physiologic events leading to first ovulation in the female rat is summarized in Figure 8-6. As mentioned previously, these events can be understood as time points on a developmental clock, and represent expression of critical puberty genes. Among the most important of these events are three "activational periods of the hypothalamic-pituitary unit." These events set into motion responses that coordinate activity of various reproductive tissues culminating in first ovulation.

8

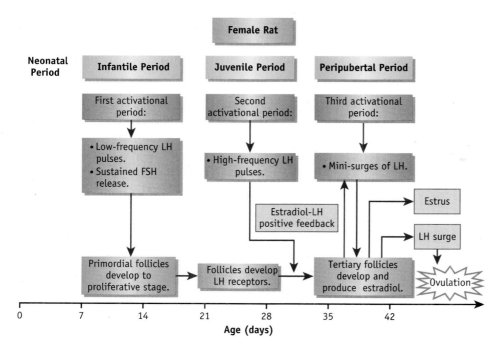

FIGURE 8-6

Mechanism of puberty onset in the female rat. Events highlighted in red, yellow, and green correspond to changes in hypothalamic-pituitary activity, patterns of gonadotropins, and changes in ovarian activity, respectively. Arrows indicate cause-effect relationships between events.

The first activational period occurs during the infantile stage of development. This period is characterized by elevated concentrations of FSH and variable concentrations of LH in blood. These patterns of gonadotropins have been attributed to disorganized activity of GnRH-secreting cells. The resulting pattern of GnRH release induces a sustained release of FSH, but only low-frequency pulses of LH. The elevated concentrations of FSH stimulate development of some primordial follicles, moving them into a pool of follicles that will develop further (i.e., a proliferative pool).

Much of the juvenile period of development is characterized by low circulating concentrations of LH and FSH. This is due to the fact that GnRH release by the hypothalamus is restrained during this period of development. This restraint is likely attributed to a dominance of inhibitory neuronal inputs regulating GnRH-secreting neurons; that is, a central restraint. A second activational period occurs at the end of the juvenile period. At this time, the influence of inhibitory inputs diminishes and the effects of excitatory inputs increase. This results in activation and synchronization of GnRH neurons, which causes a high-frequency mode of pulsatile LH secretion. Interestingly, this pattern of LH secretion is confined to the afternoon. The occurrence of high-frequency LH pulses has tremendous physiologic significance. The elevation in LH resulting from this pattern of secretion coincides with appearance of LH receptors in the ovary. An increase in LH release, coupled with the ability of follicles to respond to LH, results in enhanced development of the ovarian follicles that entered the proliferative pool during the infantile period. Follicle development during the juvenile period is also stimulated by FSH, other pituitary hormones (prolactin and growth hormone), as well as neurotransmitters produced by neurons that innervate the ovaries. One of the more significant consequences of this increased follicular development is an increase in production of estradiol by the ovaries. During the juvenile period, an estradiol-LH positive feedback system has begun to develop. This means that estradiol (produced by the ovaries) feeds back on the hypothalamus and to further enhance GnRH release. This results in the appearance of "mini-surges" of LH during the afternoon. These mini-surges of LH further enhance production of estradiol by the ovaries.

The third and final activational period marks the transition between the juvenile and peripubertal periods, and involves maturation of the estradiol-LH positive feedback system. During the juvenile period, a population of GnRH neurons begins developing the capacity to release increased amounts of GnRH in response to elevated concentrations of estradiol, causing mini-surges of LH. Once these neurons become fully responsive to estradiol, they gain the capacity to elicit a full surge of LH in response to elevated estradiol concentrations. By the time this occurs, ovarian follicles have become fully developed and are maximally responsive to gonadotropins. The massive increase in LH that characterizes the LH surge induces rupture of pre-ovulatory follicles and release of oocytes.

In addition to inducing a pre-ovulatory surge of LH, the rise in estradiol caused by follicle maturation induces estrus behavior. In this way sexual receptivity is synchronized with ovulation, thereby enhancing the chance of a fertile mating.

Before we discuss puberty in other mammalian species, it is important to mention that the negative feedback actions of ovarian steroids do not seem to play an important role in the sexual development of rodents. The increase in LH secretion that is pivotally important in pubertal development in rats is attributed to changes in neuronal inputs that activate and synchronize activity of GnRH neurons. In domestic ungulates the prepubertal increase in LH secretion is attributed to a decrease in response to the negative feedback actions of estradiol. The female rat expresses a similar change in response to estradiol negative feedback, but this occurs after (not before) first ovulation.

Domestic Ungulates

Much of what has been learned about the endocrine mechanisms governing puberty onset in rodents also applies to domestic ungulates. In each case, onset of puberty depends on four necessary conditions: 1) the ability of the hypothalamic-pituitary unit to produce high basal levels of LH (i.e., high-frequency pulses), 2) the ability of the ovaries to develop pre-ovulatory follicles and produce high levels of estradiol in response to elevated LH secretion, 3) the ability of the hypothalamic-pituitary unit to elicit a pre-ovulatory surge of LH in response to high levels of estradiol, and 4) the ability of pre-ovulatory follicles to ovulate in response to an LH surge. Of these four conditions, the first is the last to be achieved, and therefore appears to be the event that ultimately determines the timing of puberty onset. As noted in the previous discussions, the prepuberal rise in basal LH secretion in rats is due primarily to steroid-independent mechanisms, in particular removal of the central restraint that suppresses GnRH release. In domestic ungulates, steroid-dependent mechanisms play a major role in keeping LH pulse frequencies low during the prepubertal period. Specifically, the hypothalamic-pituitary unit becomes less sensitive to estradiol negative feedback within the last few weeks of the prepubertal period, allowing LH pulse frequencies to increase and stimulate follicle growth to the pre-ovulatory stage.

8

BOX 8-1   Focus on Fertility: The Gonadostat Hypothesis

The so called gonadostat hypothesis is the earliest attempt to explain onset of puberty in the rat. It was developed during the 1930s and gained wide support until it was refuted in the early 1980s. Ironically, the hypothesis has been proven to be applicable to species other than the one in which it was originally developed.

According to the gonadostat hypothesis, the increase in LH pulse frequency that is necessary for gonadal maturation results from a resetting of the hypothalamic-pituitary system (the gonadostat) to the negative feedback actions of gonadal steroids (testosterone in males and estradiol in females). The earliest evidence supporting this idea comes from observations regarding age-related changes in response to gonadodectomy in rats. Removal of the gonads results in formation of "castration cells" in the anterior pituitary gland. Castration cells reflect the removal of the negative feedback actions of gonadal steroids on pituitary cells that produce gonadotropins. Formation of castration cells is prevented if animals are provided with injections of gonadal steroids. The fact that the dose of estradiol required to prevent formation of these cells in immature female rats is only 1 percent of that required to produce the same response in adults was interpreted to mean that the young animal is much more sensitive to the negative feedback actions of estradiol on the pituitary gland. Research in the 1960s and 1970s confirmed that the sensitivity of the gonadostat to the negative feedback actions of testosterone and estradiol decreased as rats approached puberty (Figure 8-7). However, a careful analysis of LH patterns in developing rats

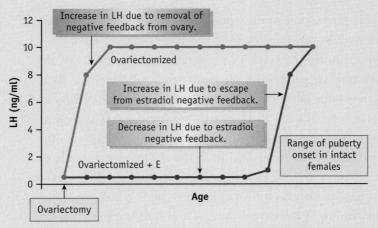

Figure 8-7 Experimental paradigm used to test the gonadostat theory of puberty onset in several species of mammals. Ovariectomy before puberty (red line) results in an increase in LH due to removal of negative feedback signals from the ovary. Estradiol replacement therapy prevents such an increase (blue line) demonstrating that this hormone is responsible for most if not all of the negative feedback. Note that LH levels increase in estradiol-treated animals as they approach the age at which puberty normally occurs (green box). This observation has been interpreted to mean that females escape the negative feedback actions of estradiol before attaining puberty.

revealed that the decrease in response to negative feedback occurs *after* onset of puberty. The prepubertal increase in LH in rats appears to be due to removal of a gonadal steroid independent (most likely centrally mediated) inhibition of GnRH release.

In spite of the fact that the gonadostat hypothesis proved to be inadequate to explain onset of puberty in rats, it did provide a useful conceptual framework for studying puberty in a variety of mammals. As it turns out, the hypothesis appears to be valid for sheep, cattle, and pigs, but not for primates. The history of the gonadostat hypothesis illustrates that a scientific hypothesis can be wrong but extremely important in advancing our understanding of a phenomenon.

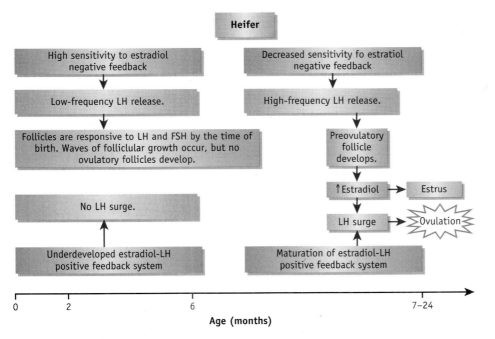

FIGURE 8-8

Mechanism of puberty onset in the heifer. Events highlighted in red, yellow, and green correspond to changes in hypothalamic-pituitary activity, patterns of gonadotropins, and changes in ovarian activity, respectively. Arrows indicate cause-effect relationships between events.

The heifer is a good model for understanding puberty onset in domestic ungulates because the physiologic events associated with sexual maturation have been thoroughly characterized and this information has been used to manipulate age at puberty in production of beef and dairy cattle (Figure 8-8). Sheep and swine have also been used extensively in studies of puberty onset. The time course of events leading to the onset of puberty is well characterized in heifers. A similar pattern of developmental events applies to the ewe lamb and gilt, but the ages at which these events occur vary according to how rapidly the animal develops sexually.

Antral follicles first appear on the ovaries of calves before birth and become responsive to gonadotropins between 2 and 4 weeks of age. Exogenous LH and FSH have been shown to induce development of pre-ovulatory follicles by 4 weeks of age, and injections of LH that mimic the LH surge induce ovulation of such follicles. The hypothalamic-pituitary unit of cattle is intact and functional by the first few weeks of life. In calves the hypothalamus releases GnRH in a pulsatile manner by 2 weeks of age. Moreover, pulses of LH have been detected in peripheral circulation by this age. The ability to respond to the stimulatory feedback action of estradiol appears to be fully developed by 5 to 6 months of age. Based on this evidence it seems that the heifer calf has the potential to become pubertal by 6 months of age. How is it that most heifers do not reach puberty until much later (12 months)? The answer to this question can be developed by understanding the regulation of pulsatile LH release in the developing heifer.

The pattern of pulsatile LH in calves changes between birth and puberty. During the infantile period (0 to 2 months), LH pulses begin to appear in the circulation and the number begins to increase. This may be due to organization and activation of GnRH-secreting neurons and the pituitary gland gaining the ability to respond to GnRH. The frequency of LH pulses increases and then declines during the early prepubertal period (2 to 5 months), and remains low throughout the late prepubertal period (5 to 10 months). The increase in LH may be due to a lack of feedback inhibition by the ovary. At this stage of development, ovarian follicles are beginning to develop and produce very little estradiol. The subsequent decline in LH concentrations is likely due to the negative feedback actions of estradiol which is being produced by developing follicles. This negative feedback prevails throughout the late prepubertal period and sustains a low-frequency pattern of LH secretion. During the peripubertal period (between 10 months of age and puberty), LH pulse frequency increases to a level exceeding that of the early prepubertal period. This has been attributed to a decrease in sensitivity to the negative feedback actions of estradiol.

The conclusion that the low frequency pattern of LH during the prepubertal period is attributed to a high sensitivity of the hypothalamic-pituitary unit to estradiol negative feedback is based on two important observations. First, ovariectomy of prepubertal heifers results in an increase in LH pulse frequency. Second, administration of estradiol prevents the effects of ovariectomy on pulsatile LH secretion. The conclusion that the prepubertal increase in pulsatile LH release is due to a loss of sensitivity to estradiol negative feedback is based on the observation that ovariectomized heifers given estradiol exhibit an increase in pulsatile LH secretion at about the time of puberty onset in intact heifers. These observations do not exclude the possibility that a steroid-independent mechanism (similar to removal of central restraint) might also contribute to the prepubertal increase in pulsatile LH secretion in heifers and other ungulates.

What are the consequences of the escape from estradiol negative feedback? The resulting increase in LH pulse frequency sets into motion a series of events that culminate in estrus and ovulation. In order to understand how this works it is necessary to keep in mind that the ovaries of prepubertal animals are expressing waves of follicular growth. In other words, proliferative pools of follicles develop to various stages. Some of these may become tertiary follicles, but they ultimately undergo atresia rather than develop into pre-ovulatory follicles. However, if a large tertiary follicle encounters a high-frequency pattern of LH, it develops into a pre-ovulatory follicle. The details of this process will be discussed in the next two chapters. The pre-ovulatory follicle produces high levels of estradiol and causes circulating concentrations of this steroid hormone to increase. At a particular threshold concentration, estradiol induces estrus as well as an LH surge. The LH surge induces ovulation of the pre-ovulatory follicle.

There is little doubt that the escape from estradiol negative feedback is an important prerequisite for onset of puberty in heifers. However, this physiologic change does not appear to be a developmental event as much as a means by which various environmental cues influence reproductive activity. Changes in sensitivity to estradiol negative feedback also mediates the effects of season, nutrition, and lactation on reproductive activity of domestic ungulates. Under certain circumstances (e.g., high plane of nutrition), heifer calves attain puberty as early as 6 months of age. Thus it appears that the inhibition of LH release by estradiol negative feedback can be overcome. Such a mechanism may serve as a means to allow the female to reproduce when environmental conditions are favorable.

Primates

Unlike rodents and many other mammals, primates experience a long interval between birth and puberty. For example, in human females, puberty isn't initiated until the second decade of life. Another important difference between most primates and other mammals is the fact that in the vast majority of primate species, females do not express a heat period (estrus). Although there may be fluctuations in sexual activity in these females they appear to be sexually receptive at all times. Whereas first estrus is useful in estimating age at puberty in non-primate species, first menstruation (menarche) is the most noticeable external change associated with sexual maturation in female primates. Unlike estrus, menstrual flow is not tightly coupled with ovulation. Ovulation occurs approximately 14 days after the initiation of menstruation during the normal menstrual cycle of adults.

FIGURE 8-9

Mechanism of puberty onset in the human female. Events highlighted in red, yellow, and green correspond to changes in hypothalamic-pituitary activity, patterns of gonadotropins, and changes in ovarian activity, respectively. Arrows indicate cause-effect relationships between events.

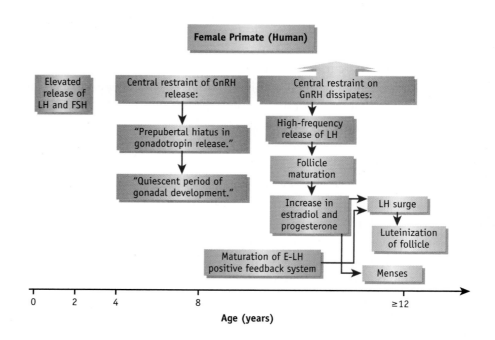

8

Figure 8-9 summarizes the major events leading to onset of puberty in the female primate. The key to understanding the onset of puberty in the females of these species is the fact that an increase in the pulsatile release of LH is an important necessary condition that sets into motion a cascade of events which lead to menarche and ovulation. In fact, this is similar to what occurs in female rodents. Recall that in the female rat, the occurrence of mini-surges of LH during the afternoon stimulates follicle growth to the pre-ovulatory stage. Another similarity between rats and primates is that prior to the prepubertal increase in LH, GnRH secretion is held in check by a central restraint.

Ovarian development in humans and other primates begins early in gestation. By 16 to 20 weeks of pregnancy the number of germ cells in the primate ovary reaches a peak. Primordial follicles appear at this time and soon give rise to primary follicles. Such development is not dependent on gonadotropins. Later in gestation (after the second trimester), the fetal ovaries develop FSH receptors. This permits FSH to induce development of primary follicles into antral follicles. Although the fetal and prepubertal ovaries contain antral follicles, they undergo atresia rather than develop into preovulatory follicles. The size of the ovaries increases between infancy and adulthood, primarily due to an increase in number of antral follicles as well as an increase in mass of the medullary stromal tissue. Steroid hormone production by the ovaries parallels follicular development. By 8 to 10 years of age, production of estradiol by the human ovaries is comparable to that of adult women, and induces development of secondary sex traits (e.g., pubic hair, and breast development). By 12 to 13 years of age, most girls undergo menarche, which is indicative of the fact that the ovaries have produced sufficient amounts of estradiol and progesterone to induce proliferation of the uterine endometrium. The female primate is rarely fertile at menarche. In most cases, first ovulation doesn't occur until at least 6 months after first menstruation. Regular, fertile menstrual cycles may not occur until several years later in humans.

In primate females, attainment of puberty appears to be dependent on development of a high-frequency pattern of pulsatile gonadotropin secretion. The hypothalamic-pituitary portal system develops early in pregnancy allowing GnRH to stimulate release of LH and FSH. Concentrations of these gonadotropins in fetal blood increase throughout the first half of gestation, and then decline by the end of gestation. In humans, circulating concentrations of LH and FSH increase to adult levels during the first 2 years of life, but then decline until late in the prepubertal period. This period of low gonadotropin secretion is known as the quiescent period of gonadal development, or the prepubertal hiatus in gonadotropin secretion. In humans this occurs between 4 and 11 years of age. The low concentrations of LH and FSH prevent ovarian follicles from developing to the pre-ovulatory stage.

The physiologic basis of low gonadotropin secretion during the prepubertal period of primates has been elucidated by several important experiments. As noted earlier, suppression of gonadotropin release can be attributed to steroid-dependent and steroid-independent mechanisms. In the case of primates, steroid-independent mechanisms appear to be more important. The strongest evidence for this conclusion is the fact that removal of the ovaries at 1 week of age in rhesus monkeys failed to abolish the prepubertal hiatus in gonadotropin secretion. This is consistent with the idea that the inhibition of gonadotropin secretion during this period is due to inhibitory inputs that do not depend on actions of ovarian steroids.

There is evidence to suggest that steroid-dependent mechanisms play a role in prepubertal regulation of gonadotropin secretion. It is clear that the negative feedback actions of estradiol develop by the end of fetal development and are likely present throughout the prepubertal period. Although such effects cannot account for the prepubertal hiatus in LH and FSH secretion, they may be of significance before development of the central restraint as well as after removal of this steroid-independent inhibition.

Late in the prepubertal period (11 to 12 years of age), the hypothalamic-pituitary unit "re-awakens" and gonadotropin secretion increases. This response is diurnal in nature, consisting of a nocturnal increase in the pulsatile

8

release of LH and FSH. In humans, the increase in pulsatile release of LH is associated with rapid-eye movement (REM) sleep. There is consensus that this increase in gonadotropin secretion is due to a decline in central inhibition of GnRH as well as a decreased sensitivity to the negative feedback actions of estradiol.

The consequence of increased gonadotropin secretion is enhanced development of ovarian follicles. The increase in pulsatile LH secretion is associated with appearance of large, antral follicles and elevated release of estradiol by the ovaries. As noted earlier, this increase in estradiol induces development of secondary sex traits and enhances growth of the genital organs. At the same time, the estradiol-LH positive feedback system within the hypothalamus becomes mature. The extremely high levels of estradiol produced by a pre-ovulatory follicle induce and LH surge, which in turn can induce ovulation. The first LH surge typically fails to induce a fertile ovulation. In most cases, the ovulated follicle is not completely luteinized resulting in a short menstrual cycle. Nevertheless, the abrupt decline in estradiol and progesterone following the first LH surge results in sloughing of the uterine endometrium (menarche).

Regulation of Puberty in Males

Compared to the female, our understanding of puberty onset in the male is limited. Whereas puberty in females involves abrupt changes in activity of the reproductive system (estrus, LH surge, and ovulation), the physiologic changes associated with sexual maturation in males can be best described as subtle and gradual. The primary event that sets into motion the onset of puberty in the male appears to be an increase in the pulsatile secretion of LH. The immediate consequence of this rise in LH secretion is development of steroidogenic pathways within the testes. This results in an increase in testosterone production which is the primary stimulus for spermatogenesis. These principles are best understood for the bull calf (Figure 8-10).

The sexual maturation of bull calves can be divided into four stages: infantile, prepubertal, transitional, and pubertal. During the infantile (birth to 3 months of age) stage, the testes contain cells that are precursors for Leydig and Sertoli cells. As in the heifer calf, frequency of LH pulses is low, but begins to increase by 2 to 3 months of age. This increase in LH release has been attributed to initiation of pulsatile GnRH release along with increased pituitary responsiveness to GnRH. During the prepubertal period (3 to 6 months of age), LH pulse frequency continues to increases, but then declines. The drop in LH concentrations coincides with an increase in

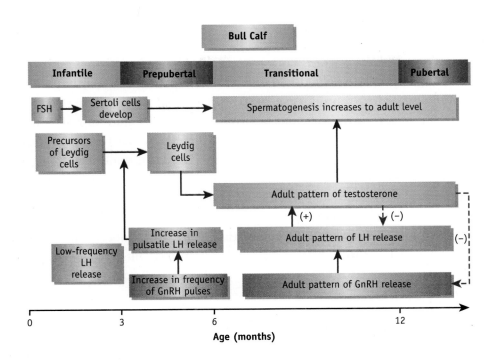

FIGURE 8-10

Mechanism of puberty onset in the bull calf. Events highlighted in red, yellow, and green correspond to changes in hypothalamic-pituitary activity, patterns of gonadotropins, and changes in testicular activity, respectively. Arrows indicate cause-effect relationships between events.

testosterone secretion by the testes. The increase in testosterone production is due to LH-induced development of Leydig cells. The inverse relationship between LH and testosterone is due to the negative feedback effects of testosterone on LH release. Concentrations of FSH are also elevated during the prepubertal period, due to a lack of negative feedback actions of testicular hormones. FSH induces the development of Sertoli cells, which produce estradiol and inhibin. As concentrations of these two testicular hormones increase, they feed back negatively on the pituitary gland to suppress FSH release. Therefore, FSH levels decline during the prepubertal period. The development of Sertoli cells and the production of testosterone by Leydig cells mark the transition of the testes from the prepubertal to the pubertal state. The presence of Sertoli cells and testosterone initiates spermatogenesis. During the transitional stage (6 to 12 months of age), pulsatile secretion of LH remains low and stable and reflects the negative feedback relationship between testosterone and LH. Concentrations of testosterone increase gradually during this period, reflecting a gradual maturation of the testes. Functionally speaking, the endocrine relationships between the testes and the hypothalamic-pituitary unit are mature by the end of the prepubertal period. The characterizing feature of the transitional period is the gradual maturation of the testes, culminating in an adult pattern of spermatogenesis; that is, sperm production sufficient for impregnating a female.

TIMING OF PUBERTY ONSET

Having reviewed the major theories describing how puberty onset occurs, it now seems appropriate to consider one of the more fundamentally important questions regarding reproductive biology: How is the age at puberty determined? In other words, what mechanisms control the timing of events that lead to onset of puberty, or what determines when the puberty alarm sounds? The bulk of research addressing this question has been done with females. The previous discussion of puberty onset in females highlights the importance of an increase in the pulsatile release of LH in determining age at puberty. In light of this important concept, our question about the timing of puberty onset becomes one of what regulates the timing of the prepubertal increase in LH secretion. To address this question, we turn to our earlier consideration of thermodynamics and reproduction. Recall that reproduction is an energy-dependent process. Thus it should be no surprise that the mechanisms controlling onset of puberty are also energy dependent. There is an abundance of information to support this claim. In the 1950s, animal scientists first noted an inverse relationship between prepubertal growth rate and age at puberty. Similar observations have been documented in numerous species including humans. Today it is generally accepted that the timing of puberty onset is largely dependent on the generation of metabolic signals that indicate that the animal has attained a metabolic status that will support successful reproduction.

SUMMARY OF MAJOR CONCEPTS

- Puberty can be defined as attainment of the ability to reproduce, and includes all of the morphologic, physiologic, and behavioral changes associated with this ability.

- Age at puberty is a major determinant of the lifetime reproductive success of an individual, but there are significant biological costs associated with early sexual maturation.

- Theoretically speaking, the process of sexual maturation can be viewed as involving a "developmental clock," that governs the sequential expression of key regulatory genes that lead to onset of puberty. The rate at which this process occurs is genetically predetermined, but whether or not expression of these genes culminates in onset of puberty depends on permissive signals that convey information about the animal's internal and external environments.

- Of the genes regulating sexual maturation, those controlling the pulsatile secretion of LH appear to be particularly important. Onset of a high-frequency mode of pulsatile LH secretion initiates a cascade of events that culminates in full maturation of the gonads.

- The prepubertal increase in pulsatile LH secretion is attributed to both steroid-independent (loss of central restraint) and steroid-dependent (escape from negative feedback actions of gonadal steroids) mechanisms. The relative importance of these two types of mechanisms varies with species.

DISCUSSION

1. Bruce, age 12, is the only boy in his seventh-grade class to sport a mustache and sideburns. Would it be correct to conclude that Bruce has reached puberty? Why or why not? Explain, in theoretical terms, why Bruce's male classmates lack an adult pattern of facial hair.

2. The feral Soay sheep inhabit the islands of the St. Kilda archipelago off the coast of Great Britain and live under extremely harsh environmental conditions. The Soay sheep are seasonal breeders; rutting season and mating occurs during the autumn, just before the onset of harsh winter weather, whereas lambing occurs during the milder weather of the spring. During rut, males reduce their feeding time and increase their physical activity in an attempt to find and secure mates. This behavior limits opportunities to build up stores of metabolic fuel that are vital to survival during the winter months. Under these circumstances, one might hypothesize that natural selection would favor a slow rate of sexual maturation so that rams would not engage in rut as juveniles. Interestingly, males typically attain puberty by 7 months of age (at the onset of the rutting season), when they are only one-third of their adult size. The risks of such a strategy are readily apparent. The mortality rate among Soay males is extremely high during the winter months that follow rut; sometimes reaching 99 percent. Explain how such a presocial sexual maturity ensures reproductive success of individual rams in the face of such environmental conditions?

3. Imagine that you are the first reproductive biologist to study sexual maturation in the female alpaca. You already know when these animals attain puberty, but you know nothing about the hormonal control of sexual maturation in this species. Based on your knowledge of puberty in other female mammals, what are some of the most important research questions you would seek to address for this species?

8

4. Describe an experiment that would allow you to distinguish between steroid-dependent and steroid-independent inhibition of LH secretion in a prepubertal female mammal.

5. Describe several ways (treatments) that would induce early puberty onset in male and female mammals. Describe how these treatments would work to induce puberty.

REFERENCES

Andrews W.W., J.P. Advis and S.R. Ojeda. 1981. The maturation of estradiol negative feedback in female rats: evidence that the resetting of the hypothalamic "gonadostat" does not precede the first preovulatory surge of gonadotropins. *Endocrinology* 109:2022–2031.

Day, M.L. and L.H. Anderson. 1998. Current concepts on the control of puberty in cattle. *Journal of Animal Science* 76 (Suppl. 3):1–15.

Foster, D.L. 1994. Puberty in the sheep. In E. Knobil and J.D. Neill, *The Physiology of Reproduction Vol. 2.,* Second Edition. New York: Raven Press, pp. 411–451.

Ojeda, S. R. and H.F. Urbanski. 1994. Puberty in the rat. In E. Knobil and J.D. Neill, *The Physiology of Reproduction Vol. 2.,* Second Edition. New York: Raven Press, pp. 363–409.

Plant, T.M. 1994. Puberty in primates. In E. Knobil and J.D. Neill, *The Physiology of Reproduction Vol. 2.,* Second Edition. New York: Raven Press, pp. 453–485.

Ramirez, V.D. and S.M. McCann. 1963. Inhibitory effect of testosterone on testicular growth in rats. *Endocrinology* 76:412–417.

Ramirez, V.D. and S.M. McCann. 1963. Comparison of the regulation of luteinizing hormone (LH) secretion in immature and adult rats. *Endocrinology* 72:452–464.

Schillo, K.K., J.B. Hall, and S.M. Hileman. 1992. Effects of nutrition and season on the onset of puberty in the beef heifer. *Journal of Animal Science* 70:3994–4005.

Sisk, C. and D.L. Foster. 2004. The neural basis of puberty and adolescence. *Nature Neuroscience* 7:1040–1047.

Steele, R.E. and J. Weisz. 1974. Changes in sensitivity of the estradiol-LH feedback system with puberty in the female rat. *Endocrinology* 95:513–520.

Dynamics of Testicular Function in the Adult Male

CHAPTER OBJECTIVES

- Describe the major functions of the testes.

- Describe the regulation of testicular hormone production.

- Describe the process of spermatogenesis.

- Describe the regulation of spermatogenesis.

- Describe the morphology of the spermatozoon.

MORPHOLOGY OF THE TESTIS

In this chapter we will explore the major functions of the testis and relate these functions to the structure of this gonad. With this purpose in mind, it is convenient to divide the testis into two major compartments: tubular (within the seminiferous tubules) and interstitial (between the tubules; Figure 9-1). The two testicular compartments are separated by the basement membrane of the tubule. This layer of connective tissue encompasses **peritubular myoid cells,** which are joined by punctate junctions to form a continuous sheet of cells. Connective tissue cells known as fibrocytes are typically seen on the interstitial surface of the tubular membrane. The major function of the tubular compartment is production of sperm cells (spermatogenesis). The interstitial compartment is concerned primarily with production of testosterone. Although these regions have different structural and functional features, they are related; that is, testosterone enters the tubular compartment to regulate spermatogenesis.

The anatomic separation of the two testicular compartments has physiologic significance. When various dyes are injected into the blood, they diffuse out of capillaries in the interstitial testis and equilibrate rapidly and freely within the surrounding tissue and lymphatics. In contrast, the penetration of such dyes into the basal compartment of the tubule is much slower. Interestingly, neither of these dyes penetrate beyond the basal region into the so-called adluminal compartment. These results have been interpreted to suggest that a **blood-testis barrier** exists between the interstitial and tubular compartments of the testis. The fact that dyes penetrate the basal compartment suggests that the peritubular boundary (basement membrane and peritubular cells) do not account for this barrier. The barrier appears to be attributed to the tight junctions that link together adjacent Sertoli cells. With the exception of some extremely small molecules, any substance that enters the adluminal and central

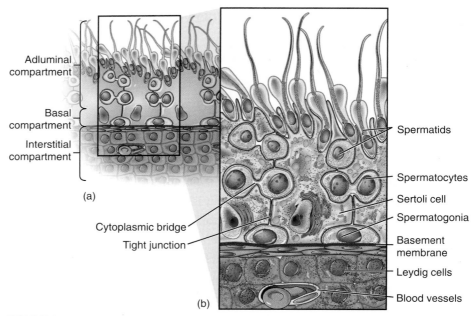

Adluminal compartment

Basal compartment

Interstitial compartment

(a)

Cytoplasmic bridge

Tight junction

Spermatids

Spermatocytes

Sertoli cell

Spermatogonia

Basement membrane

Leydig cells

Blood vessels

(b)

FIGURE 9-1

Schematic illustrations of the major compartments of the testis showing perspectives from a light microscope (a) and an electron microscope (b). The organ can be divided into tubular and interstitial compartments. The tubular compartment consists of a basal segment, lying below the tight junctions between Sertoli cells, and the adluminal portion, which lies above these junctions. Note that the Sertoli cells envelope the developing sperm cells that reside in layers which correspond to different generations arising from cell divisions. As these cells develop they remain connected by intracellular bridges.

compartments of the tubule must first enter and then be secreted by the Sertoli cells. Secretions from the Sertoli cells account for the bulk, if not all, of the fluid that flows from the seminiferous tubules. Such secretion is active and is opposed by considerable hydrostatic pressure and diffusion gradients. The composition of this fluid is different from blood because not all of the blood constituents that enter the basal compartment of the testis are taken up and/or secreted by Sertoli cells.

Based on the previous description of the testis, it is possible to further subdivide the testis into four compartments: intravascular (consisting of blood), interstitial (containing lymphatics and Leydig cells), basal intratubular (consisting of the region between the basement membrane of the tubule and the band of tight junctions connecting Sertoli cells), and adluminal intratubular (between the band of tight junctions and the lumen of the tubule).

OVERVIEW OF TESTICULAR FUNCTIONS

As noted in the previous section, the testes serve two important physiologic functions: spermatogenesis and synthesis/secretion of testosterone and other hormones. The processes underlying these functions are regulated by two endocrine feedback loops, each of which involves a pituitary gonadotropin and a testicular hormone (Figure 9-2). One of these feedback loops involves luteinizing hormone (LH), which acts on the Leydig cells to regulate production of testosterone, which, in turn, exerts a negative feedback action on LH release. The other involves follicle-stimulating hormone (FSH), which acts on Sertoli cells to induce release of inhibin, which exerts a negative feedback effect on FSH release. The production of sperm cells by the testes is dependent on LH and FSH. Each of these hormones influences spermatogenesis by acting on the Sertoli cells. The effects of LH are indirect and are mediated by testosterone, whereas the effects of FSH are direct. Spermatogenesis is a complex process consisting of three major phases: proliferation, meiosis, and differentiation **(spermiogenesis).** During proliferation, stem cells known as spermatogonia undergo a series of mitotic divisions. The proliferation of these cells serves the purpose of providing a large and continuous supply of precursor cells that can ultimately develop into sperm cells. Following proliferation, spermatogonia undergo meiosis and are referred to as spermatocytes. The haploid cells resulting from meiosis are spermatids, cells that undergo morphologic changes to differentiate into spermatozoa.

9

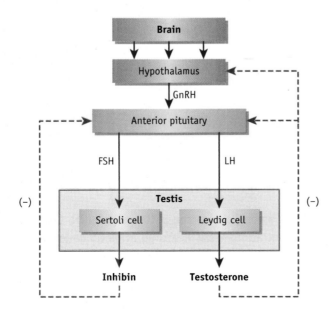

FIGURE 9-2

Schematic illustration of the major hormonal interactions within the hypothalamic-pituitary-testicular system. Homeostasis is maintained by two negative feedback relationships between testicular hormones and gonadotropins.

9

ENDOCRINE REGULATION OF THE TESTES

One of the hallmarks of puberty in the male is the establishment of adult patterns of gonadotropin secretion, which results from the development of negative feedback relationships between the testes and the hypothalamic-pituitary system. These feedback loops ensure that circulating concentrations of gonadotropins and testicular hormones fluctuate within the narrow physiologic ranges required to sustain male reproductive functions. In order to understand how these systems work, it is necessary to learn about the mechanisms regulating release of gonadotropins and testicular hormones.

Regulation of LH and FSH Release by GnRH

Release of LH and FSH is controlled by gonadotropin-releasing hormone (GnRH), which is produced by neurosecretory cells that are loosely distributed within the anterior portion of the hypothalamus. These cells release GnRH into a portal vascular system that carries this peptide hormone to the anterior lobe of the pituitary gland where it induces release of LH and FSH from cells known as gonadotrophs. GnRH is released in pulses that evoke corresponding pulses of LH (Figure 9-3). The relationship between release of GnRH and FSH is less clear. Injections of GnRH induce FSH release and most FSH pulses are preceded by GnRH pulses. However, patterns of LH and FSH are not always similar and some FSH pulses are not preceded by pulses

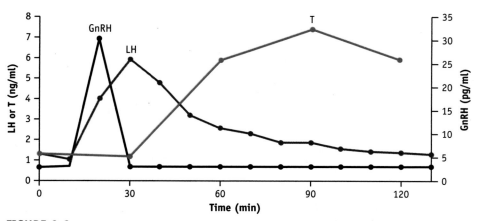

FIGURE 9-3

Temporal associations among circulating patterns of GnRH, LH, and testosterone in a ram. GnRH concentrations were measured in blood collected from hypophysial portal vessels, whereas LH and testosterone concentrations were determined in blood collected from the jugular vein. Note that an increase in testosterone is preceded by a pulse of LH, which is preceded by a pulse of GnRH (Data from Rhim et al., 1993).

of GnRH. The frequency of GnRH and LH pulses is typically four to six pulses per 24-hour period, but this varies considerably among individual males as well as with physiologic status.

Regulation of Testicular Hormones by LH and FSH

The only known effects of LH and FSH in mammals occur within the gonads. In adult males, LH is both a necessary and sufficient condition to sustain a normal level of spermatogenesis. As noted earlier, this effect is mediated entirely by stimulating production of testosterone by Leydig cells. This relationship is manifested by the close temporal association between circulating patterns of these two hormones (Figure 9-3). In general, pulses of LH are followed by corresponding increases in testosterone concentrations. This effect is due to the fact that LH enhances the rate-limiting step in the biosynthesis of testosterone (Figure 9-4). Briefly, LH binds to its receptor located on

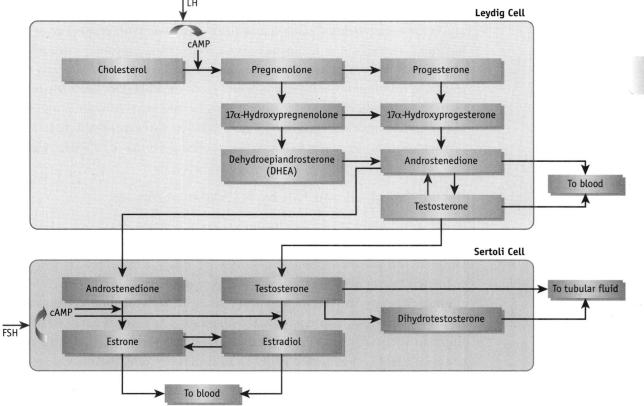

FIGURE 9-4

Schematic illustration summarizing testosterone synthesis in the testis. LH enhances the rate-limiting step in testosterone synthesis. In addition to testosterone, the adult testis produces small amounts of estrogen as well as the more potent androgen, DHT. Production of estrogen and DHT involve LH and FSH acting on Leydig and Sertoli cells, respectively.

the membranes of Leydig cells. This stimulates production of cyclic AMP (cAMP) an intracellular messenger that induces intracellular changes to promote transport of cholesterol to the mitochondria where it is converted to pregnenolone by an enzyme known as desmolase. Pregnenolone is the rate-limiting precursor for testosterone and other steroid hormones. One of the more important effects of LH on Leydig cells is to enhance synthesis of a **steroidogenic acute regulatory (StAR) protein,** which facilitates transport of cholesterol across the mitochondrial membrane.

The most important physiologic effects of FSH occur during the early part of postnatal sexual development. During this time FSH stimulates proliferation of Sertoli cells. This effect depends on the ability of FSH to stimulate aromatase, an enzyme that promotes the conversion of testosterone to estradiol. An absence or deficiency of FSH during this critical period of development results in infertility. Thus FSH plays a critical role in initiation of sperm production. In contrast, a deficiency FSH occurring during adulthood does not result in subnormal fertility suggesting that this gonadotropin is not necessary for maintaining spermatogenesis once it has been initiated. With respect to steroid hormone synthesis, the ability of FSH to convert testosterone to estradiol in Sertoli cells is reduced once Sertoli cells mature. Therefore, production of estradiol by the adult testes is extremely low. However, FSH does cause the release of inhibin, another hormone produced by Sertoli cells. Inhibin is a peptide hormone and appears to have little or no effect on sperm production. Its major physiologic role appears to be a negative feedback regulator of FSH release.

Regulation of LH and FSH by Testicular Hormones

In the adult male, testicular hormones exert negative feedback actions on gonadotropin release. There are no known positive feedback relationships between the pituitary gland and testes in males. LH release is primarily regulated by the negative feedback action of testosterone. This conclusion is supported by results of classic ablation-replacement experiments (Figure 9-5). In this experiment, bilateral castration (removal of both testes) of bulls resulted in an increase in pulsatile LH release. This demonstrates that the testes produce some factor(s) that exert(s) an inhibitory effect(s) on LH release. When castrated bulls are treated with testosterone, pulsatile LH release decreases and resembles patterns observed in normal bulls. This response supports the idea that testosterone accounts for most, if not all, of the negative feedback signal produced by the testes. Other experiments show that castration causes an increase in FSH release, but that this effect is not prevented by testosterone replacement. However, injections of inhibin have been shown to inhibit FSH release in castrated males.

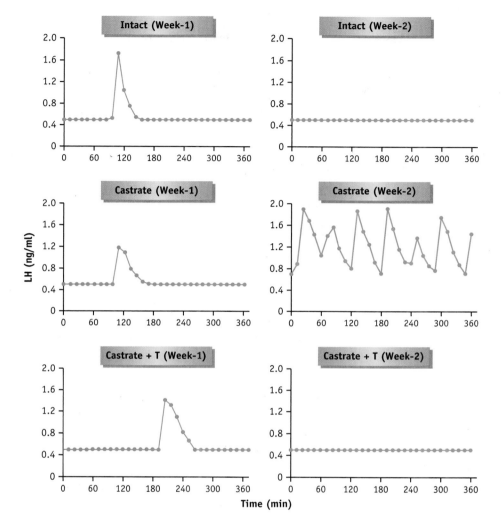

FIGURE 9-5

Patterns of LH in bulls that remained intact (top row), were castrated (middle row) or were castrated and treated with daily injections of testosterone (bottom row) 1 week before and 2 weeks after removal of the testes. The fact that castration caused an increase in LH together with the fact that testosterone replacement therapy reduced LH concentrations to precastration levels demonstrates that testicular testosterone provides a major negative feedback signal regulating LH release (Data from Imwalle and Schillo, 2002).

Hormones that exert negative feedback actions on release of LH and FSH can act on the hypothalamus, anterior pituitary gland, or both tissues. The inhibitory effects of testosterone on LH secretion are brought about via effects at both levels. Testosterone has been shown to suppress GnRH release as well as suppress the responsiveness of the pituitary gland to GnRH. In contrast inhibin seems to suppress FSH release by acting exclusively at the level of the pituitary gland in most mammals.

Hormonal Control of Spermatogenesis

As noted previously, testosterone and other androgens are the primary regulators of sperm production in the adult male. A deficiency in these hormones and (or) a deficiency of androgen receptors results in infertility. It is well accepted that androgens sustain spermatogenesis by acting on Sertoli cells, the only somatic cell type that has direct contact with developing germ cells in

FIGURE 9-6

Schematic illustration showing the importance of testosterone and ABP in regulating spermatogenesis.

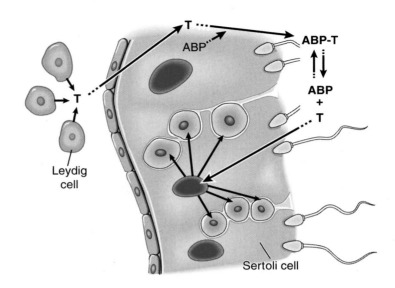

the male. It is beyond the scope of this chapter to provide a detailed account of how androgens regulate spermatogenesis. However, data from years of research support the hypothesis that the androgen receptors of Sertoli cells regulate various genes that play key regulatory roles in spermatogenesis. More specifically, there is consensus that the androgen receptor plays important roles during the meiosis and differentiation phases of spermatogenesis.

Although LH is sufficient to sustain spermatogenesis, FSH plays, an important role as well. One of the most important effects of FSH is to stimulate production of **androgen-binding protein (ABP)** by Sertoli cells (Figure 9-6). This protein has a low affinity for androgens, but exists in high concentrations. Thus ABP sequesters testosterone ensuring that large amounts of the hormone are continuously available to support spermatogenesis.

SPERMATOGENESIS

When studying spermatogenesis it is helpful to keep in mind that the production of spermatozoa by the testes is continuous. In other words, there is a steady flow of sperm cells entering the seminferous tubules of both testes throughout adulthood. This is markedly different from gametogenesis in females where release of oocytes from the ovary occurs periodically (during the ovulatory phase of the estrous cycle). According to some estimates the production rate of sperm cells ranges between 300 and 600 per gram of testis per second. This continuous production of spermatozoa by the testes is attributed to two features of seminiferous tubule activity. First, within a particular section of seminiferous tubule, spermatogonia, located in the basal intratubular compartment, give rise to a new generation of developing germ cells on a periodic basis. Second, the generation of germ cells is not synchro-nized along the length of the tubule, so spermatogenesis begins at different times in different locations of the seminiferous tubules. In order to envision the stages of spermatogenesis, one would have to trace the development of a tubular stem cell into a spermatozoon over time within a particular segment of a tubule. The time required for completion of spermatogenesis varies among species. In most mammals a period of 2 to 3 months is required.

The production of sperm cells begins during embryogenesis when primor-dial germ cells migrate to and infiltrate the gonad, eventually becoming sper-matogonia. During pubertal development these germ cells begin to develop within the seminiferous tubule. As they develop they migrate from the basal to adluminal region. The most immature germ cells (spermatogonia) are found along the basement membrane of the tubule, whereas the most developed (spermatids) are found in the central region of the tubule. As the germ cells develop they remain connected via so-called intercellular bridges (see Fig-ure 9-1). A group of connected cells is known as a cohort and can consist of as many as 50 cells. Such connections permit communication among the cells and may serve as a means to synchronize development of a germ cell cohort.

Proliferation

Spermatogenesis is typically divided into three major phases (Figure 9-7). The proliferation phase serves the purpose of generating large numbers of sper-matogonia via mitosis. The process begins when a type A_1 spermatogonium di-vides to generate two type A_2 spermatogonia. This is followed by additional cell divisions each of which results in a different type of spermatogonium. The number of mitotic divisions varies among species. In rats, there are six divisions resulting in the generation of a maximums of 64 clones, known as primary spermatocytes. The actual number of primary spermatocytes produced during the proliferative phase is usually less than 64, due to cell death. Several types of spermatogonia can be identified during proliferation. For example in the rat, the first three divisions result in type A spermatogonia, whereas the fourth and fifth give rise to type I (intermediate) and type B spermatogonia, respectively. These cells are distinguished based on various morphologic features. At some point during the proliferation of type A spermatogonia, mitosis ceases in one of the resulting daughter cells, and the cell then serves as a stem cell that can initiate a new cycle of proliferation. In rats, mitosis ceases with the production of type A_4 spermatogonia. This prevents depletion of stem cells thus ensuring that spermatogenesis is continual. It is important to note that as spermatogonia develop into primary spermatocytes they move from the basal to the adluminal

FIGURE 9-7

Schematic illustration summarizing the process of spermatogenesis in the seminiferous tubules of the testis. Note that the development of sperm cells involves mitosis, meiosis, and cell differentiation and that these processes occur in different compartments of the tubule.

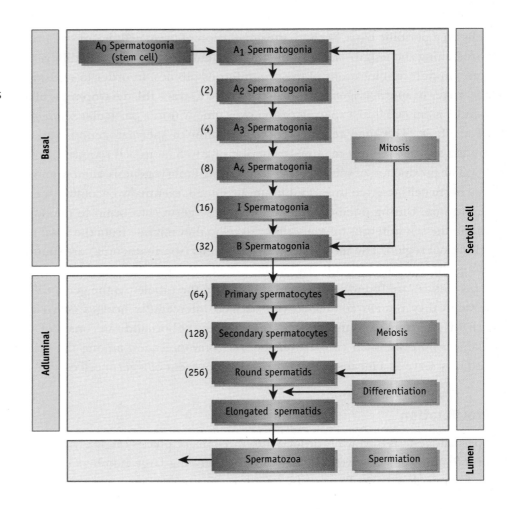

9

compartment of the tubule. By the end of proliferation, they have pushed their way through the tight junctions that join adjacent Sertoli cells. Once the primary spermatocytes emerge into the new environment of the adluminal compartment, they begin meiosis, the second phase of spermatogenesis.

Meiosis

As you may recall, meiosis occurs in two stages. In spermatogenesis, primary spermatocytes are undergoing meiosis I, which gives rise to secondary spermatocytes. Secondary spermatocytes undergo meiosis II to give rise to spermatids. Upon formation, a primary spermatocyte enters interphase, or the resting phase between cell divisions. It then progresses through the first meiotic prophase, which consists of leptotene, zygotene, pachytene, and diplotene phases. This is the longest phase of spermatogenesis, accounting for as much as 30 percent of the time required for a spermatogonium to develop into a spermatozoon. Thus the life span of primary spermatocytes is longer than those of spermatogonia, secondary spermatocytes, or spermatids. During prophase I, the DNA of chromosomes replicates such that each chromosome consists of two chromatids joined at a centromere. Homologous pairs of chromosomes pair up and align along the equatorial region of the cell to form tetrads. At this point the chromatids of the pairs can cross over resulting in a random exchange of DNA fragments between homologous chromosomes. During metaphase I, the homologous chromosomes migrate to opposite poles along the spindle. At this stage the chromosomes consist of two chromatids joined at a centromere. During telophase I, two secondary spermatocytes are formed, each one containing a set of homologous chromosomes. During meiosis II, the chromosomes align on the equatorial region, separate, and then migrate to opposite ends of the cells. Eventually each of the secondary spermatocytes divides resulting in two spermatids each of which contains one haploid set of chromosomes. Meiosis II progresses rapidly so the lifespan of a secondary spermatocyte lasts only 1 or 2 days. At this stage of development the spermatids are spherical and referred to as early spermatids. If you are keeping track of the number of cells generated by spermatogenesis, you would realize that a maximum of 256 early spermatids can result from 64 primary spermatocytes (or from a single type A_1 spermatid).

Differentiation

The completion of meiosis marks the end of the so-called genetic phases of spermatogenesis (i.e., proliferation and meiosis). The DNA of spermatids and spermatozoa does not replicate and gene expression is minimal in these haploid cells. The final phase of sperm cell production involves morphologic changes that result in the transformation of a round spermatid into an elongated spermatozoon, which has the ability to "swim" and fuse with an oocyte (Figure 9-8). Some reproductive physiologists view the spermatozoa as a "self-propelled package of DNA," and refer to the differentiation phase as the packaging phase. Another commonly used term to describe this phase is spermiogenesis. It is important to understand that although sperm cells have a unique shape and function, they contain the same cellular organelles as any other cell type. Differentiation refers to the re-arrangement of these organelles in ways that correspond to the function of spermatozoa.

The round spermatids look like ordinary epithelial cells. However, there are noticeable changes in their structure soon after they form. One of the more noticeable changes is the condensation of chromatin in the nucleus, resulting in a denser nucleus that loses its ability to react with certain types of

9

FIGURE 9-8

Schematic illustration depicting spermiogenesis; that is, transformation of a spermatid into a spermatozoon. This process involves the rearrangement of organelles and genesis of specialized structures that allow sperm cells to become motile and eventually penetrate an oocyte.

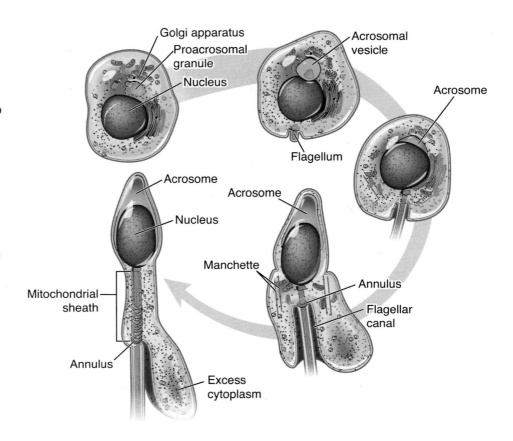

9

stains. Another early change involves the Golgi apparatus. Small vesicles of this organelle fuse to create a single acrosomal vesicle containing a single acrosomal granule. The vesicle, with its enclosed granule, migrates toward the nucleus and eventually fuses with it before spreading over it. While this is occurring, the two centrioles migrate to the side of the nucleus opposite to that which is being covered by the acrosomal vesicle. The centriole that is closest (proximal) to the nucleus becomes oriented radially, whereas the other (distal) centriole becomes oriented perpendicular to the more proximal one.

Once the **acrosome** and centrioles assume their new positions, the shape of the cell begins to change. The distal centriole begins to develop the axoneme (thread) that will become the tail (flagellum) of the sperm cell. The proximal centriole forms the neck, which is the connecting piece that joins the tail to the head of the spermatid. As the axoneme grows it pushes cytoplasm away from the nucleus. While the tail grows, the nucleus of the spermatid begins to elongate and the cytoplasm moves distally relative to the nucleus. At this stage of differentiation, a ring-shaped structure (annulus) appears at the base of the tail. This marks the point of attachment of the tail to the head of the cell. As the spermatid takes on a new shape, the endoplasmic reticulum regresses and disappears. During the time the nucleus and tail elongate, a strange system of microtubules known as the manchette assembles to form a sleeve-like structure enveloping the nucleus of the spermatid posterior to the acrosomal cap. Some of these microtubules will eventually form the postnuclear cap. During the final stages of differentiation mitochondria migrate away from the nucleus and gather in the distal cytoplasm, which will eventually form the midpiece of the sperm cell. The mitochondria are assembled around the base of the flagellum arranged in a helical fashion.

Differentiation culminates with the release of spermatozoon from the Sertoli cells. Various types of junctions between Sertoli cells and developing germ cells exist during spermatogenesis and are probably involved with moving the developing germ cells toward the luminal region of the tubule. By the time spermatids have completed differentiation, these contacts are terminated and spermatozoa are released into the lumen of the seminiferous tubule. This process is known as **spermiation** and can be understood as being analogous to ovulation of the oocyte in females.

Morphology of Spermatozoon

As noted previously in this chapter, the sperm cell can be viewed as a system for delivering paternal DNA to the female oocyte. Such a specialized function requires a specialized structure (Figure 9-9). The major components include a head and a tail. The interphase between the head and tail is known

9

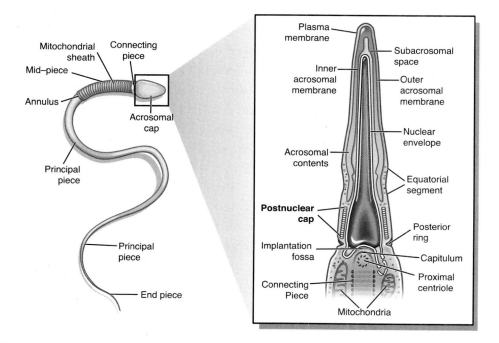

Mitochondrial sheath
Connecting piece
Mid–piece
Annulus
Acrosomal cap
Principal piece
Principal piece
End piece

Plasma membrane
Subacrosomal space
Inner acrosomal membrane
Outer acrosomal membrane
Acrosomal contents
Nuclear envelope
Equatorial segment
Postnuclear cap
Posterior ring
Implantation fossa
Capitulum
Connecting Piece
Proximal centriole
Mitochondria

FIGURE 9-9

Major structural features of the mammalian spermatozoon. Note that the enlarged view of the head region represents a longitudinal section, revealing the cell nucleus, acrosomal membranes, and plasma membrane, as well as structures involved with attachment of the tail to the head.

9

as the connecting piece. In this region, the base of the tail (capitulum) fits into a so-called implantation socket located at the base of the sperm head. The tail is typically divided into a midpiece, principal piece, and end piece. A prominent mitochondrial sheath encompasses the midpiece of the tail.

The shape of the sperm head varies among species. For example, it is hook-shaped in rodents and flattened in humans and domestic livestock species. In all cases, sperm heads are no more than a few μm long and 1 to 2 μm thick. A section through the head reveals a core of highly condensed mass of DNA and protein (chromatin) housed in the membrane of the cell's nucleus. Unlike other cells, the sperm has very little cytoplasm. As noted earlier, the acrosome forms a cap around the sperm nucleus. The inner membrane of this cap is separated from the nuclear membrane by an extremely narrow subacrosomal space. The entire head of the sperm (for that matter the entire sperm cell) is encased in the cell membrane. However, the cell membrane of the sperm cell fuses with the nuclear membrane at the base of the head to form the posterior ring. The area between the posterior limits of the acrosome and the posterior ring houses a dense lamina called the postacrosomal sheath. This postacrosomal region is where the sperm cell attaches to the oocyte during fertilization.

It is beyond the scope of this discussion to describe in detail the ultrastructure of the sperm tail and how this allows the cell to move. However, it is possible to develop a general appreciation for the morphology of this important organelle. The connecting piece of the tail consists of nine segmented columns that form an articulated segment, which make this section highly flexible and permits the tail to move from side to side. These columns taper in the middle piece and overlap with nine coarse fibers that run the length of the flagellum. The connecting piece contains few organelles, whereas the midpiece is covered by a sheath of mitochondria arranged in a helix. The principle piece consists primarily of the 9 coarse outer fibers which surround smaller fibers. These smaller fibers consist of two central microtubules surrounded by nine, evenly spaced doublet microtubules. The entire principal piece is encased in a fibrous sheath made up of circumferentially arranged fibers. These fibers become thinner as the sperm tail narrows and eventually end at the terminal piece. The terminal piece consists only of the 9+2 arrangement of microtubules. In addition to providing rigidity, the microtubules that run the length of the flagellum are responsible for sperm motility. Briefly, the doublet fibers slide past one another, causing the flagellum to move from side to side in a whip-like fashion. This process requires ATP, which is generated by the mitochondria located in the middle piece.

SPERMATOGENIC CYCLE

So far our analysis of spermatogenesis has been limited to a description of how primitive spermatogonia develop into spermatozoa. Although such knowledge is useful, it does not explain how release of spermatozoa is sustained at fairly constant rates in fertile males. As noted earlier, once males attain puberty, the release of spermatozoa from the testes is continuous, unless the individual experiences illness or environmental changes that influence sperm production. Understanding the physiologic basis for this continuous gamete production is arguably one of the more difficult tasks students encounter in reproductive physiology courses. One reason for this difficulty is related to the fact that such understanding requires the ability to envision the day-to-day changes that take place within each of the four to five concentric layers of seminferous epithelium. If you were to examine numerous sections of seminiferous tubule you would come to recognize that particular cell types are only seen in the presence of certain other cell types. For example you might notice that type A spermatogonia that are beginning their first mitotic divisions are always found with spermatocytes in the early and late stages of meiosis and both young (round) and mature (elongated) spermatids. Meticulous examination of the tubules would reveal several different patterns of cell associations. If you were to observe the same section of tubule over consecutive days a particular sequence of these patterns would emerge. This repeating sequence of patterns over time is known as the spermatogenic cycle. The particular patterns of cellular associations have been defined as the stages of this cycle. The length of this cycle varies among species. The boar's cycle is one of the shortest, lasting only 8 days. In humans, rats, and bulls the spermatogenic cycle is approximately 2 weeks in length. It is important to understand that the length of the spermatogenic cycle is not the same as the amount of time required for a generation type A spermatogonia to develop into a generation of spermatozoa. By the time a particular generation of spermatogonia undergoes proliferation, meiosis, and differentiation, the epithelial tissue in which it resides will experience several spermatogenic cycles. For example, in the bull the cycle lasts 13.5 days and 4.5 cycles pass between the initiation of proliferation and spermiation. Thus spermatogenesis takes 61 days in the bull.

Figure 9-10 summarizes the concepts described in the previous paragraph, which is a schematic illustration of the spermatogenic cycle of the bull. You can develop an understanding of this cycle by referring to this figure and following the fate of a group of type A spermatogonia, which appear in the first layer of tissue during the end of stage III. This generation of

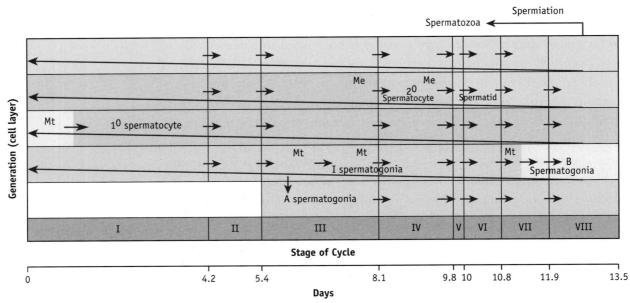

FIGURE 9-10

Schematic illustration of the spermatogenic cycle. The horizontal axes of the figure include a timeline corresponding to days on which the seminiferous epithelium is observed and the different stages of spermatogenesis (i.e., patterns of cellular associations), which are indicated by Roman numerals. The vertical axis represents the concentric layers of cells (generations) that make up the tubular epithelium. The lower part of the figure represents the basal region of the tubule, whereas the upper part of the figure represents the luminal region of the tubule. The different cells are indicated by different colors and shading patterns. Note that the stages of the cycle are depicted by the columns, each of which has a different vertical pattern. To follow the developmental path of a particular type A spermatogonium, begin at the basal layer at stage III and follow the arrows.

type A spermatogonia remains visible in this layer through stage VIII. Toward the end of the final stage, these cells migrate to the second layer of cells. The fate of this particular generation of cells is not depicted on the figure. However, their fate will be the same as that of the previous generation of type A spermatogonia that have migrated into the second layer (layer two, stage I). Once type A spermatogonia have migrated to the second layer, they remain inactive until the end of stage III. At this time they undergo mitosis, resulting in an increased number of cells in the layer. These are the type I spermatogonia. It is important to note that not all of the type A spermatogonia undergo mitosis. Some of these cells migrate back to the first layer and replenish the pool of stem cells. This explains why both type A spermatogonia and type I spermatogonia are observed in layers one and two, respectively, during stage IV. Now let's return to our cohort of type I spermatogonia.

By the middle of stage VII, these cells divide and give rise to a generation of type B spermatogonia. At this stage, the cells move to the third layer. The fate of this particular cohort is not shown in Figure 9-10. However, we can follow what happens to them by examining the previous cohort of type B spermatogonia. Once these cells enter the third layer of cells they undergo mitosis to form primary spermatocytes (layer 3; stage I). As discussed in an earlier section, primary spermatocytes undergo the first meiotic division. Because this step requires a considerable amount of time, primary spermatocytes can be observed for a good portion of the cycle. At some point during this process, the primary spermatocytes migrate to the next layer of cells. The fate of this particular set of primary spermatocytes is not shown, but the development of the previous generation of primary spermatocytes can be seen by referring to the next layer (layer 4, stage I). Once meiosis I is completed (layer four; stage IV), secondary spermatocytes develop. However they do not exist for long because meiosis II occurs rapidly (layer 4; stage IV to V). Secondary spermatocytes are observed only during stage IV. As noted earlier, completion of meiosis results in formation of round spermatids. Differentiation of spermatids accounts for a significant portion of the spermatogenic cycle. Some of this occurs while the cells occupy layer four. As the spermatids develop into spermatogonia, they migrate toward the lumen and enter the fifth layer of cells. The fate of the early spermatids of layer four is not illustrated. They migrate to the next layer and share the fate the previous generation of spermatids (layer 5). As differentiation continues, the spermatids elongate and develop tails. Eventually the resulting spermatogonia separate from the Sertoli cells (layer 5; stage VIII). If you study Figure 9-10 and the accompanying text, you will soon develop an appreciation for the dynamics of the spermatogenic cycle. Try to focus on understanding the concept instead of memorizing the cellular associations that characterize each stage of the cycle. Unless one is attempting to diagnose particular defects in spermatogenesis the ability to identify particular stages of the cycle is of limited value.

SPERMATOGENIC WAVE

The process of spermatogenesis is not synchronized throughout all sections of the tubules. Therefore, one would expect to see different stages of spermatogenesis in different portions of the tubule at a particular time. Indeed this is the case. Interestingly, as one moves along the seminiferous tubule one observes a repeating pattern of spermatogenic stages. Entire segments of tubule may at the same stage, but adjacent sections are typically at the next or previous stages. This is spatial pattern of spermatogenic stages is referred to as the spermatogenic wave. We do not understand how this wave is generated. However, it seems plausible that it is the result the way in which spermatogenesis is first established at puberty.

BOX 9-1 | **Focus on Fertility: Artificial Insemination**

Fundamental knowledge concerning the reproductive physiology of males and females provides the basis for several techniques that have had profound effects on management of fertility of farm animals, wild animals, and humans. Such methods are commonly referred to as **assisted reproduction technologies** and include artificial insemination, embryo transfer, in vitro production of embryos, and cloning. Among these technologies artificial insemination is the oldest and most commonly used. Artificial insemination involves collecting semen from males and depositing it into the female reproductive tract for the purpose of fertilizing an oocyte.

Artificial insemination consists of two major components: 1) collecting, analyzing, and processing semen, and 2) introduction of semen into the female reproductive tract. The major incentives for developing and using artificial insemination include 1) dissemination of desirable genetic material, 2) overcoming infertility, and 3) reducing the spread of venereal diseases. This technology is used commonly in livestock production, and there are many technical reports and manuals that provide detailed descriptions of semen collection, analysis, processing, and handling. Much of the genetic progress in milk production in dairy cows has been attributed to the extensive use of artificial insemination in the dairy industry. This has been accomplished by identifying sires that carry genes for high production and then making their semen available to the majority of dairy farmers. The technique is also used as a routine means to overcome some types of infertility in

humans and employed in some endangered mammalian species as a means to counter genetic depression (reduced genetic diversity) in free-living species that have experienced fragmentation and to enhance captive breeding programs in zoological parks.

Semen can be collected from males by one of three methods: 1) artificial vagina, 2) electroejaculation, and 3) digital manipulation. An artificial vagina consists of an insulated cylinder open at one end and equipped with a receptacle (plastic test tube or bottle) at the other. Although the construction varies for different species, most designs consist of an inner liner and an outer casing that creates a middle compartment filled with warm water. Artificial vaginas are commonly used to collect semen from bulls and stallions (Figure 9-11). Each of these animals can be trained to mount a dummy while a technician positions the artificial vagina around the penis. Once intromission is achieved the male ejaculates and the semen drains into the collection receptacle. Electroejaculation is used commonly in bulls, rams, and goat bucks. This technique requires an electroejaculator, which consists of a power supply that delivers current to electrodes embedded in a phallus-shaped rectal probe. The probe is positioned in the rectum over the pelvic urethra. Power increases gradually until the pelvic muscles involved in the ejaculatory process begin to contract. The ejaculate is collected into a test tube that is held at the tip of the penis by a technician. Digital manipulation is the preferred method

Figure 9-11 Collection of semen from a stallion. After becoming sexually stimulated by an estrual mare, the stallion mounts a dummy and a technician directs the penis into an artificial vagina. The pressure and warm temperature provided by the artificial vagina induces the stallion to ejaculate, and the ejaculated semen collects into a plastic bottle attached to the tip of the device.

of semen collection in boars. This approach requires the boar to mount a dummy. While the boar mounts, a technician grasps the glans penis using a gloved hand and then applies manual pressure to mimic the sensation provided by the sow's cervix during copulation. As the boar ejaculates, semen is collected into an insulated container held at the tip of the penis.

Once semen is collected, it must be evaluated to determine if it is suitable for insemination. Semen analysis involves assessment of sperm motility, morphology, and concentration. Motility is typically assessed in a subjective manner. An aliquot of ejaculate is placed on a glass slide and examined under a microscope using low power. If the sample displays rapid swirling patterns, motility is usually adequate (i.e., more than 75 percent of the cells are motile). Morphology of sperm is usually assessed by microscopic examination. A small drop of semen is smeared on a slide, allowed to dry,

treated with a stain, and examined under high magnification. Abnormalities can be primary (major) consisting of abnormally shaped heads or secondary (minor) consisting of bent or abnormal tails. Bovine semen should have fewer than 30 percent abnormalities to be considered satisfactory. In some species, it is a common practice to calculate the concentration of sperm cells in an ejaculate. This involves dilution of a sample of semen and counting the number of sperm cells in a given volume using a hemocytometer.

There are two major ways semen is processed following collection. Fresh semen may be diluted (extended) and then directly deposited into females. In species such as cattle, semen is typically extended, frozen, and then stored at low temperatures (cryopreservation). Whether or not fresh semen is used depends on management practices and the ease with which semen can be frozen and thawed without reducing quality. Semen extenders consist of antibiotics to prevent bacterial growth, buffers to prevent changes in pH resulting from metabolic activity of sperm cells, carbohydrates to provide energy for sperm cells, and membrane stabilizers such as egg yolk. Once the semen has been extended it is divided into small aliquots (0.5 mL) placed in thin plastic straws, which are frozen and stored in liquid nitrogen until used.

Insemination procedures vary greatly among species. Major differences involve when to inseminate a female and where to deposit the semen into the female's reproductive tract. With the exception of the queen (female cat), semen is typically deposited transcervically into the uterus. This requires the use of a cannula or insemination gun, which can be passed through the vagina and cervix so that the tip of these devices protrudes a short distance into the body of the uterus. In large species such as cattle placement of the cannula is facilitated by manipulating the cervix *per rectum* (Figure 9-12).

9

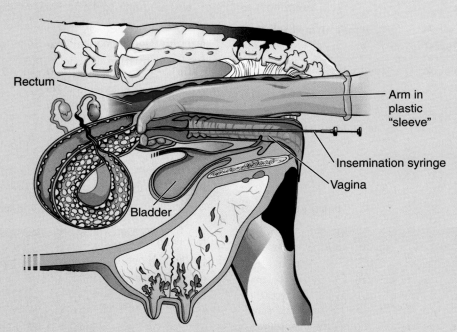

Rectum

Arm in
plastic
"sleeve"

Insemination syringe

Vagina

Bladder

Figure 9-12 Schematic illustration of artificial insemination in a cow. One hand is used to gently grasp the cervix through the rectum, whereas the other hand guides an insemination syringe through the lumen of the cervix. The tip of the syringe holds a plastic straw containing semen. Once the tip of the syringe is positioned properly in the uterine body the plunger of the syringe is depressed and semen is injected from the straw.

The timing of insemination relative to ovulation is critical to achieving high conception rates with artificial insemination. For example, in cattle the best conception rates are achieved when cows are inseminated between 6 and 24 hours after onset of estrus. This illustrates the need for accurate detection of estrus.

SUMMARY OF MAJOR CONCEPTS

- The major functions of the testes are to produce spermatozoa and hormones that regulate spermatogenesis.
- Testosterone is the major testicular hormone and its synthesis is regulated by LH.
- Spermatogenesis involves three major stages including 1) proliferation of spermatogonia to produce spermatocytes, 2) meiosis to produce spermatids, and 3) differentiation to produce spermatozoa.
- Testosterone is the primary regulator of spermatogenesis and its production is dependent on LH.

DISCUSSION

1. Following hemicastration (removal of one testicle), the remaining testicle hypertrophies and eventually becomes almost twice as large as normal. Explain these results in terms of the negative feedback relationship between testosterone and LH.

2. Some forms of infertility in men are attributed to a failure of the pituitary to release LH. In these cases, normal fertility can be induced by fitting these men with small pumps that deliver GnRH in a pulsatile fashion. Explain how this treatment works.

3. Males that suffer from a fever or an episode of heat stress (i.e., a rise in body temperature) experience infertility, but this appears to be a biphasic response. In other words, the male will produce nonviable spermatozoa immediately after the stress and then again several weeks later (depending on the species). Based on your understanding of the spermatogenic cycle, explain this phenomenon.

REFERENCES

Amory, J.K. and W. Bremner. 2001. Endocrine regulation of testicular function in men: implications for contraceptive development. *Molecular and Cellular Endocrinology* 182:175–179.

Holdcraft, R.W. and R.E. Braun. 2004. Hormonal regulation of spermatogenesis. *International Journal of Andrology* 27:335–342.

Imwalle, D.B. and K.K. Schillo, 2002. Castration increases pulsatile luteinizing hormone release, but fails to diminish mounting behavior in sexually experienced bulls. *Domest. Anim. Endocrinol.* 22:223–235.

Rhim, T-H, D. Kuehl, and G.L. Jackson. 1993. Seasonal changes in the relationships between secretion of gonadotropin-releasing hormone, luteinizing hormone, and testosterone in the ram. *Biology of Reproduction* 48:197–204.

Setchell, B.P. 1982. Spermatogenesis and spermatozoa. In C.R. Austin and R.V. Short, *Reproduction in Mammals: 1. Germ cells and fertilization.* Cambridge: Cambridge University Press, pp. 63–101.

Sharpe, R.M. 1994. Regulation of spermatogenesis. In E. Knobil and J.D. Neill, *The Physiology of Reproduction Vol. 1. Second Edition.* New York: Raven Press, pp. 1363–1434.

Walker, W.H. and J. Cheng. 2005. FSH and testosterone signaling in Sertoli cells. *Reproduction* 130:15–28.

9

Dynamics of Ovarian Function in the Adult Female: Ovarian Cycles

CHAPTER OBJECTIVES

- Introduce the concept of an ovarian cycle.

- Describe the basic characteristics of ovarian cycles in mammals.

- Compare and contrast the ovarian cycles of several types of mammals.

THE OVARIAN CYCLE AND REPRODUCTIVE ACTIVITY OF FEMALES

Our analysis of pubertal development in the female mammal revealed that ovarian activity is dynamic. In other words, on a given day the ovaries of developing females express different combinations of ovarian structures. Part of this phenomenon is due to waves of follicle growth. During the prepuberal period, large vesicular follicles develop, but then undergo atresia. Puberty occurs when a follicle reaches the pre-ovulatory stage and then goes on to ovulate and form a corpus luteum. Once an individual becomes pubertal, this sequential pattern of ovarian activity (follicle growth-ovulation-corpus luteum formation) is continually repeated, and is known as the ovarian cycle. Before we consider the details of ovarian cycles, it may be helpful to note that in cases of multiple births more than one follicle reaches the pre-ovulatory stage and ovulates. In these cases, more than one corpus luteum develops. For simplicity, we will focus on cases involving single ovulations.

Figure 10-1 shows ovarian cycles associated with the pregnant and nonpregnant state. Females that do not conceive express a repeating pattern of follicle growth, ovulation, development of a corpus luteum, and regression of a corpus luteum. If the individual conceives after ovulation, regression of the corpus luteum is delayed until parturition. Following birth, the female will resume ovarian cycles that are characteristic of the nonpregnant state. In most cases, resumption of ovulation and occurrence of regular ovarian cycles is delayed for a period of time following parturition.

Follicular and Luteal Phases

The period of the ovarian cycle during which a follicle develops to the preovulatory stage is known as the follicular phase. During this phase

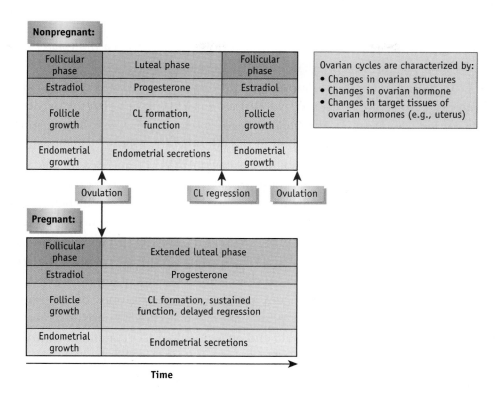

Nonpregnant:

Follicular phase	Luteal phase	Follicular phase
Estradiol	Progesterone	Estradiol
Follicle growth	CL formation, function	Follicle growth
Endometrial growth	Endometrial secretions	Endometrial growth

↑ Ovulation   ↑ CL regression   ↑ Ovulation

Ovarian cycles are characterized by:
• Changes in ovarian structures
• Changes in ovarian hormone
• Changes in target tissues of ovarian hormones (e.g., uterus)

Pregnant:

Follicular phase	Extended luteal phase
Estradiol	Progesterone
Follicle growth	CL formation, sustained function, delayed regression
Endometrial growth	Endometrial secretions

→ **Time**

FIGURE 10-1

Major phases of ovarian cycles in nonpregnant and pregnant females. During the nonpregnant state (top) a female alternates between follicular and luteal phases. During pregnancy the length of the luteal phase is prolonged. These phases are characterized by changes in the type of ovarian steroid produced, type of ovarian structures, and morphology of the uterine endometrium.

of the cycle, the corpus luteum regresses and the dominant structure on the ovaries is a pre-ovulatory follicle. The follicular phase ends at ovulation and is followed by the luteal phase. During the luteal phase, a corpus luteum is the dominant ovarian structure. Although follicle growth occurs during the luteal phase, and large tertiary follicles develop, none of the follicles developing during this period reaches the pre-ovulatory stage. The end of the luteal phase is marked by regression of the corpus luteum; that is, luteolysis.

The cycle of ovarian activity is reflected by fluctuating patterns of ovarian hormones, namely estradiol and progesterone. During the follicular phase, estradiol concentrations in blood increase due to elevated production of this hormone by the dominant follicle. During the luteal phase, the corpus luteum produces large amounts of progesterone. In the nonpregnant female, this sequence of estradiol-ovulation-progesterone is repeated and coincides with the aforementioned pattern of follicular phase-ovulation-luteal phase. Pregnancy can be viewed as an extended luteal phase. In other words, if a pregnancy is established after ovulation, progesterone concentrations remain elevated until birth.

The cyclic release of estradiol and progesterone results in cyclic changes in activity of a variety of tissues that have receptors for these hormones. Effects on the genital tract are of particular importance. In general, the estrogenic

10

portion of the ovarian cycle prepares the tract for receiving spermatozoa and fertilization, whereas the progesterone-dominated portion of the cycle prepares the tract for receiving and nurturing the embryo.

In most mammals, ovarian steroids produce profound effects on female reproductive behavior. "Estrus" was first used in 1900 to describe the particular period when mammalian females become sexually receptive. The term is derived from the Latin adaptation of the Greek word "oistros," which means frenzy. During estrus, females exhibit behaviors that promote sexual contact with males. Such behaviors are most prevalent soon before or at the time of ovulation and are caused by the high levels of estradiol. Because the behaviors associated with estrus are readily observable, and occur at regular intervals corresponding to the follicular phase of the ovarian cycle, species that express this pattern of sexual behavior are said to express estrous cycles. In the absence of pregnancy, females will express estrous behavior at regular intervals, whereas pregnant females will not exhibit estrus. Derivations of the word "estrus" are used describe other phases of the estrous cycle. **Anestrus** refers to a nonbreeding period when estrus and ovulation do not occur and the female resists males' attempts at mating. The estrous cycle is traditionally divided into four phases: **proestrus, estrus, metestrus,** and **diestrus** (Figure 10-2). Proestrus refers to the early portion of the follicular phase and includes events that cause an animal to come into heat. As noted previously estrus refers only to the time when the female is willing to copulate with a male, and corresponds

10

FIGURE 10-2

Major stages of the estrous cycle. The follicular phase is commonly divided into a proestrus and estrus period, whereas the luteal phase is divided into metestrus and diestrus.

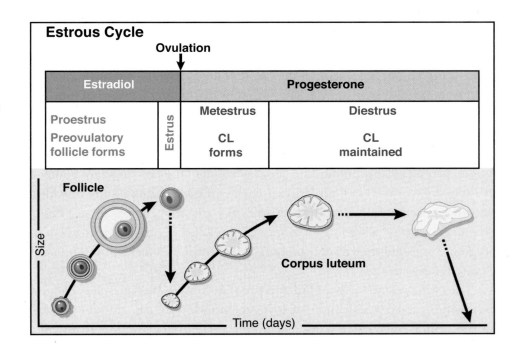

to the latter portion of the follicular phase. During metestrus (also known as diestrus 1), a period corresponding to the early luteal phase, ovarian steroids prepare the genital tract for receiving the newly fertilized ovum. Diestrus (also known as diestrus 2) refers to the remaining portion of the luteal phase. During this phase the reproductive tract is in a state that can maintain pregnancy. If fertilization has not occurred, this phase is followed by a new proestrus period, marking the beginning of a new cycle.

Sexual receptivity of some mammals is not confined to a particular portion of the ovarian cycle (Figure 10-3). In most species of primates, females engage in sexual activity throughout the cycle. Because these individuals do not express repeating patterns of sexual activity, it is inappropriate to refer to their ovarian cycles as estrous cycles. In these cases, the name of the ovarian cycle is based on other outward signs of changing ovarian activity; that is, menstruation. Menstruation (also known as menses) refers to the sloughing of the endometrium accompanied by bleeding. As noted earlier, fluctuations in estradiol and progesterone concentrations associated with the follicular and luteal phases exert important effects on the genital tract. For example, these hormones work together to promote changes in the uterine endometrium that support pregnancy. The drop in progesterone concentrations during the transition between the follicular and luteal phases of the ovarian cycle causes cells of endometrium to slough off and is accompanied by bleeding. In the nonpregnant state, menstruation occurs at regular

10

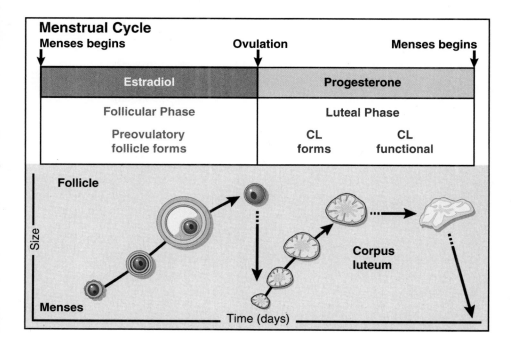

FIGURE 10-3

General characteristics of a menstrual cycle. Females that express this type of ovarian cycle do not exhibit estrus. The cycle begins and ends with the onset of menses. Note that menstrual cycles consist of a follicular and a luteal phase, which are approximately the same lengths.

intervals associated with the transition between the follicular and luteal phases. Species that exhibit this are said to express menstrual cycles. The vast majority of mammalian species that exhibit menstrual cycles are primates.

Ovarian Cycles of Various Species of Mammals

Among mammalian species there is considerable variation in lengths of ovarian cycles. Table 10-1 summarizes the major characteristics of ovarian cycles in several species for the nonpregnant state. The estrous cycles of domestic ungulates such as the sheep, pig, cow, and horse are approximately 3 weeks in length. In each case the length of the follicular phase is shorter than that of the luteal phase. The menstrual cycles of higher primates last approximately 4 weeks and lengths of the follicular and luteal phases are approximately the same (14 days).

Rodents such as rats and mice have very short estrous cycles, due to a truncated luteal phase. In these species, females exhibit estrus and ovulate once every 4 to 5 days with follicular and luteal phases each lasting 2 to 3 days. Interestingly, the length of the luteal phase is lengthened to 11 to 12 days if female rodents are mated to infertile males or if the cervix is stimulated mechanically. Because this pattern of luteal development and progesterone

10

TABLE 10-1 Ovarian cycles of different mammalian species

Species	Classification (Estrus)	Classification (Ovulation)	Length of Cycle (Days)	Follicular Phase (Days)	Luteal Phase (Days)
Human	Menstrual cycle	Spontaneous	24–32	10–14	12–15
Cattle	Polyestrous	Spontaneous	20–21	2–3	18–19
Swine	Polyestrous	Spontaneous	19–21	5–6	15–17
Sheep	Seasonally polyestrous	Spontaneous	16–17	1–2	14–15
Horse	Seasonally polyestrous	Spontaneous	20–22	5–6	15–16
Rabbit	Polyestrous	Induced	1–2 (non-mated) 14–15 (+infertile male)	1–2	0 (non-mated) 13 (+infertile male)
Rat	Polyestrous	Spontaneous	4–5 (non-mated) 13–14 (+infertile male)	2	2–3 (non-mated) 11–12 (+infertile male)
Dog	Monoestrous - polyestrous	Spontaneous	84–252	18	63

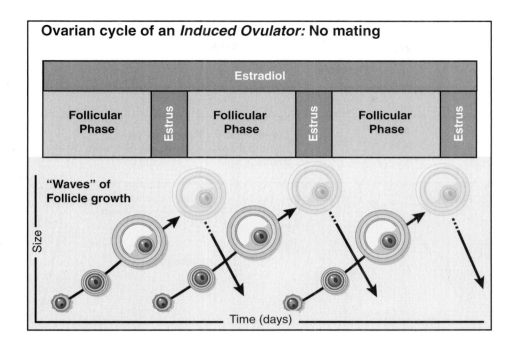

Ovarian cycle of an *Induced Ovulator:* No mating

FIGURE 10-4

General characteristics of the ovarian cycle of an induced ovulator in the absence of mating. This type of female ovulation depends on copulation. Therefore, females not exposed to males exhibit consecutive follicular phases during which a preovulatory follicle develops, induces estrus, but then undergoes atresia.

production is similar to that seen in the pregnant female, the lengthened luteal phase is referred to as **pseudopregnancy.**

The data for the rabbit may be confusing at first glance. In the absence of mating the rabbit doe has a very short ovarian cycle consisting of only a follicular phase and estrus; that is, a large tertiary follicle develops, but undergoes atresia instead of ovulating (Figure 10-4). This cycle of follicle growth and atresia continues so long as the doe in not mated. The pattern of follicle growth and regression is commonly referred to as a wave of follicle growth or follicular wave (Figure 10-4). Unlike the previous examples, where ovulation occurs spontaneously at the end of each follicular phase, rabbit does will not ovulate unless they copulate. Animals with this sort of ovulatory mechanism are known as induced ovulators, meaning that they exhibit a luteinizing hormone (LH) surge and ovulation only if the cervix is stimulated. When a doe mates with an infertile buck (or if its cervix is mechanically stimulated), ovulation is induced and corpora lutea will develop and produce progesterone for 13 days; that is, pseudopregnancy (Figure 10-5). Other induced ovulators include the cat, ferret, black bear, camel, and llama.

Figure 10-6 compares the mechanisms of ovulation for induced ovulators and spontaneous ovulators. In both cases high levels of estrogen, produced by large preovulatory follicles, induce sexual receptivity in the female. In spontaneous ovulators, the high levels of estrogen act on the hypothalamus and pituitary gland to induce an LH surge, which then causes ovulation. In

10

FIGURE 10-5

General characteristics of the ovarian cycle for an induced ovulator when mating occurs. Mating during estrus induces an LH surge, which causes ovulation and development of a corpus luteum. The length of the luteal phase is the same for fertile (pregnancy) and nonfertile (pseudopregnancy) matings.

Ovarian cycle of an *Induced Ovulator:* Mating

Copulation

Ovulation

Estradiol		Progesterone
Follicular Phase	Estrus	Luteal Phase = Pegnancy or pseudopregnancy (in case of sterile mating)

Follicle

Size

Corpus luteum

Time (days)

FIGURE 10-6

Comparison of ovulatory mechanisms in induced and spontaneous ovulators. In each case elevated concentrations of estradiol induce estrus. However, the LH surge in induced ovulators is caused by stimulation of the cervix (a), not estradiol (b) as is the case with spontaneous ovulators.

Classification of Ovarian Cycles Based on Mechanism of Ovulation:

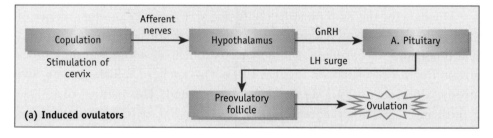

(a) Induced ovulators

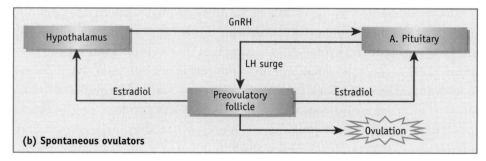

(b) Spontaneous ovulators

induced ovulators, the LH surge is induced by a neural reflex; that is, stimulation of the cervix during copulation stimulates afferent nerves, which induce a surge of GnRH release from the hypothalamus.

Animals that exhibit estrous cycles can be further classified based on the number of times they express cycles in a year (see Table 10-1). Nonseasonal breeders such as cattle and swine will express estrous cycles continuously throughout the year and are referred to as **polyestrous** animals. Seasonal breeders such as sheep, goats, and horses express estrous cycles continuously, but only during a particular time of year (the breeding season). Each year these seasonal breeders experience an anestrous period during which they fail to exhibit ovarian cycles. These are the seasonally polyestrous mammals. The pattern of estrus in canids is rather unique. Wild canids such as wolves and coyotes express only one estrus annually and are classified as **monoestrous** species. In contrast, most breeds of domestic dogs express estrus biannually, but it is not uncommon for some bitches (especially the smaller breeds) to express several cycles in a given year. It is important to note that this classification scheme applies to both spontaneous and induced ovulators.

COMPARATIVE ANALYSIS OF OVARIAN CYCLES IN SEVERAL TYPES OF MAMMALS

As noted in the previous sections, the ovarian cycles of mammals share some important characteristics, namely that ovulation is sandwiched between periods of dominance by the preovulatory follicle (producing high levels of estradiol) and the corpus luteum (producing high levels of progesterone). In the next two chapters we will study the mechanisms of follicle growth, ovulation, and the development and regression of the corpus luteum. Before we enter such discussions, it is important to develop an understanding of how the various reproductive hormones interact to regulate these events. Your understanding of these regulatory mechanisms will be facilitated if you keep in mind the following concepts:

10

- Regulation of ovarian activity depends on communication between the ovaries and the hypothalamic-pituitary unit.

- Pituitary gonadotropins (LH and [follicle-stimulating hormone [FSH]) are the primary regulators of follicles and the corpus luteum, including secretion of hormones by these ovarian structures.

- Hormones produced by the ovaries feed back on the hypothalamic-pituitary unit to inhibit or enhance release of LH and FSH.

- Endocrine interactions between the ovaries and the hypothalamic-pituitary unit result in distinct patterns of ovarian and pituitary hormones, which characterize each phase of the ovarian cycle.

In view of these concepts, our approach to understanding the mechanisms controlling ovarian cycles will be to consider each of the major phases of the ovarian cycle. We will first examine patterns of reproductive hormones, and then consider regulation of these hormone patterns as well as the physiologic implications of them.

Regulation of Ovarian Cycles of Rodents

The ovarian cycle of the rat is the most thoroughly documented one among mammalian species. As noted earlier, the domestic rat has been the primary subject of medical research since the early twentieth century. Much of what we know about reproductive biology was first discovered in rats. Today this animal continues to be the focus of most research dealing with reproductive physiology.

Rats that do not copulate express estrus once every 4 to 5 days (Figure 10-7). Proestrus, estrus, metestrus (diestrus 1), and dietestrus (diestrus 2) each lasts approximately 1 day. When studying estrous cycles, the time of ovulation serves as a useful reference point that divides events of the cycle into those leading to and those resulting from this important event. In rats, ovulation occurs approximately 8 to 10 hours after the beginning of estrus. Figure 10-7 shows patterns of the major reproductive hormones throughout the rat estrous cycle. These patterns reflect concentrations of hormones measured at 2-hour intervals throughout the 4-day cycle. Due to this infrequent mode of blood sampling, pulsatile patterns are not evident.

10

FIGURE 10-7

Circulating patterns of various reproductive hormones throughout the 4-day estrous cycle of the rat.

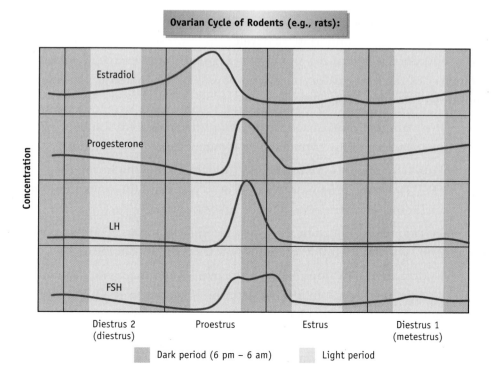

Estradiol

The major physiologic features of estradiol patterns can be summarized as follows:

- Concentrations are lowest during estrus, but then increase gradually throughout the cycle, reaching peak levels during proestrus.
- The pattern is attributed to the pattern of follicle growth. By the time estrus begins, a group of follicles becomes committed to a preovulatory pool. These follicles were members of a larger group of follicles that initiated growth 16 days earlier.
- Follicles selected for ovulation grow throughout the estrous cycle and attain maximal size during proestrus. The increase in estradiol concentrations between estrus and proestrus reflects the follicles' enhanced capacity to produce this hormone.
- Once follicles ovulate, they lose their capacity to produce estradiol. Consequently, concentrations of estradiol fall abruptly during proestrus and reach minimal levels by the morning of estrus.

Progesterone

Patterns of progesterone during the cycle are similar to those of estradiol in some respects and different in others.

- Concentrations begin to increase during diestrus and peak for the first time during metestrus.
- The source of progesterone during this part of the cycle is the corpora lutea that develop following ovulation. As noted earlier, the life span of these structures is short and they lose their ability to produce progesterone by early metestrus.
- A second peak of progesterone occurs during proestrus. Unlike the previous peak, the source of progesterone in this case is the preovulatory follicles, and the stimulus for this response is the LH surge that occurs during proestrus. The precipitous drop in progesterone seen during late proestrus is attributed to the reduced capacity of follicles needed to produce this steroid following ovulation.

Luteinizing Hormone

The profile of LH concentrations shown in Figure 10-7 does not reveal all of the physiologically important features of LH secretory patterns. When LH

10

concentrations are measured in sequential blood samples collected at more frequent intervals (e.g., once every 10 to 15 minutes) over several hours, it becomes clear that LH concentrations fluctuate in a pulsatile manner throughout the cycle.

With respect to the LH profile presented in Figure 10-7, the most physiologically relevant features are:

- Concentrations of LH remain low until the afternoon of proestrus, when there is a surge of LH.

- The LH surge is caused by the positive feedback actions of estradiol. As estradiol concentrations increase they enhance LH secretion, which in turn stimulates release of more estradiol from pre-ovulatory follicles.

- The stimulatory effects of estradiol on LH secretion involve both enhanced GnRH release from the hypothalamus and increased responsiveness of the anterior pituitary gland to GnRH.

- It is unclear whether the LH surge is caused by a single constant pulse of LH or a series of high-frequency LH pulses.

- The physiologic effects of the LH surge include rupture of pre-ovulatory follicles (ovulation), completion of oocyte maturation, and transformation of follicular cells into luteal cells (luteinization).

The mechanism mediating the positive feedback action of estradiol deserves special attention because it is somewhat unique among mammals. Three conditions seem to be necessary for an LH surge to be induced in the rat.

- Neuronal mechanisms responsible for evoking this response.

- A surge of estradiol.

- A surge of progesterone.

The neuronal mechanism that mediates the positive feedback effects of estradiol on LH release appears to be controlled by an endogenous circadian rhythm. In other words, this mechanism becomes operative only during a 1- to 2-hour period each day. The timing of the operative period is dependent on the daily photoperiod. For example, in rats maintained on a daily cycle of 12 hours of light and 12 hours of dark, this so-called critical period will occur between 2:00 and 3:45 p.m. The reason the LH surge occurs only during proestrus is due to the fact that high levels of estradiol and progesterone are required to evoke this response. Moreover, the proestrus surge of progesterone appears to prevent release of surge amounts of LH during the days following proestrus (estrus, diestrus, and metestrus).

The physiologic effects of LH are not limited to those associated with the pre-ovulatory surge of this hormone. LH also plays an important role in development of pre-ovulatory follicles.

The major stimulus for the increased secretion of estradiol by follicles between diestrus and proestrus is the high-frequency pattern of pulsatile LH release.

Follicle-Stimulating Hormone

The most important characteristics of FSH patterns during the estrous cycle include the following:

- A surge of FSH occurs concomitantly with the LH surge during proestrus.
- The elevated concentrations of estradiol and progesterone constitute the major stimuli for inducing this surge, and the neuronal mechanisms that bring about this response are probably the same as those controlling the LH surge.
- The primary surge of FSH is followed by a secondary surge during early estrus.
- The secondary surge of FSH is likely brought about by an abrupt decrease in inhibin, a peptidergic ovarian hormone (produced by the granulosa cells of follicles) that serves as the major negative feedback signal controlling FSH release.
- Concentrations of FSH are minimal during diestrus and metestrus due to the combined inhibitory feedback effects of estradiol, progesterone, and inhibin.

The main physiologic effect of FSH is to stimulate development of secondary follicles to the tertiary stage. We will discuss details of follicle growth and development (folliculogenesis) in the next chapter. For the moment it is only important to understand that the secondary surge of FSH promotes growth of secondary follicles such that they enter a growing pool of follicles on estrus. The subsequent exposure of these follicles to consecutive surges of LH and FSH associated with the next four to five cycles further enhances growth of these follicles such that they become pre-ovulatory follicles. Thus, as mentioned earlier, follicles that ovulate during a particular cycle are selected during an estrus that occurred 20 days earlier.

10

Prolactin

The most significant features of prolactin patterns during the nonmated state can be summarized as follows (Figure 10-8):

- The pattern of prolactin during the rat estrous cycle is very similar to that of LH; that is, a proestrus surge followed by low concentrations during the remainder of the cycle.
- The major stimulus for the prolactin surge is the rising concentrations of estradiol between metestrus and early proestrus.
- Estrogen induces the prolactin surge by acting directly on the anterior pituitary gland as well as on the hypothalamus.
- Like the LH surge, the timing of the prolactin surge is subject to regulation by a circadian clock.
- The physiologic effects of the prolactin surge of proestrus appear to be minimal.

In nonmated female rats, the corpora lutea persist for only a brief interval. In contrast, mating or mechanical stimulation of the cervix prevents regression of corpora lutea. As noted earlier in this chapter, a fertile mating extends the lifespan of the corpora lutea for 20 to 22 days, whereas mating with a sterile male or mechanical stimulation of the cervix extends the

10

FIGURE 10-8

Circulating patterns of prolactin in non-mated and mated rats. Following mating, females exhibit two nocturnal surges of prolactin. This pattern of prolactin release extends the lives of the corpora lutea.

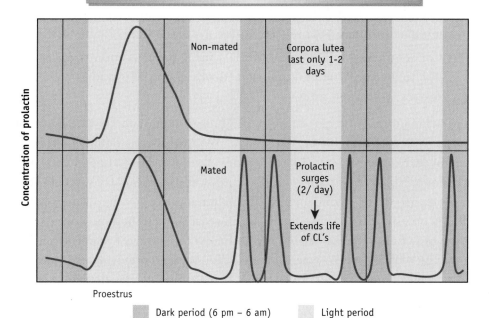

Ovarian Cycle of Rodents (e.g., rats): Pseudopregnancy and Pregnancy

Non-mated

Corpora lutea last only 1-2 days

Concentration of prolactin

Mated

Prolactin surges (2/ day)

Extends life of CL's

Proestrus

Dark period (6 pm – 6 am)   Light period

lifespan of corpora lutea for 12 to 14 days (pseudopregnancy). This extended luteal function is caused by unique changes in prolactin secretion (Figure 10-8). The specific nature of this effect in the pseudopregnant rat can be summarized as follows.

- During pseudopregnancy, female rats express two daily surges of prolactin for approximately 2 weeks.
- The daily prolactin surges peak during the early (5 to 7 p.m.) and late (3 to 7 a.m.) dark periods.
- The end of pseudopregnancy is marked by an abrupt termination of the prolactin surges, which appears to be caused by the reduction in progesterone attributed to regressing corpora lutea.
- The physiologic significance of the mating-induced prolactin surges deals with maintenance of pregnancy. These high levels of prolactin are necessary for corpora lutea to produce progesterone, which is essential for maintaining pregnancy.

Regulation of the Ovarian Cycles of Induced Ovulators

Induced or reflex ovulators are widespread among the orders of mammals. However, this type of ovulation seems to predominate in Insectivores, Lagomorphs, and Carnivores. At least 22 species of mammals have been documented to be induced ovulators, and another 25 species may ovulate in this manner. As noted earlier, ovulation in these species occurs only after copulation. However, ovulation can be induced experimentally by mechanical stimulation of the cervix. The general mechanism of induced ovulation was discussed in an earlier section of this chapter (Figure 10-6). You should recognize this as a neuroendocrine reflex arc. During mating various stimuli (e.g., tactile) are detected by various receptor organs. This generates neuronal signals that are conducted to the central nervous system via afferent neurons. Integration of these various inputs occurs within the hypothalamus to evoke a hormonal response. In the case of induced ovulation, the hormonal response is a surge of LH that induces ovulation of ovarian follicles.

Induced ovulation has been studied most in the rabbit doe. However, the domestic queen is probably the induced ovulator with which most people are familiar. Figure 10-9 summarizes the ovarian cycle of the queen. The ovarian cycle of the unmated queen consists only of proestrus and estrus. Proestrus refers to the period separating consecutive estrous periods and is the time during which preovulatory follicles develop. This is most likely brought about by a high-frequency pattern of pulsatile LH release. As follicles

10

FIGURE 10-9

Circulating patterns of progesterone and estradiol in the queen before and after mating. In the absence of mating the queen will express a 9-day estrus once every 17 days. Upon stimulation of the cervix the female will ovulate resulting in formation of corpora lutea that will persist for 60 to 65 days.

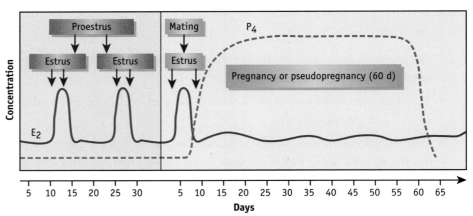

Ovarian Cycle of the Queen:

NOTE: Estrus lasts about 9 days and occurs at approximately 17-day intervals.

reach the preovulatory stage and they produce large amounts of estradiol and induce behavioral estrus. If the queen does not copulate, the large follicles will become atretic and estradiol production will cease. By the time estrus ends, a new crop of follicles is recruited to enter the preovulatory stage. The duration of estrus in the queen is approximately 9 days and occurs once every 17 days in the nonmated state. Notice that concentrations of progesterone remain extremely low during the ovarian cycles of nonpregnant queens.

Like the rat, mating induces dramatic changes in the ovarian cycle of induced ovulators. Mating or stimulation of the cervix induces a surge of LH. The LH surge induces ovulation, and luteinization of ruptured follicles resulting in formation of corpora lutea. The corpora lutea persist for approximately 2 months, whether or not the mating is fertile. In the case where pregnancy does not occur the queen is said to be in a state of pseudopregnancy.

We know less about the role of FSH in regulation of the feline estrous cycle than we do in regulation of the ovarian cycles of rodents, higher primates, and domestic farm animals. In rabbits, two surges of FSH occur postcoitus. The first is smaller and rises at a slower rate than the LH surge, and is probably due to a rise in gonadotropin-releasing hormone (GnRH) brought about by cervical stimulation. The secondary surge of FSH occurs 12 to 24 hours after mating and is the result of a sudden drop in inhibin caused by follicle rupture and ovulation. Presumably this response is the stimulus for initiating development of a new crop of preovulatory follicles.

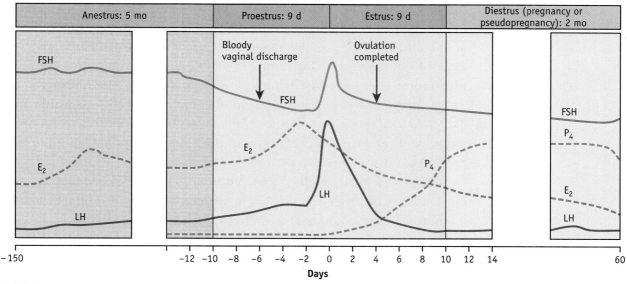

Ovarian Cycle of the Bitch:

| Anestrus: 5 mo | Proestrus: 9 d | Estrus: 9 d | Diestrus (pregnancy or pseudopregnancy): 2 mo |

FIGURE 10-10

Circulating patterns of LH, FSH, estradiol, and progesterone in the bitch at various times relative to estrus. During most of the year, the bitch remains in an anestrous state during which gonadotropin levels remain at basal levels and regulate waves of follicle growth. Onset of high-frequency LH pulses during the proestrus period stimulates follicle growth to the pre-ovulatory stage. Pre-ovulatory follicles release large amounts of estradiol which induces estrus and an LH surge. The LH surge lasts several days and induces ovulation. The resulting corpora lutea are viable and produce progesterone for 60 to 65 days.

The Ovarian Cycles of Canids

The ovarian cycles of canids are rather unique among mammals. First, as noted earlier, most wild canids and many of the larger breeds of the domestic dogs do not experience more than one or two cycles each year. Thus the bitch experiences a long (5 months) period of anestrus. Another unique feature of the bitch's estrous cycle is that the diestrus period is indistinguishable between pregnant and nonpregnant states. Figure 10-10 summarizes highlights of the ovarian cycle of the bitch.

Toward the end of anestrus, circulating concentrations of LH begin to increase due to onset of a high-frequency mode of pulsatile LH secretion. This pattern of LH release persists throughout proestrus and culminates in a surge of LH, which marks the onset of estrus. Concentrations of LH remain low following the LH surge. As in the previous two examples, the high-frequency pattern of LH enhances development of preovulatory follicles and the LH surge induces ovulation of these follicles.

Concentrations of ovarian steroids are low during anestrus. Concentrations of estradiol fluctuate in accordance with waves of follicle growth, but do not reach levels that induce estrus or a pre-ovulatory surge of LH. The onset of high-frequency LH pulses during the final stages of anestrus prevent large antral follicles from becoming atretic and stimulate them to the pre-ovulatory stage of development. This results in elevated concentrations of estradiol, which induce an LH surge as well as behavioral estrus. The LH surge lasts several days (24 to 96 hours) and induces multiple ovulations within 2 to 3 days. As follicles ovulate, they lose their ability to produce large amounts of estradiol. Therefore, concentrations of this steroid hormone fall leading to the end of estrus.

As corpora lutea develop from ruptured follicles they produce increasing amounts of progesterone. These structures persist for approximately 2 months in dogs regardless of the reproductive state of the bitch. Like in rodents and induced ovulators, persistence of corpora lutea in the absence of pregnancy is referred to as pseudopregnancy.

The Ovarian Cycles of Domestic Ungulates

The estrous cycles of the major livestock species (cattle, sheep, swine, and horses) have been extensively studied. By far the greatest emphasis has been on the cycles of the cow and ewe. The cow has received attention because of the economic importance of reproductive management in the dairy and beef industries. The ewe has been the focus of a tremendous number of studies because the smaller body size of sheep makes this species more suitable to some difficult experimental techniques and its seasonal reproductive activity makes it a good model for studying seasonal control of reproductive activity. Tremendous insight into the regulation of ovarian cycles in domestic ungulates can be gained by focusing on either the cow or the ewe. We shall focus our analysis on the ewe because information on the neuroendocrine control of its estrous cycle is more complete than that of the cow. Moreover, in a later chapter we will use the sheep as a model to study the effects of season on reproductive activity.

The estrous cycle of the ewe averages 17 days in length. There appears to be little variation in this trait among sheep. Approximately 95 percent of the cycles last between 16 and 19 days. Like the rat, the ovarian cycle of the sheep consists of a follicular phase and a luteal phase. These phases can be subdivided into proestrus, estrus, diestrus, and metestrus (Figure 10-11). Proestrus and estrus make up the follicular phase, whereas metestrus and diestrus make up the luteal phase. During proestrus, one or two preovulatory follicles are

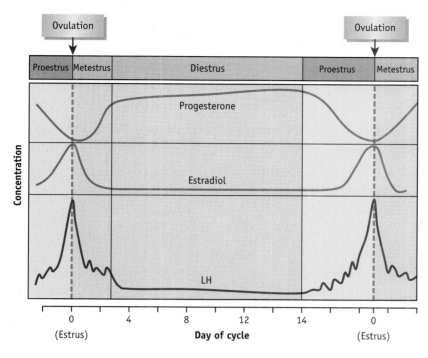

FIGURE 10-11

Circulating patterns of progesterone, estradiol, and LH in the ewe throughout the estrous cycle.

developing. Ovulation occurs during the latter part of estrus. Metestrus refers to the time during which the corpus luteum is forming. The corpus luteum is fully functional during diestrus, and the regression of the corpus luteum marks the end of this period and the beginning of proestrus.

Our approach to analyzing the ovine estrous cycle will be similar to the one we used for the estrous cycle of the rat. The pattern of each reproductive hormone will be considered separately. Each of the following discussions refers to hormone patterns depicted in Figure 10-11.

Progesterone

By now it should be clear that circulating concentrations of progesterone reflect activity of the corpus luteum. More specifically, concentrations of progesterone are directly proportional to both the size and number of corpora lutea. Thus concentrations of progesterone are low during proestrus and estrus due to the lack of functional corpora lutea. Within the first 3 days following ovulation, concentrations of this steroid are extremely low. However, between days 3 and 8 progesterone concentrations increase as the ruptured follicles develop into corpora lutea and gain the capacity to produce large amounts of this steroid. Concentrations of progesterone increase gradually throughout the luteal phase of the cycle (between days 8 and 14). The end of the luteal phase is marked by an abrupt decrease in progesterone concentrations. This occurs over a 1- to 2-day period.

10

The pattern of progesterone during the ovine estrous cycle is the result of the following regulatory mechanisms.

- The rise in progesterone levels during metestrus is a consequence of ovulation and subsequent transformation of tissue from ovulated follicles into luteal tissue. These events are mediated by the LH surge.
- The elevated concentrations of progesterone during the luteal phase are sustained by the luteotrophic actions of LH.
- The drop in progesterone marking the end of the luteal phase is due to the death of luteal tissue brought about by increased release of prostaglandins $F_{2\alpha}$ ($PGF_{2\alpha}$) by the uterus.

Estradiol

Our previous analyses of ovarian cycles firmly established the relationship between estradiol concentrations and follicle activity. These same principles apply to the ovarian cycle of the sheep. Specifically, circulating concentrations of estradiol reflect both the number and size of follicles. With this in mind the patterns of estradiol associated with the ewe's estrous cycle can be readily understood. The principal rise in estradiol occurs during the 2- to 3-day follicular phase (proestrus), beginning after the aforementioned drop in progesterone concentrations and peaking at the onset of estrus. The existence of other peaks in estradiol concentrations is unclear. A secondary rise in this hormone may occur on day 4 of the cycle. This is consistent with what occurs in the cow. Cows experience a wave of follicle development soon after ovulation, which results in an increase in estradiol during diestrus. Some cows experience another follicular wave and corresponding increase in estradiol concentrations during the middle of the luteal phase.

The pattern of estradiol concentrations during proestrus and estrus are strictly the result of follicular activity which can be summarized as follows.

- The increase in estradiol concentrations during proestrus reflects growth and activity of a preovulatory follicle, which begins during late diestrus.
- The decrease in estradiol concentrations during estrus are due to ovulation.
- Increases in estradiol concentrations during other times of the cycle are attributed to additional waves of follicle growth, which end with follicular atresia rather than ovulation.

LH

One of the more dramatic features of the estrous cycle is the pre-ovulatory surge of LH. In the ewe, the LH surge is tightly coupled with the onset of

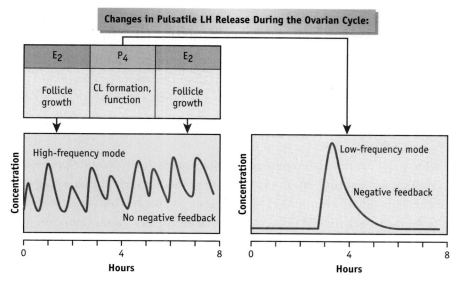

Changes in Pulsatile LH Release During the Ovarian Cycle:

FIGURE 10-12

Variation in the pulsatile pattern of LH release during the estrous cycle of the ewe. During the follicular phase, LH is released in a high-frequency pattern due to the absence of progesterone negative feedback. Elevated levels of progesterone during the luteal phase suppress the pulsatile secretion of LH, resulting in a low-frequency pattern.

estrus. This massive increase (50- to 100-fold increase) in LH concentrations begins at the onset of estrus, peaks 4 to 8 hours later and ends within 12 hours of its initiation. During other phases of the estrous cycle, LH is released tonically. Average concentrations increase during diestrus, then decrease and remain low during most of metestrus. When concentrations of progesterone drop, tonic levels of LH begin to increase and continue to rise throughout proestrus. The rate of increase rises abruptly with the onset of the LH surge.

The day-to-day changes in LH patterns shown in Figure 10-11 are caused by changes in the pulsatile secretion of this hormone. Figure 10-12 illustrates pulsatile patterns of LH for the follicular and mid-luteal phases. The increase in LH concentrations during proestrus is due to the onset of a high-frequency pattern of release. The lower concentrations of the luteal phase reflect a low-frequency pattern of release. You might have noticed that the amplitude of LH pulses is also different between phases of the cycle. Generally, pulse amplitude is a function of pulse frequency. More specifically, a lower pulse frequency usually results in higher pulse amplitude.

Ovarian steroids play a primary role in regulating patterns of LH during the estrous cycle. In particular, the positive and negative feedback effects of estradiol and progesterone regulate both the pattern of GnRH release from the hypothalamus and the responsiveness of the anterior pituitary gland to GnRH. Highlights of these regulatory mechanisms are listed below.

- The high levels of estradiol on proestrus induce the preovulatory surge of LH by increasing GnRH release and by enhancing responsiveness of the pituitary gland to GnRH.

- The high levels of progesterone found during the luteal phase block the positive feedback actions of estradiol on LH release.

10

- The low-frequency pattern of LH, characteristic of the luteal phase, is caused by the negative feedback actions of progesterone on hypothalamic release of GnRH.

- The high-frequency pattern of LH, characteristic of proestrus and diestrus, is due to the absence of a negative feedback signal; i.e., progesterone.

FSH

Patterns of FSH have been more difficult to assess due to unique problems associated with measuring this hormone in the blood. It is clear that a surge of FSH occurs coincident with the LH surge during estrus. It is also well accepted that a secondary peak in FSH occurs 24 hour after the first peak. Other increases in FSH might occur during the luteal phase, but these have not been consistently identified. However, careful monitoring of FSH patterns in cows reveals that increases in FSH concentrations occur at the onset of follicular waves.

Regulation of FSH secretion has not been studied to the same extent as the regulation of LH secretions in sheep. However, it is clear that ovarian hormones play significant roles in controlling FSH release. The following aspects of this control are noteworthy.

- It is generally agreed that the preovulatory surge of FSH is controlled by the same mechanisms as those controlling the LH surge; that is, the high concentrations of estradiol present during proestrus appear to be induce the FSH surge, and luteal phase concentrations of progesterone block this effect.

- The positive feedback effects of estradiol on FSH release are mediated by enhanced release of GnRH. It remains unclear if estradiol enhances pituitary response to GnRH.

- The secondary surge of FSH (and other increases that precede follicular waves) is due to reductions in ovarian hormones that exert negative feedback actions on FSH. Estradiol and inhibin are among the most important of these hormones.

- Throughout the estrous cycle there is an inverse relationship between concentrations of FSH and concentrations of estradiol and inhibin, hormones produced by follicles.

Prostaglandin F$_{2\alpha}$

One of the most important events regulating the estrous cycle is luteolysis. Recall that luteal phase concentrations of progesterone prevent the high-frequency pattern of LH release as well as the preovulatory surge of LH. Therefore, ovulation cannot occur until luteal regression. In Chapter 12, we will explore the details of the mechanisms controlling this important event. For the time being it is only important to understand that luteolysis is induced by $PGF_{2\alpha}$. During the luteal phase, release of this hormone by the uterine endometrial cells is minimal. However, between days 12 and 14 large surges of $PGF_{2\alpha}$ appear in the circulation due to increased production by the uterus. The major consequence of this response is regression of luteal tissue which results in a rapid decrease in progesterone concentrations. The mechanisms controlling $PGF_{2\alpha}$ release and luteolysis will be examined in detail in Chapter 12.

There is consensus among reproductive biologists that the mechanisms regulating luteolysis in the other major livestock species are similar to the one described for the sheep. It is also likely that a similar mechanism exists in rodents. We know little about the control of luteolysis in induced ovulators and canids. In addition, luteolytic mechanisms in higher primates have been elusive until recently.

Ovarian Cycles of Higher Primates

The task of understanding the menstrual cycles of higher primates is facilitated by our analysis of estrous cycles in other species. Much of what we know about regulation of ovarian activity in livestock, rodents, and induced ovulators applies to regulation of menstrual cycles. In both types of ovarian cycles, ovulation is a convenient reference point with which to begin. The major difference between estrous and menstrual cycles is not as much biological as much as how the cycle is described. More specifically the terminology used to describe the stages of the cycle. With respect to estrous cycles, estrus is typically referred to as day 0. In contrast, day 0 of the menstrual cycle is typically defined as the day menstruation begins (Figure 10-13). The menstrual cycle is often divided into two phases, based on activity of the ovaries; that is, follicular and luteal phases. However, the cycle is also described based on the activity of the uterine endometrium. The follicular phase consists of a 5-day period of menses and a 9-day proliferative phase. During the luteal phase, the uterus is said to be in the secretory phase. Menses refers to the sloughing of blood and endometrial tissue. This is followed by the proliferative phase during which the endometrium increases in thickness. This is largely due to formation of glands in the endometrial mucosa. During the secretory phase, the endometrium attains maximal thickness and uterine glands begin to secrete fluids, which help create an environment favorable to pregnancy. At this time small spirally shaped arteries proliferate and extend through the submucosa to supply blood to the endometrial glands. The physiologic connections between the two ovarian phases and

10

FIGURE 10-13

Circulating patterns of LH, FSH, estradiol, and progesterone during the menstrual cycle.

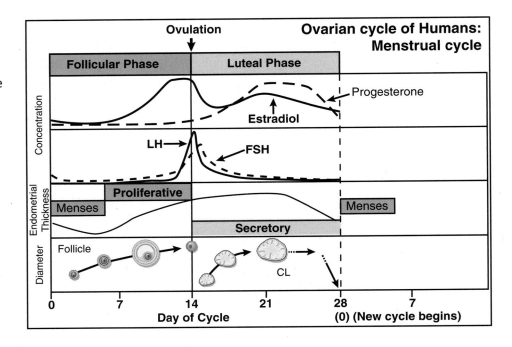

the three uterine phases are the ovarian hormones. The proliferative phase of the uterus is due to the high levels of estradiol found during the follicular phase, whereas the secretory phase is due to the high levels of progesterone and estradiol characteristic of the luteal phase. Menses is induced by the sudden drop in progesterone, which marks the end of the luteal phase. This decrease in progesterone induces constriction of the spiral arteries which results in ischemia and then necrosis of the endometrium. Dead tissue and blood then slough off into the lumen of the uterus. One shouldn't develop the notion that these phases of uterine activity are unique to animals with menstrual cycles. The uteri of other mammals undergo similar changes in association with changes in ovarian steroid secretion. However, the extensive endometrial sloughing and bleeding does not occur in these nonprimate species.

Our discussion of the menstrual cycle will rely heavily on what is known about the human. However, many of the details concerning regulation of the menstrual cycle have been elucidated based on research done with Old World monkeys, in particular the Rhesus monkey. Figure 10-13 shows patterns of the major reproductive hormones throughout the menstrual cycle of women. As you will soon learn, the mechanisms responsible for these patterns are similar to those described for the aforementioned species.

Progesterone

The pattern of progesterone during the menstrual cycle should be quite familiar by this point. Briefly, concentrations remain low during the follicular phase and are elevated during the luteal phase. Progesterone concentrations begin to increase within 3 days after ovulation, reach peak levels by the middle of the luteal phase, and then wane thereafter. Progesterone levels reach minimal concentrations 14 days following ovulation. The primary source of progesterone is the corpus luteum. As in other species, the LH surge is responsible for inducing ovulation, luteinizing the ruptured follicle, and maintaining function of the corpus luteum. The precise mechanism controlling regression of the corpus luteum remains elusive. However, it seems clear that the corpus luteum begins to lose its ability to respond to LH during the latter half of the luteal phase. Whether or not there is a luteolytic agent inducing luteal regression is unclear. It is clear that the uterus does not produce a luteolysin as in many nonprimate species. This topic will be addressed further in Chapter 13.

Estradiol

Fluctuations in estradiol concentrations during the menstrual cycle are similar to those of the various estrous cycles described earlier. Estradiol concentrations increase during the follicular phase due to increased production by a developing preovulatory follicle. Selection of a preovulatory follicle occurs during the follicular phase. During this time the fastest growing follicle suppresses development of other antral follicles and becomes the major source of estradiol.

Concentrations of estradiol also increase during the luteal phase, but this is not due to follicular development. Rather, the source of this secondary rise in estradiol is the corpus luteum. Follicles contribute little to circulating estradiol concentrations during the luteal phase because there is very little follicular growth beyond the early antral stage during this phase.

LH and FSH

Concentrations of LH are highest during the mid-cycle LH surge. During most of the follicular and luteal phases, LH levels are minimal. Intensive blood sampling reveals that concentrations of this hormone fluctuate in a pulsatile manner throughout the cycle. A high-frequency pattern of LH pulses is present throughout the follicular phase. At this time the frequency of LH pulses is approximately one per hour. This pattern is maintained during the LH surge, but the amplitudes of pulses increases greatly at this time. During the luteal phase LH pulse frequency decreases to one pulse every few hours.

Concentrations of FSH are also highest during mid-cycle. A preovulatory surge of FSH occurs coincidently with the LH surge. Like LH, concentrations of FSH are minimal during other times of the cycle. However, unlike LH, concentrations of FSH increase during menses but then decline and remain

10

low during the follicular phase. FSH concentrations fluctuate in a pulsatile manner but these patterns have not been as thoroughly characterizes as the pulsatile patterns of LH.

Patterns of gonadotropins during the menstrual cycle are the consequence of the feedback actions of ovarian hormones on the hypothalamic-pituitary unit. Specifically,

- The low frequency of LH pulses during the luteal phase is due to the negative feedback actions of progesterone, whereas the high frequency of LH pulses characteristic of the follicular phase is due to the absence of this negative feedback.

- The preovulatory surges of LH and FSH are due to the positive feedback actions of estradiol on the hypothalamus and pituitary gland.

- The elevation in FSH during menses is due to the absence of large follicles; that is, a lack of estradiol and inhibin, or major negative feedback signals regulating FSH release.

HORMONAL REGULATION OF OVARIAN CYCLES

Although there are significant differences in hormone patterns associated with the ovarian cycles of various mammals, it is possible to draw some firm conclusions regarding how these hormones interact to regulate these cycles. Figure 10-14 summarizes the major sequence of events that occur during the follicular and luteal phases of the estrous cycles of spontaneous ovulators.

FIGURE 10-14

Sequence of hormone events occurring during the luteal and follicular phases of the ovarian cycle of spontaneous ovulators. Note the LH surge marks the transition between the follicular and luteal phases, whereas regression of the corpus luteum marks the transition between the luteal and follicular phases.

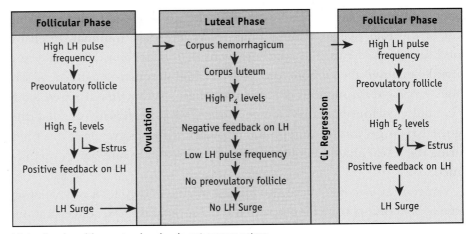

Sequence of Hormonal Events Controlling Ovarian Cycles of Spontaneous Ovulators*

Follicular Phase	Ovulation	Luteal Phase	CL Regression	Follicular Phase
High LH pulse frequency		Corpus hemorrhagicum		High LH pulse frequency
Preovulatory follicle		Corpus luteum		Preovulatory follicle
High E_2 levels → Estrus		High P_4 levels		High E_2 levels → Estrus
Positive feedback on LH		Negative feedback on LH		Positive feedback on LH
		Low LH pulse frequency		
		No preovulatory follicle		
LH Surge →		No LH Surge		LH Surge

*Note: Species with menstrual cycles do not express estrus.

SUMMARY OF MAJOR CONCEPTS

- The ovarian cycle of mammals consists of a repeating sequence of events, which includes follicular development-ovulation-corpus luteum development-corpus luteum regression.

- The ovarian cycle is typically divided into two major phases: a follicular phase, when the dominant ovarian structure is a preovulatory follicle producing high levels of estradiol and inhibin, and a luteal phase, when the dominant ovarian structure is a corpus luteum producing high levels of progesterone.

- Ovarian hormones determine the pattern of gonadotropin secretion during the ovarian cycle and gonadotropins in turn regulate development and activity of ovarian structures.

- A lack of negative feedback allows a high-frequency mode of LH secretion during the follicular phase, whereas progesterone negative feedback sustains a low-frequency mode of LH secretion during the luteal phase. High levels of estradiol induce a pre-ovulatory surge of LH at the end of the follicular phase.

- Estradiol and inhibin are the major negative feedback regulators of FSH secretion. High levels of these hormones suppress FSH during the follicular phase, but the maximal levels of estradiol induce a preovulatory surge of FSH. In some species the decline in inhibin and estradiol coincident with ovulation cause a secondary surge in FSH.

- Specifically, FSH and LH interact to govern follicle growth, whereas LH induces ovulation and luteinization of follicles and sustains luteal activity.

- In many species the regression of the corpus luteum is caused by an increase in release of PGF by the uterus.

DISCUSSION

1. Some reproductive physiologists have used the pseudopregnant rabbit as a model to study the estrous cycle. Does this make sense? Explain your answer.

2. Suppose you discover a new species of mammal and you want to determine the length of its ovarian cycle. Unfortunately all you have to work with is a small herd of females and your observation skills. Describe how you might go about estimating the length of this species' ovarian cycle.

10

3. Thinking in terms of evolution theory, what might be an advantage of induced ovulation?

REFERENCES

Freeman, M.E. 1994. The neuroendocrine control of the ovarian cycle of the rat. In E. Knobil and J.D. Neill, *The Physiology of Reproduction Vol. 2. Second Edition*. New York: Raven Press, pp. 613–658.

Goodman, R.L. 1994. Neuroendocrine control of the ovine estrous cycle. In E. Knobil and J.D. Neill, *The Physiology of Reproduction Vol. 2. Second Edition*. New York: Raven Press, pp. 659–709.

Hotchkiss, J. and E. Knobil. 1994. The menstrual cycle and its neuroendocrine control. In E. Knobil and J.D. Neill, *The Physiology of Reproduction Vol. 2. Second Edition*. New York: Raven Press, pp. 711–749.

Johnston, S.D., M.V. Root Kustritz, and P.N.S. Olson. 2001. *Canine and Feline Theriogenology*. Philadelphia: Saunders.

Ramirez, V.D. and W. Lin Soufi. 1994. The neuroendocrine control of the rabbit ovarian cycle. In E. Knobil and J.D. Neill, *The Physiology of Reproduction Vol. 2. Second Edition*. New York: Raven Press, pp. 585–611.

Stevenson, J.S. 2007. Clinical reproductive physiology of the cow. In R.E. Youngquist and W. R. Threlfall, *Current Therapy in Large Animal Theriogenology 2*. St. Louis: Saunders, pp. 258–270.

10

Dynamics of Ovarian Function: Folliculogenesis, Oogenesis, and Ovulation

CHAPTER OBJECTIVES

- Describe phases of follicle development.

- Describe mechanisms regulating follicle development.

- Describe synthesis of estradiol by follicles.

- Describe development of the oocyte.

- Describe the mechanism of ovulation.

OVERVIEW

The ovarian cycle facilitates fertilization of an oocyte via four important processes:

- development of an oocyte that can be fertilized (oogenesis),
- development of a preovulatory follicle that will respond to an LH surge (**folliculogenesis**),
- release of the oocyte contained within the follicle (ovulation), and
- preparation of the reproductive tract for transport of gametes.

Although these processes can be studied separately, they are intimately related. For example, follicular cells support oogenesis, whereas ovulation requires development of a mature follicle. In addition, the hormones produced by follicular cells influence the motility of the reproductive tract which is important in gamete transport. In this chapter we will explore the mechanisms whereby follicles and oocytes develop and ovulate.

EARLY DEVELOPMENT OF THE OOCYTE AND FOLLICLE

Oogenesis and folliculogenesis are intimately associated (Figure 11-1). Each of these processes begins during sexual differentiation, when primordial germ cells infiltrate the embryonic gonad. The final stages of oogenesis and folliculogenesis occur in the adult during the follicular phase of the ovarian cycle. Oogenesis begins in the fetal ovary when the primordial germ cells enter the gonad and become oogonia. These cells proliferate via mitosis during fetal development. Toward the end of pregnancy, all of the oogonia begin meiosis and become primary oocytes. Further development of these cells is arrested during the first meiotic prophase. Thus at the time of birth the ovaries

FIGURE 11-1

Overview of oogenesis (top) and folliculogenesis (bottom). Both processes begin during the embryonic period and continue throughout much of adulthood. Note that all follicles contain a primary oocyte. Complete maturation of the oocyte requires ovulation and fertilization.

Overview of Oocyte and Follicle Development

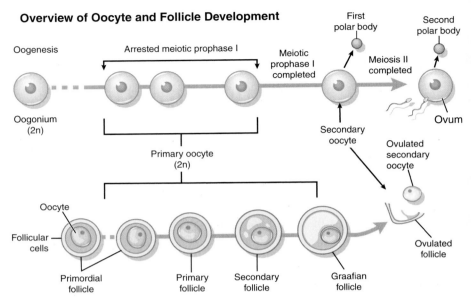

Folliculogenesis

contain all the oocytes they will ever have. In most mammalian species, completion of the first part of meiosis occurs at ovulation, whereas the second meiotic division is completed upon fertilization. Unlike spermatogenesis, oogenesis in the adult does not increase the number of gametes. The first and second meiotic divisions produce the first and second polar bodies, which are small cells that are not viable and eventually regress.

Follicle development begins about the time oogonia enter the first meiotic prophase. During the early stages of meiosis, the primary oocyte becomes surrounded by a single layer of follicular cells. These cells produce a thin basement membrane, which surrounds the oocyte. At this point the oocyte, along with the follicular cells and basement membrane, is known as a primordial follicle. It is believed that follicular cells generate a meiosis inhibitory factor, which keeps the oocyte in an arrested state of development until ovulation. Development of the oocyte ceases soon after the follicular cells condense around the oocyte. Gap junctions form between the oocyte and the follicular cells that surround it. Inhibition of oocyte maturation depends on a cell-to-cell communication facilitated by these gap junctions. Resumption of oocyte development occurs near the time of ovulation and is related to disruption of gap junctions brought about by the preovulatory surge of LH.

Primordial follicles can remain in this state of suspended animation for many years. For example, human females do not begin ovulating until the end of the first decade of life and continue expressing ovarian cycles until 50 years of age or more.

FOLLICULOGENESIS

A primordial follicle has three possible fates. First, it can remain quiescent. Some primordial follicles never resume development. Second, the follicle can undergo atresia. This is what happens to the vast majority of follicles. Third, the follicle can resume development. There are two possible fates for follicles that resume development. Most undergo atresia. A few develop to the fully mature state and ovulate.

The amount of time required for a follicle to grow to a preovulatory size is much longer than a single ovarian cycle. For example, in humans (Figure 11-2) this process takes almost 1 year. Follicle growth is typically divided into three major phases: preantral, antral, and preovulatory. The preantral phase refers to the period of development when the follicle lacks an antrum (primordial, primary, and secondary follicles). This phase begins when a primordial follicle enters a pool of developing follicles, becomes a primary follicle, and then develops multiple layers of granulosa cells to become a secondary follicle. The preantral phase is the longest of the three phases, but less is known about this stage than the others. The antral phase of follicular growth begins when a secondary follicle begins to develop a fluid-filled antrum to become a tertiary follicle. During

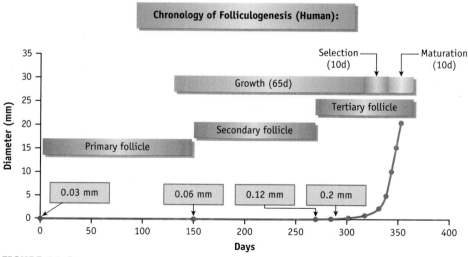

FIGURE 11-2

Chronology of folliculogenesis in the human. Follicles begin to develop when a primordial follicle develops a layer of granulosa cells around the oocyte and becomes a primary follicle. However, appreciable growth does not occur until the primary follicle develops multiple layers of granulosa cells and becomes a secondary follicle. As granulosa cells proliferate, some follicles develop an antrum and become tertiary follicles (selection). Eventually one tertiary follicle will undergo rapid growth and develop into a preovulatory follicle (maturation).

this phase of development, follicular cells proliferate and produce increasing amounts of follicular fluid, which causes the size of the follicle to expand. The end of this phase is not as clearly demarcated as the transition between the pre-antral and antral phases and depends on the prevailing hormonal conditions. For most tertiary follicles, the end of the antral phase is marked by atresia. The few tertiary follicles that do not become atretic enter the preovulatory phase. This phase is known as selection. During this final phase, one follicle experiences rapid growth and becomes the ovulatory follicle (maturation).

Preantral Phase

As noted earlier, the preantral phase of folliculogenesis begins with the growth of primordial follicles. During this phase, follicular diameter increases from 20 to 400 μm due largely to growth of the primary oocyte. Although the oocyte is very active during this phase, it remains in the first meiotic prophase. Some of the more significant activities include RNA synthesis, development of a dense nucleolus in the nucleus, and loading of the cytoplasm with organelles (Golgi apparatus and endoplasmic reticulum). Such changes are prerequisites for later stages of oocyte development that occur following ovulation.

Changes in follicular cells also contribute to follicle growth during the preantral phase. The granulosa cells surrounding the oocyte divide and form multiple layers. The cells that are in contact with the oocyte become the corona radiata and secrete a glycoprotein that becomes part of the zona pellucida, a membrane that surrounds the oocyte. Cytoplasmic processes project from the innermost layer of granulosa cells through the zona pellucida and form gap junctions with the oocyte membrane. This creates a system through which the granulosa cells provide various low-molecular weight substrates (e.g., amino acids and nucleotides) to the oocyte. These compounds are the building blocks for important macromolecules such as proteins and nucleic acids. A basement membrane develops and encases the outermost layer of granulosa cells. Loosely organized layers of spindle-shaped cells develop around the layer of granulosa cells and become known as thecal cells. Unlike the granulosa cell layer, the layer of thecal cells becomes richly supplied with capillaries.

Antral Phase

Resumption of follicle development occurs sporadically and incompletely during prenatal and neonatal life. During these periods a few follicles may develop to the antral stage, but they soon regress. As females approach puberty, initiation of follicle growth becomes more common. In the adult

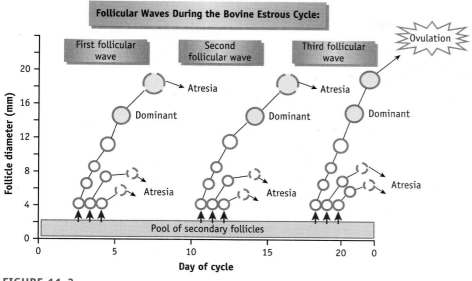

FIGURE 11-3

Patterns of follicle growth in a cow during the estrous cycle. Most cows experience a wave of follicular growth during the early and late portions of the estrous cycle. Some cows experience a follicular wave during the middle of the cycle. In each case the wave begins when a group of tertiary follicles emerge from a pool of secondary follicles. Each of these follicles enlarges but all but one become atretic. The follicle that continues to grow is known as the dominant follicle. The only dominant follicle to ovulate is the one formed during the follicular wave marking the transition between consecutive cycles. The other dominant follicles undergo atresia.

11

female, primordial follicles enter a pool of growing follicles at a steady trickle. At regular intervals, a group of follicles enters the antral stage and these follicles grow in wave-like patterns. Briefly, a follicular wave is the pattern of growth expressed by a follicle during the antral phase of development. Typically the diameter of an antral follicle increases until it ovulates or undergoes atresia. The number of follicle waves varies among and within species. In primates, growth of antral follicles is restricted to the follicular phase of the menstrual cycle. In contrast the cow, the ewe and the mare can express more than one follicular wave during a particular estrous cycle (Figure 11-3). For example, two- or three-wave estrous cycles predominate in various herds of cattle. In cows, a follicular wave consists of the emergence of a group of antral follicles, approximately 4 mm in diameter, followed by growth of all follicles and then continued growth of only the largest follicle. The wave that ends with ovulation is typically referred to as the **ovulatory wave,** whereas waves that do not lead to ovulation are known as **anovulatory waves.**

FIGURE 11-4

Summary of follicle growth in humans. The same pattern of development applies to other species of mammals. In humans, follicles are grouped into eight classes based on diameter.

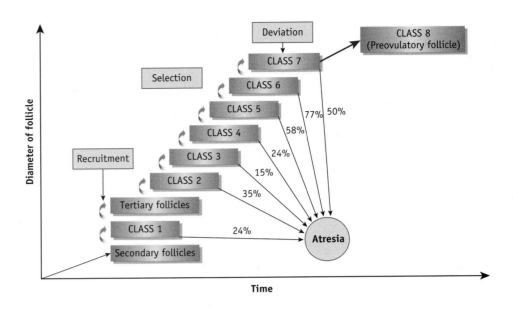

Selection of the Ovulatory Follicle

One of the most fundamentally important questions associated with folliculogenesis is, how is the preovulatory follicle selected from the vast pool of follicles? Figure 11-4 summarizes the events leading to selection of the ovulatory follicle in humans. It appears that the same sequence of events applies to other species as well.

Events leading to the selection of the preovulatory follicle begin during the preantral phase of follicular development. At any given time, there is a group of secondary follicles that have completed preantral growth and developed receptors for LH and FSH. If concentrations of these gonadotropins are sufficient, secondary follicles will develop into antral follicles. This process is known as **emergence** or **recruitment.** Recruitment of a group of follicles prevents additional follicles from entering the pool of antral follicles.

The fate of the future dominant follicle is sealed during emergence. Even though antral follicles are at similar stages of development at the time of emergence, there is variation in their *rates* of development. The follicle developing at the fastest rate will be the one that ultimately becomes the dominant follicle and ovulates. In cattle, this future dominant follicle emerges from the pool of secondary follicles at least 6 to 7 hours before the other follicles when it is only 3 to 4 mm in diameter. This phase of development is commonly referred to as **selection;** that is, when the largest follicle emerges from the pool of tertiary follicles. Antral follicles grow in parallel such that the largest follicle maintains a slightly larger diameter than other follicles. In cattle, the diameter of the future dominant follicle remains approximately 0.5 mm larger than that of the next largest follicle.

During the antral phase, both the granulosa and theca cells proliferate, but there is little if any change in the size of the oocyte. The thecal cells form two distinct layers: a richly vascularized inner layer of endocrine cells (theca interna) separated from a connective tissue layer (theca externa) by a fibrous capsule. As the granulosa cells proliferate they produce follicular fluid, which accumulates between the cells and eventually displaces them, forming a fluid-filled antrum. A layer of granulosa cells known as the cumulus oophorus surrounds the oocyte, which is suspended by a thin column of granulosa cells.

Preovulatory Phase

The end of the antral phase of folliculogenesis is marked by an event known as **deviation** (Figure 11-4). During deviation, the largest follicle continues to grow and smaller follicles experience a reduction or cessation of growth. It is at this stage when a dominant follicle can be clearly identified. In cattle, deviation occurs within 2.5 days after emergence of the future dominant follicle. Growth of the dominant follicle is rapid during the preovulatory phase. The tremendous increase in follicle diameter is due primarily to expansion of the volume of follicular fluid. In addition, production of estradiol by follicular cells is reaching a maximum level during this time. The preovulatory phase ends with ovulation.

Regulation of Folliculogenesis

Follicle growth can be divided into two phases based on how the process is regulated; that is, gonadotropin-independent and gonadotropin-dependent (Figure 11-5). Preantral follicle growth is not dependent on gonadotropins, whereas LH and FSH direct follicle growth during the antral and preovulatory phases.

Gonadotropin-Independent Follicle Growth

The preantral phase of follicle growth is not disrupted by hypophysectomy (removal of the pituitary gland). Thus it appears that development of primary and secondary follicles does not depend on gonadotropins. Although the mechanisms responsible for development of follicles during the preantral phase are not well understood, it is clear that this phase of development is orchestrated by a variety of hormones produced in the ovary and acting via paracrine, neurocrine, and autocrine mechanisms. It is noteworthy that receptors for LH and FSH do not appear on follicular cells until the end of the preantral phase; that is, on secondary follicles. In fact, the appearance of these receptors is a prerequisite for emergence and subsequent follicle growth.

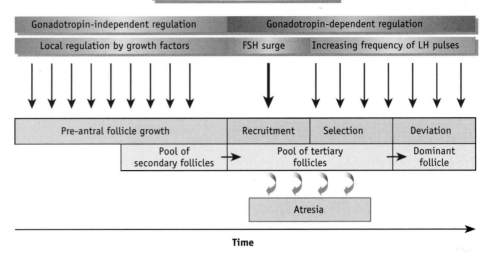

FIGURE 11-5

Overview of the regulation of follicle growth in mammals. Growth of pre-antral follicles does not depend on LH or FSH. Rather this phase of growth is orchestrated by locally produced growth factors acting in a paracrine manner. Recruitment of secondary follicles requires a surge of FSH. Subsequent growth of tertiary follicles depends on the pulsatile secretion of LH. Follicles that become dominant are the ones that are recruited first and therefore mature before the others.

Gonadotropin-Dependent Follicle Growth

Follicle growth during the antral and preovulatory phases is regulated largely by the actions of FSH and LH. In general, FSH plays a pivotal role in follicle development between emergence and deviation, whereas LH is most important during postdeviation growth (Figures 11-5 and 11-6).

ROLE OF FSH An increase in FSH is required to induce recruitment of follicles (Figure 11-6). A surge of FSH precedes and initiates the occurrence of a follicular wave in cows, horses, and sheep. In primates, the gradual increase in FSH concentrations observed during the early follicular phase is sufficient to induce recruitment of antral follicles. The growth of all recruited follicles continues as concentrations of FSH decline following the surge and this growth appears to require FSH.

Concentrations of FSH decline for several days following the peak of the FSH surge. This decrease in FSH continues through selection and deviation and appears to play a critical role in regulating these processes. This is supported by the observation that injections of FSH prevent or delay deviation in cows.

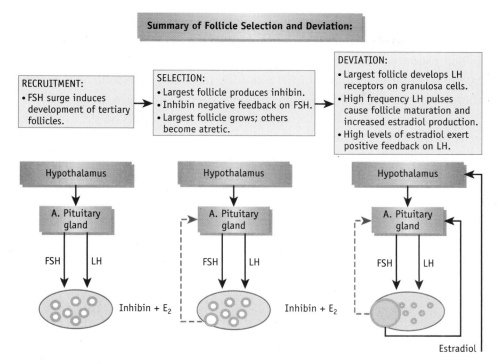

FIGURE 11-6

Endocrine interactions between the ovary and hypothalamic-pituitary axis during the recruitment, selection, and deviation of follicles during the ovarian cycle.

The mechanism causing the decrease in FSH concentrations following the peak in FSH are complex. Initially, the decrease may be due to depletion of releasable pools of FSH in pituitary gonadotropes. In addition, negative feedback signals generated by growing follicles play a major role in regulating this response. In cows, antral follicles develop an FSH-depressing ability within 1 day after emergence and contribute to the drop in FSH for about 2 days. Between 2 and 3 days after emergence the largest follicle develops an enhanced capacity to suppress FSH. As a result the concentration of this gonadotropin falls below that required to sustain the growth of smaller follicles. Apparently, the largest follicle is more sensitive to FSH than the smaller follicles and continues to develop even in the presence of lower FSH concentrations. The negative feedback signal mediating the inhibitory effect of the follicles on FSH release appears to include both estradiol and inhibin. Follicular production of these hormones increases greatly as follicles develop. The enhanced sensitivity of the largest follicle to FSH may be attributed to its larger surface area, and thus a greater number of FSH receptors, and/or intrafollicular factors (e.g., estradiol), which enhance responsiveness to FSH. In summary, the largest follicle induces deviation by suppressing concentrations of FSH while retaining the ability to respond to decreasing concentrations of FSH. The fact that the largest follicle grows while suppressing development of smaller follicles serves as the basis for calling the largest follicle the **dominant** follicle.

11

ROLE OF LH LH plays an important role in maturation of the dominant follicle following deviation (Figures 11-6 and 11-7). Continued growth of this follicle following deviation is an LH-dependent process. This idea is supported by the fact that suppression of LH following deviation stalls follicular growth, thereby preventing follicles from reaching a preovulatory size. In cows, the effects of LH are mediated by the appearance of LH receptors on the granulosa cells of the largest follicle between 2 and 4 days after emergence, the time when deviation begins. Before this stage, LH receptors are found only in the thecal cells (Figure 11-7). The appearance of LH receptors in granulosa cells

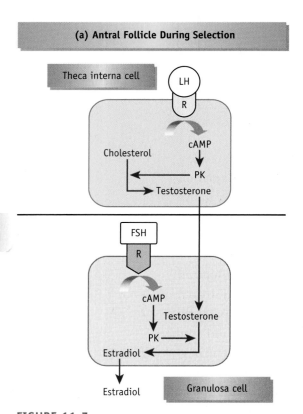
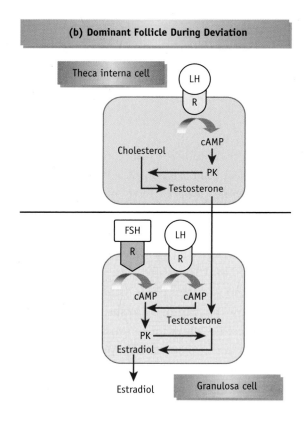

FIGURE 11-7

Overview of prevailing theory explaining regulation of estradiol synthesis by ovarian follicles. Production of estradiol requires the actions of LH and FSH acting on the theca interna and granulosa cells, respectively (a). Each of these hormones acts on receptors that promote synthesis of cyclic AMP (cAMP), which then activates protein kinase (PK) an ezyme that regulates activity of key enzymes in steroidogenesis. In order for deviation to occur, a follicle must develop LH receptors on its granulosa cells (b). This allows the cells to remain active and continue steroidogenic activity in the face of decreasing FSH concentrations. Follicles that do not develop this ability become atretic.

permits these cells to respond to both LH and FSH (Figure 11-7). In light of the fact that FSH concentrations decrease during deviation, it appears that LH becomes an important regulator of follicular growth during this period. It is likely that the combined effects of LH and FSH account for the rapid growth and enhanced activity of the dominant follicle. As the largest follicle attains dominance, production of estradiol increases dramatically. The enhanced production of estradiol by the dominant follicle and the appearance of LH receptors on its granulosa cells are necessary conditions for ovulation. However, these events are not sufficient conditions for ovulation. Ovulation also requires the estradiol-induced LH surge.

The fate of the dominant follicle depends on the prevailing hormonal environment. If the dominant follicle develops at a time when there is no corpus luteum and progesterone concentrations are low, then the elevated concentrations of estradiol induce an LH surge, which causes the dominant follicle to ovulate. Follicle waves that reach a crest in the presence of a corpus luteum will not ovulate due to the fact that high levels of progesterone block the positive feedback effects of estradiol on LH. Thus neither the LH surge nor ovulation can occur in the presence of a corpus luteum. In these cases, the dominant follicle undergoes atresia and regresses.

BOX 11-1   Focus on Fertility: Superovulation and Embryo Transfer

11

Embryo transfer was the next assisted reproduction technology developed after artificial insemination. Although the use of this technology is not as widespread as and often less successful than artificial insemination it is employed routinely in several species of livestock, species of wildlife, and humans. Embryo transfer can be used to accelerate proliferation of genetic material from females, minimize spread of reproductively transmitted diseases, salvage genetic material of valuable individuals, and facilitate development of new lines or breeds of livestock. The general paradigm for successful embryo transfer in cattle is shown in Figure 11-8. Today embryos are routinely recovered nonsurgically by flushing them from the uterine

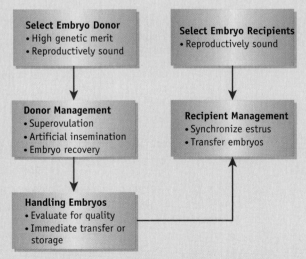

Figure 11-8 Major steps for an embryo transfer paradigm commonly used in cattle and other species.

horns several days after fertilization (when they are at the blastocyst stage of development). They can be recovered and immediately transferred to recipients, but this requires that the recipients be at the same stage of the ovarian cycle as the donors (7 days after estrus). It is also possible to subject embryos to cryopreservation and store them for transfer at later times. Embryos can also be prepared by in vitro fertilization, which requires mixing of spermatozoa and an oocyte in a Petri dish and allowing the embryo develop to the blastocyst stage before transfer.

The success of embryo transfer is heavily dependent on the ability to produce and recover viable embryos from donor females. In order to enhance the possibility of achieving this goal, donors are typically subjected to hormone treatments that superstimulate the ovaries to induce multiple ovulations (Figure 11-9). This technique is called **superovulation.** The basic concept underlying this method is that the number of follicles that undergo deviation and become dominant can be increased by boosting the number of follicles that are recruited. Enhancement of recruitment is accomplished by administering FSH during the late luteal phase, the time when follicles are recruited for the next cycle. In most cases, donor cows are treated with $PGF_{2\alpha}$ toward the end of FSH treatments in order to induce estrus and ovulation at predictable times. This facilitates use of artificial insemination. Response to superovulation regimens are quite variable producing anywhere between 0 and 20 (or more) ovulations. The average response is nine ovulations.

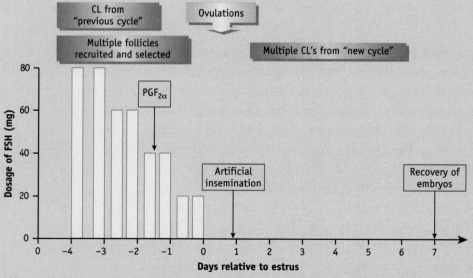

Figure 11-9 Paradigm for inducing superovulation in cattle. Administration of FSH to donor cows during the latter portion of the luteal phase increases the number of follicles that are recruited, selected, and undergo deviation. Note that in order to begin FSH treatments four days prior to estrus, it is necessary to monitor the ovarian cycles of donors. Administration of $PGF_{2\alpha}$ serves the purpose of inducing regression of the corpus luteum at a predetermined time such that onset of estrus and ovulation can be predicted. This step facilitates use of artificial insemination, but can be omitted and the donor allowed to come into heat and ovulate spontaneously.

ENDOCRINE ACTIVITY OF FOLLICLES

In addition to regulating folliculogenesis and oogenesis, the gonadotropins interact to regulate the endocrine activity of follicles (see Figure 11-7). During follicle selection, LH acts on theca interna cells to promote conversion of cholesterol to testosterone. Recall that this effect is similar to that in the Leydig cells of the testes. Testosterone is released into the extracellular fluid and can be taken up into the blood to enter the general circulation, or diffuse across the follicular wall and be taken up by granulosa cells. In the granulosa cells, testosterone is the substrate for the aromatase enzyme, which converts it to estradiol. This enzyme is regulated by FSH. Estradiol is released by the granulosa cells into the follicular fluid. A small portion of this hormone can diffuse out of the follicle and enter the blood.

The regulation of steroid hormone production by the follicle changes during deviation. Recall that the dominant follicle is the first one to develop LH receptors on granulosa cells. At this point the granulosa cells are stimulated by the combined effects of LH and FSH. This causes rapid growth of the follicle and elevated secretion of estradiol and inhibin. Note that the effects of LH and FSH on estradiol synthesis are additive due to the fact that both hormones act by increasing synthesis of cAMP. It is also important to point out that the increased production of estradiol and inhibin by the dominant follicle exert negative feedback effects on FSH release. The resulting decrease in FSH prevents recruitment of new follicles and is ultimately responsible for the slowed growth and atresia of the nondominant follicles. The decrease in FSH has no inhibitory effect on the dominant follicle due to the ability of its granulosa cells to respond to LH.

OOGENESIS

As noted earlier, development of the oocyte begins before birth, but then ceases until the time of ovulation. Figure 11-1 summarizes the major aspects of oogenesis as related to follicle development, ovulation, and fertilization. Figure 11-10 summarizes highlights of oogenesis with reference to mitosis and meiosis. Once the primordial germ cells enter the gonad of female embryos, they differentiate into oogonia. These cells undergo a series of mitotic divisions before entering meiosis. By the time of birth the oogonia enter the first meiotic prophase and are then referred to as primary oocytes. This phase of meiosis is completed only to the diplotene stage. Although the primary oocyte enlarges greatly at this point, it exists in this arrested state (resting phase) of development until ovulation. An important implication of this phenomenon

11

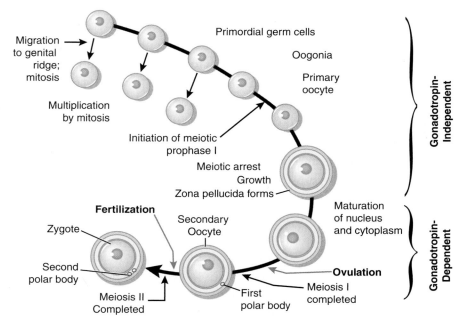

FIGURE 11-10

Overview of oogenesis in humans and domestic ungulates. The process begins early in embryogenesis with the migration of primordial germ cells to the genital ridge. Once these cells reside in the ovary they expand their numbers via a series of mitotic divisions and eventually become oogonia. Oogonia begin meiosis, and become primary oocytes upon completion of first meiotic prophase. These oocytes will not develop further unless they are housed in follicles that undergo recruitment, selection, and deviation. The primary oocytes of dominant follicles will mature and complete meiosis I if exposed to an LH surge. Thus the secondary oocyte appears only after ovulation. Completion of meiosis requires fertilization.

is that female mammals achieve a maximum number of oocytes at some point during the fetal stage of development (a peak number of 7 million by midgestation in humans). Thereafter, the number of oocytes decreases progressively due to the loss of follicles by atresia. The growth of the primary oocyte is passive; that is, not dependent on follicular cells. As the oocyte grows it becomes enveloped by the zona pellucida, a translucent, jelly-like membrane consisting of mucopolysaccharides and proteins. This membrane forms once the oocyte is surrounded by a layer of cuboidal granulosa cells. The zona pellucida is not solid. Rather it contains microscopic canals through which microvilli from the adjacent granulosa cells extend and terminate in indentations of the oocyte. This structural arrangement seems to provide a means for transport of metabolites from granulosa cells to the oocyte and persists until ovulation.

Removal of Meiotic Arrest by the LH Surge:

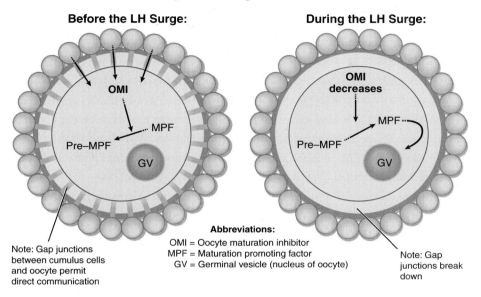

FIGURE 11-11

Schematic illustration showing how the LH surge removes the blockade of meiosis in primary follicles. By inducing degeneration of gap junctions between granulosa cells and the oocyte, LH prevents oocyte maturation inhibitor (OMI) from entering the oocyte, thereby permitting maturation promoting factor (MPF) to be produced and act on the oocyte nucleus to trigger completion of meiosis I.

Up until the point of ovulation, development of the oocyte is gonadotropin-independent. The first meiotic division is completed at ovulation and appears to be induced by the LH surge. Completion of this phase results in formation of a secondary oocyte and a first polar body. The polar body is a small offspring cell that has substantially less cytoplasm than the secondary oocyte. It eventually degenerates.

The mechanism whereby the LH surge liberates the primary oocyte from meiotic arrest is quite elegant and involves structural changes in the relationship between the oocyte and cells of the cumulus oophorus that send cytoplasmic projections through the zona pellucida (Figure 11-11). Before the LH surge, gap junctions between the cumulus cells and the oocyte permit direct chemical communication between the cells. The cumulus cells produce oocyte maturation inhibitor (OMI), which enters the cytoplasm of the oocyte and inhibits activation of another protein known as maturation promoting factor (MPF). The LH surge destroys the gap junctions between the cumulus cells and oocyte, thus lowering concentrations of OMI. The decrease in OMI allows MPF to be produced and this protein acts on the nucleus of the oocyte to induce completion of the first meiotic division.

Immediately after completion of the first meiotic division, the secondary oocyte enters the second meiotic prophase. Completion of meiosis depends on fertilization by a spermatozoon. The second meiotic division results in formation of an ootid and a second polar body.

OVULATION

Not only is ovulation a pivotal event in the ovarian cycle, it is arguably the most important rate-limiting process in the overall reproductive fitness of a species. As noted in previous discussions, ovulation serves as a convenient point of demarcation that separates the cycle into two major portions: follicular and luteal phases. The rupture of the dominant, preovulatory follicle and release of its oocyte is also a critical physiologic event that is necessary for the female gamete to become available for fertilization. Two functionally independent events occur in association with ovulation: rupture of the follicular wall and luteinization. The former is concerned with release of the oocyte from the ovulatory follicle, whereas the latter is concerned with transformation of follicular cells into cells of the corpus luteum. Both of these processes are triggered by the pre-ovulatory surge of LH.

During the past 75 years several hypotheses have been put forth to explain how the LH surge brings about rupture of the ovulatory follicle. By the early 1960s, it was generally assumed that ovulation was the result of increasing intrafollicular pressure brought about by contraction of the smooth muscle cells located in the ovarian stroma. This hypothesis is no longer tenable. There is little evidence to support the idea that smooth muscle cells exist in the connective tissue surrounding follicles, let alone that these cells contract during ovulation. Moreover, pressure within the follicle does not increase before ovulation. Ovulation appears to be the result of a weakening of the follicular wall under the force of modest but steady intrafollicular pressure, which is primarily due to the hydrostatic pressure of capillaries in the theca interna.

Anatomic Changes in the Follicle Before and During Ovulation

An understanding of the mechanism of ovulation requires familiarity with the histology of the follicular wall (Figure 11-12). The selective growth of the preovulatory follicle causes the outer tissue layer of the follicule to come into close apposition with the surface epithelium and underlying capsule (tunica albuginea) of the ovary. The site is known as the apex of the follicle, and marks the location of ovulation. When the follicle wall in this region begins to weaken, a **stigma** forms and designates the site at which the follicular wall will rupture.

The apical follicular wall has a complex structure consisting of five tissue layers. As we study the ovulatory process, keep in mind that each of these layers must be breached in order for the oocyte to be released from the follicle. The first, or outermost layer, is the single layer of cuboidal epithelial cells known as the germinal epithelium. The second layer is the tunica albuginea,

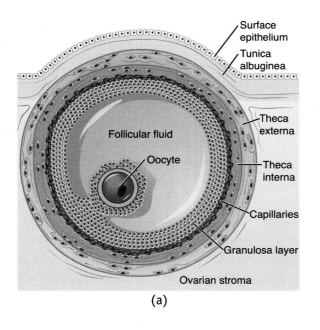

Surface epithelium
Tunica albuginea

Follicular fluid

Oocyte

Theca externa

Theca interna

Capillaries

Granulosa layer

Ovarian stroma

(a)

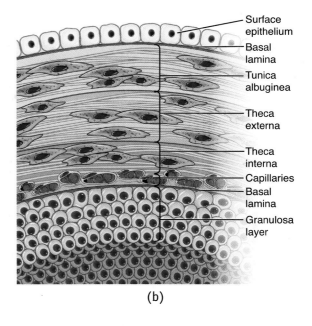

Surface epithelium
Basal lamina
Tunica albuginea

Theca externa

Theca interna
Capillaries
Basal lamina
Granulosa layer

(b)

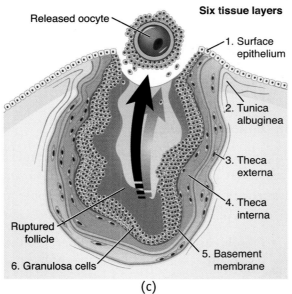

Released oocyte

Six tissue layers

1. Surface epithelium

2. Tunica albuginea

3. Theca externa

4. Theca interna

5. Basement membrane

6. Granulosa cells

Ruptured follicle

(c)

11

FIGURE 11-12

Drawings of a pre-ovulatory follicle showing tissue layers that comprise the follicle wall. The oocyte is supported by a column of granulosa cells known as the cumulus oophorus (a). Release of the oocyte from the follicle (ovulation) requires breakdown of six tissue layers including the surface epithelium, tunica albuginea, theca externa, theca interna, follicle basement membrane, and granulosa cells (b). Ovulation begins with formation of the stigma, a localized area of degeneration through which the oocyte eventually escapes (c).

a connective tissue sheath consisting of fibroblasts and collagen fibers. This layer is between 5 and 7 cells deep and envelopes the entire ovary to delineate its integrity. The theca externa forms the third layer of the follicle wall. This forms the follicle's own layer of connective tissue and includes several layers of fibroblasts and collagen fibers. It may be difficult to distinguish a border between the tunica albuginea and theca externa due to the fact that the two layers mesh together. The fourth layer is the theca interna. This layer consists of two layers of elongated cells that contain numerous mitochondria, lipid droplets, and a well-developed smooth endoplasmic reticulum, which are organelles involved in production of steroid hormones. The theca interna contains most of the capillaries supplying blood to the follicle. The fifth, and innermost tissue layer consists of granulosa cells. These cells are attached to a basal lamina which separates the theca interna from the granulosa layer. The granulosa layer is avascular because the capillaries of the theca interna do not penetrate the basement membrane. On average, the granulosa layer is five to seven cells thick. However, additional layers of cells exist at the cumulus oophorus, a pedestal of granulosa cells that support the oocyte. Gap junctions between adjacent granulosa cells create a functional syncytium. Gap junctions also exist between the oocyte and the cells of the corona radiata. This organization of cells coordinates activity of granulosa cells and permits direct communication between and oocyte and this layer of follicular cells.

The Ovulatory Process

Complete rupture of the ovulatory follicle and release of the oocyte requires between 10 and 40 hours, depending on the species. Changes in the ultrastructure of the follicular wall can be observed during the final hour before ovulation. The most important of these changes include the following:

- Fibroblasts of the tunica albuginea and theca externa change from a quiescent, resting to an active, proliferating state, and begin to dissociate from one another.
- Connective tissue of the tunica albuginea and theca externa at the apex of the follicle becomes more loosely organized and less tenacious.
- Cuboidal cells of the germinal epithelium develop vacuoles and become necrotic.
- Theca interna cells remain unchanged but some of the capillaries contain coagulated blood causing petechia on the surface of the follicle.
- Granulosa cells accumulate lipid in lipid droplets, reflecting increased synthesis of progesterone.

Additional changes occur within the apex of the follicle a few minutes before ovulation. The most noticeable is that this region of the follicle wall bulges outward, forming the stigma. This event is associated with the following changes in ultrastructure:

- Epithelial cells on the surface of the follicle slough off.
- Cells of the theca interna and granulose layers dissociate and migrate to the base of the stigma.
- The thin layer of highly degraded collagenous tissue that remains at the apex narrows to less than 20 percent of its original thickness.

Ovulation itself requires hydrostatic pressure within the antrum of the follicle. The pressure is modest and does not change throughout the ovulatory process. Rupture of the follicle is attributed to this sustained pressure and the weakening of the follicle wall. Intrafollicular pressure is the hydrostatic pressure of capillaries located within the theca interna. This pressure drops once the follicle ruptures.

Regulation of the Ovulatory Process

It is generally agreed that the major prerequisite for ovulation is degradation of collagenous layers of the theca externa and tunica albuginea at the apex of the ovulatory follicle. However, information about the biochemical events responsible for this response is incomplete. Figure 11-13 summarizes aspects of the ovulatory mechanism for which there is general agreement among reproductive biologists.

As noted earlier, events leading to ovulation are initiated by the LH surge. Receptors for this gonadotropin exist in both the thecal and granulose cells as well as the surface epithelium of the ovary. Thus the mechanism whereby LH induces ovulation is likely to involve a direct effect on each of these cell types. It also appears that the fibroblasts of the tunica albuginea, and theca externa, and the endothelial cells of the theca interna are also involved with ovulation, but regulation of these cells is likely to be dependent on signals generated by cells that respond directly to LH. The overall effect of the LH surge is to terminate expression of genes that govern activity of the pre-ovulatory follicle, and to induce expression of genes that regulate ovulation. Effects on the granulosa cells may be most important in regulating ovulation. Two of the most important ovulation-regulating genes are the gene for the progesterone receptor and the cyclooxygenase-2, an enzyme regulating synthesis of prostaglandins. These signaling pathways act independently to induce changes that promote ovulation and luteinization. With respect to ovulation, some of the more important changes include 1) activation and/or synthesis of proteolytic enzymes (collagenase) that degrade the extracellular matrix and 2) synthesis of regulatory factors that act on fibroblasts and

11

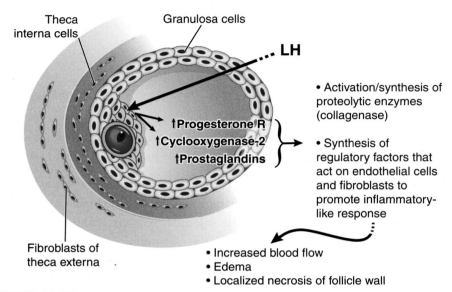

Effects of LH surge on granulosa cells are critical to ovulation

FIGURE 11-13

Schematic illustration summarizing current theories regarding how LH induces ovulation of a dominant follicle. LH acts on granulosa cells to increase expression of progesterone receptors and cyclooxygenase, a rate-limiting enzyme in the biosynthesis of prostaglandins. The increase in progesterone receptors and prostaglandin production interact to produce changes in the follicle that culminate in rupture and release of the oocyte (i.e., ovulation).

endothelial cells to induce an inflammatory-like cascade of events. These latter changes result in increased blood flow, edema, and localized necrosis of tissue.

In addition to inducing degradation of the apical follicular wall, the LH surge induces changes within the cumulus oophorus that are necessary for ovulation to occur. The LH surge acts on cumulus cells to induce production of various proteins that interact to form a biochemical matrix upon which the cumulus cells move during ovulation.

In summary, release of the oocyte from the pre-ovulatory follicle is the result of three major LH surge-induced events: 1) sustained hydrostatic pressure due to enhanced blood flow to the follicle, 2) collagenase-induced digestion of the follicle wall, and 3) separation of the oocyte from the follicular wall.

Luteinization

Within hours after the LH surge, the tissue of the ovulating follicle begins luteinization; that is, a remodeling process that leads to the formation of a corpus luteum. This process involves both morphologic and biochemical changes. Major changes include the distinct concentric structure of the follicle collapses permitting cells of the various layers to intermingle, the thecal and granulosa cells become reprogrammed to express a luteal pattern of genes, the remaining tissue undergoes rapid growth due to development of an extensive capillary network, and proliferation of various cell types.

SUMMARY OF MAJOR CONCEPTS

- Development of the follicle and oocyte occur in parallel, but the two processes are not necessarily interrelated.
- Folliculogenesis consists of a preantral, antral, and ovulatory phases.
- The preantral phase of folliculogenesis is not dependent on gonadotropins, whereas the antral ovulatory phases are gonadotropin-dependent.
- During the antral phase of folliculogenesis, follicle growth is wave-like due to three processes: recruitment, selection, and deviation.
- LH and FSH act on theca interna cells and granulosa cells, respectively, to promote estradiol synthesis.
- Oogenesis consists of major phases including: expansion of the oogonia population by mitotic divisions, a resting phase where primary oocytes remain in meiotic prophase 1, an LH surge-induce completion of the first meiotic division, and fertilization-induced completion of the second meiotic division.
- The LH surge induces ovulation (rupture of the follicle) and luteinization (transformation of the follicle to a corpus luteum).

DISCUSSION

1. One of the major functions of the testis and ovary is the production of gametes. In each case, gamete production involves both mitosis and meiosis. Make lists of similarities and differences between spermatogenesis and oogenesis.

2. Describe what you would expect to happen if you were to induce an LH surge in a cow on day 5 of her estrous cycle (i.e., after the dominant follicle of the first follicular wave has deviated). Explain your answer.

3. Daily injections of FSH given late in the cow's estrous cycle will result in multiple ovulations. Describe the physiologic mechanism responsible for this observation.

11

4. Explain how the dominant follicle induces a reduction in FSH concentrations. How is it that this reduction in FSH inhibits growth of other tertiary follicles, but does not impair growth of the dominant follicle?

REFERENCES

Farin, P.W., K. Moore, and M. Drost. 2007. Assisted reproductive technologies in cattle. In R.E. Youngquist and W. R. Threlfall, *Current Therapy in Large Animal Theriogenology 2*. St. Louis: Saunders, pp. 496–508.

Fortune, J.E. 1986. Bovine theca and granulose cells interact to promote androgen production. *Biology of Reproduction* 35:292–299.

Fortune, J.E. 1994. Ovarian growth and development in mammals. *Biology of Reproduction* 50:225–232.

Ginther, O.J. 2000. Selection of the dominant follicle in cattle and horses. *Animal Reproduction Science* 60–61:61–79.

Haughian, J.M., O.J. Ginther, K. Kot, and M.C. Wiltbank. 2004. Relationships between FSH patterns and follicular dynamics and the temporal associations among hormones in natural and GnRH-induced gonadotropin surges in heifers. *Reproduction* 127:23–33.

Pangas, S.A. and M.M. Matzuk. 2005. The art and artifact of GDF9 activity: Cumulus expansion and the cumulus expansion-enabling factor. *Biology of Reproduction* 73:582–585.

Peters, H. 1970. Migration of gonocytes into the mammalian gonad and their differentiation. *Philosophical Transactions of the Royal Society of London. B.* 259:91–101.

Richards, J.S., D.L. Russell, R.L. Robker, M. Dajee, and T.N. Alliston. Molecular mechanisms of ovulation and luteinization. *Molecular and Cellular Endocrinology* 145:47–54.

Richards, J.S., D.L. Russell, S. Ochsner and L.L. Espey. 2002. Ovulation: New dimensions and new regulators of the inflammatory-like response. *Annual Review of Physiology* 64:69–92.

Zalanyi, S. 2001. *European Journal of Obstetrics and Gynecology and Reproductive Biology* 98:152–159.

Zeleznik. A. J. 2001. Follicle selection in primates: "Many are called but few are chosen." *Biology of Reproduction* 65:655–659.

11

CHAPTER 12

Dynamics of Ovarian Function: The Corpus Luteum

CHAPTER OBJECTIVES

- Describe the pattern of progesterone release during the luteal phase of the ovarian cycle.

- Describe the structural and hormonal changes exhibited by the corpus luteum during the luteal phase.

- Describe the mechanisms regulating formation, maintenance, and regression of the corpus luteum.

OVERVIEW

The focus of this chapter is the fate of the ruptured follicle following ovulation. Within several days after ovulation the remaining follicular tissue is transformed into the corpus luteum. The mediator of this response is luteinizing hormone (LH) and the process includes both morphologic and biochemical changes (Figure 12-1). The major biochemical change involves steroidogenesis. Unlike the follicle, which synthesizes testosterone and estradiol, the corpus luteum produces only progesterone. Concentrations of progesterone in blood parallel the growth and regression of the corpus luteum (Figure 12-2). In animals that have multiple ovulations, concentrations of progesterone are directly proportional to the number of corpora lutea that form.

Progesterone plays an important role in regulating reproductive activity in mammalian females. In addition to being an important reproductive hormone, progesterone is a key intermediate in the synthesis of other steroid hormones. All cells that produce steroid hormones produce progesterone. Progesterone is the major secretory product of the corpus luteum and the placenta, but is also produced by the adrenal cortex. What little progesterone is produced in males comes from the testes and adrenal glands. The only physiologic role of progesterone discovered in males is serving as a precursor for androgen synthesis. Our concern in this chapter will be progesterone production by the corpus luteum.

The major target tissues for progesterone are the hypothalamo-pituitary system and the female genital ducts (Figure 12-1). Progesterone plays a central role in regulating the ovarian cycle. Its major actions in this regard include 1) exerting a negative feedback action on GnRH release and 2) preventing estradiol from inducing an LH surge. These effects serve to regulate the length of the ovarian cycle. In other words, ovulation cannot occur as long as progesterone remains elevated. This is due to the fact that progesterone prevents the high-frequency pulses of LH necessary to induce follicle

12

FIGURE 12-1

Schematic illustration summarizing the major effects of LH on development of a corpus luteum. In addition to inducing ovulation, the LH surge induces luteinization of the ruptured follicle. This process includes both morphological and biochemical changes. Once formed, LH is necessary for maintenance of the corpus luteum and production of progesterone, which is necessary for pregnancy and regulates LH secretion.

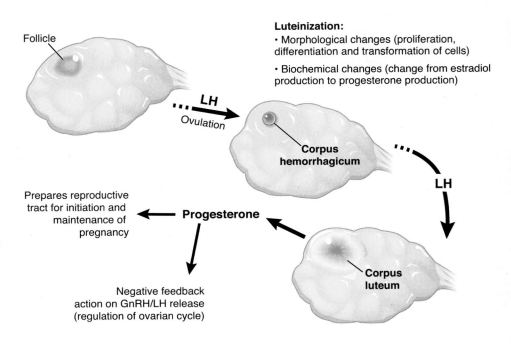

Luteinization:
• Morphological changes (proliferation, differentiation and transformation of cells)
• Biochemical changes (change from estradiol production to progesterone production)

Follicle

LH
Ovulation

Corpus hemorrhagicum

LH

Prepares reproductive tract for initiation and maintenance of pregnancy

Progesterone

Corpus luteum

Negative feedback action on GnRH/LH release (regulation of ovarian cycle)

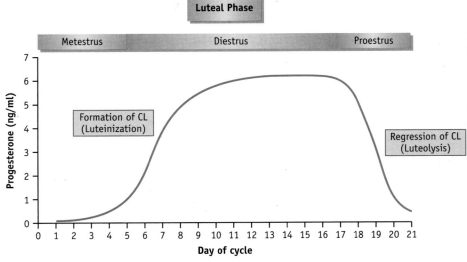

FIGURE 12-2

Timing of the formation and regression of the corpus luteum relative to patterns of progesterone during the estrous cycle of a domestic cow.

development to the preovulatory stage and prevents induction of an LH surge by estradiol.

With respect to effects on the genital ducts, progesterone's main action is to prepare the female reproductive tract for pregnancy. In the uterus, progesterone acts on the endometrium to 1) inhibit proliferation of mucosal cells, 2) stimulate secretion from glandular cells, and 3) enhance release of proteins that support early embryonic development. Progesterone also acts on the myometrium. Its major effect is to make the uterus quiescent. This action involves disruption of interactions among smooth muscle cells, interference with contractile mechanisms within smooth muscle cells, and blocking estradiol's ability to induce contractions.

GENERAL DESCRIPTION OF THE LUTEAL PHASE

The luteal phase of the ovarian cycle is defined as the period between ovulation and regression of the corpus luteum. In mammals with estrous cycles, the luteal phase encompasses metestrus and diestrus (Figure 12-2). During the first several days of the luteal phase (metestrus) what tissue remains from the ovulated follicle undergoes luteinization. As the tissue of the ruptured follicle is transformed into luteal tissue, production of estradiol ceases and progesterone production increases. During the early luteal phase, circulating concentrations of progesterone increase rapidly. The corpus luteum is fully functional by the time of diestrus, but it continues to increase in size throughout the luteal phase, releasing increasing amounts of progesterone as it grows. At the end of diestrus, the corpus luteum regresses and progesterone production ceases.

As described previously, progesterone provides the primary negative feedback signal regulating LH release in the adult female. Average concentrations of LH are low when progesterone concentrations are high (e.g., during diestrus), and high when progesterone concentrations are low (e.g., during metestrus and proestrus). This variation in LH concentrations is due to changes in the pulsatile pattern of LH release. LH is released in a high-amplitude, low-frequency mode during progesterone negative feedback. In the absence of this feedback LH is released in a low-amplitude, high-frequency mode. Progesterone suppresses the pulsatile release of LH by acting on the hypothalamus to reduce frequency of GnRH pulses. In addition to its effects on the pulsatile release of LH, progesterone inhibits the ability of estradiol to induce an LH surge. This is most likely due to effects on the anterior pituitary gland (i.e., reducing responsiveness to GnRH).

A high-frequency pattern of LH release is necessary for a follicle to develop to the preovulatory stage. Such a pattern is present before the corpus luteum develops and following luteal regression. This explains why follicular waves occur during each of these periods. The reason the dominant follicle fails to ovulate during the first wave (during metestrus) is because the rising concentrations of progesterone prevent estradiol from inducing an LH surge. You may recall in cattle that some cows have a follicular wave during diestrus, a time

12

during which LH pulse frequency is low. It is unclear how a dominant follicle can undergo deviation during conditions of low LH concentrations. Apparently, the pattern of LH that exists at this time is sufficient to promote follicle growth. Nevertheless, the dominant follicle of this wave does not ovulate because the high progesterone concentrations block induction of an LH surge.

A key concept to remember from this discussion is that progesterone plays a central role in regulating ovarian cycles. Some reproductive physiologists view progesterone as the "organizer" of the ovarian cycle. As mentioned earlier, the inter-ovulatory period is directly related to the period of time when progesterone concentrations are elevated. Thus it is possible to shorten the estrous cycle by inducing premature regression of the corpus luteum as well as extend the length of the cycle by administering progesterone. This concept is the basis of estrous synchronization techniques (see Box 12-1).

BOX 12-1   Focus on Fertility: Synchronizing Estrus and Ovulation

The ability to regulate timing of estrus and ovulation is essential for effective superovulation and embryo transfer and also enhances effectiveness of artificial insemination. One of the major factors restricting the use of artificial insemination in beef cattle is the labor required to round up and handle a herd of cows. The following example illustrates how estrus synchronization can reduce the amount of labor associated with artificial insemination. In a herd of 100 cows, an average of only 5 would express heat each day during a 3-week period. In contrast estrous synchronization methods can result in as many as 80 to 90 cows expressing heat in a 3-day period.

Development of estrus synchronization technologies has been a major focus of reproductive physiology research during the past 40 years and has occurred in five distinct phases (Table 12-1). Phases I and III (the use of progestins and PGF$_{2\alpha}$) are arguably the most important since all of the other methods have evolved from these two basic approaches.

The first attempts to control the estrous cycles of female farm animals are based on the knowledge that progesterone prevents ovulation and expression

TABLE 12-1 Phases in development of methods to synchronize estrus in cattle

Phase	Approach
I	Use of progestins.
II	Use of progestins combined with estrogens and gonadotropins.
III	Use of PGF$_{2\alpha}$.
IV	Use of progestins in combination with PGF$_{2\alpha}$.
V	Use of GnRH in combination with progestins and PGF$_{2\alpha}$.

of estrus (Figure 12-3). A common approach is to treat cyclic cows with progesterone or a progesterone-like compound (a progestin) for 14 days. There are three theoretical outcomes resulting from such a treatment: 1) creation of an artificial luteal phase (if treatment begins between luteolysis and the estrus of the next cycle), 2) extension of the luteal phase (if treatment begins when a corpus luteum

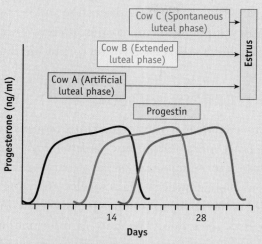

Figure 12-3 Graphic representation of how progestin can be used to synchronize estrus and ovulation in domestic cows. Randomly cycling cows would fall into one of three categories when exposed to progestin: 1) completing a cycle (cow A), 2) progressing through a cycle (cow B), and 3) beginning a cycle (cow C).

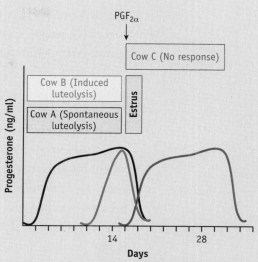

Figure 12-4 Graphic representation of how PGF$_{2\alpha}$ can be used to synchronize estrus and ovulation in domestic cows. In a herd of randomly cycling cows the hormone would be given at the end (cow A), middle (cow B) or beginning (cow C) of an estrous cycle. PGF$_{2\alpha}$ will induce a synchronized luteal regression in animals that have a functional CL (e.g., cow B). Cows experiencing spontaneous luteolysis at the time of treatment (e.g., cow A) will ovulate near the time of cows responding to PGF$_{2\alpha}$. Cows that do not have a CL at the time of treatment (cow C) will not ovulate at a time similar to that of other treated cows.

(CL) is present and the CL regresses before treatment is withdrawn), and 3) no effect on the cycle (if treatment begins within several days after estrus and coincides with the presence of a new CL). In the first case, the dominant follicle does not ovulate until the progestin is withdrawn. In the second case, the corpus luteum regresses at the expected time, but the exogenous progestin delays estrus and ovulation. In the third case, the exogenous progestin exerts no effect on the formation and regression of the corpus luteum. The important point to understand from this example is that in each animal estrus and ovulation occur within several days of progestin withdrawal.

The second fundamental approach for synchronizing estrus is based on the knowledge that PGF$_{2\alpha}$ induces regression of the corpus luteum (Figure 12-4). Injections of this hormone would be expected to induce luteolysis in any animal that has a corpus luteum at the time of injection. Since a cow has a corpus luteum for 14 days of the cycle (between days 3 and 17), there is a probability that 14 out of 21 or 67 percent of the cows would respond to PGF$_{2\alpha}$ and express estrus within 72 hours after treatment. In addition, cows receiving the injection on days 18 through 21 (14 percent) would undergo spontaneous luteolysis and also express estrus within the 72-hour time frame. Therefore, in a herd of 100 cows approximately 81 percent (67 percent + 14 percent) would come into heat within 72 hours of a single injection of PGF$_{2\alpha}$. A higher degree of synchrony is achieved if cows are given a second injection of PGF$_{2\alpha}$ 8 to 14 days after the first injection. In this way almost all of the animals will have a corpus luteum present at the time of the second injection.

12

Having reviewed the general features of the luteal phase, it is now possible to consider the corpus luteum in detail. Our discussion will be divided into three major sections: 1) development of the corpus luteum, 2) regulation of corpus luteum function, and 3) regression of the corpus luteum.

DEVELOPMENT OF THE CORPUS LUTEUM

The process by which the ruptured follicle is transformed into a corpus luteum (i.e., luteinization) involves both morphologic and biochemical changes. Morphologic changes consist of a rearrangement of the remaining follicular tissue and vascular infiltration. Biochemical changes involve changing from the ability to produce estradiol to the ability to produce progesterone. We will consider each of these processes separately. The bulk of information presented in this chapter will derived from domestic ruminants, primarily the cow and the ewe, because research on the CL of these species is extensive and has great historical relevance in the field of reproductive physiology.

Morphologic Changes

In addition to inducing ovulation, the LH surge also causes **luteinization** of the remaining follicular tissue. The major morphologic changes induced by the LH surge involve transforming the residual granulosa and theca interna cells into **large** and **small luteal cells,** respectively (Figure 12-5). Both types of luteal cells produce progesterone. In primates these cells are called granulosa-lutein and theca-lutein cells. In addition to changing the morphology of the granulosa and thecal cells, the LH surge induces a remarkable reorganization of follicular tissue. In nonprimate species, the basement membrane that separates granulosal and thecal cells disintegrates and allows the large and small luteal cells to intermingle with fibroblasts, pericytes, and endothelial cells. In primates much of the basement membrane remains and separates the two cell types. As the corpus luteum develops it also increases in size. In the ewe, there is, on average, a 16-fold increase in mass of the ovulatory tissue over several days. This growth is attributed to hypertrophy of small and large luteal cells as well as hyperplasia of small luteal cells, fibroblasts, and endothelial cells. The rate of mitosis in the developing corpus luteum is comparable to that of rapidly growing tumors. Moreover, a little over 20 percent of the mass of the CL is attributed to development of an extensive capillary plexus. The corpus luteum is one of the most richly vascularized tissues in the female and the rate of blood flow through this tissue exceeds that of all other tissues. This corresponds to its extremely high rate of oxygen consumption (per unit of mass), second only to that of the brain.

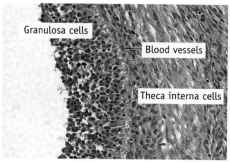
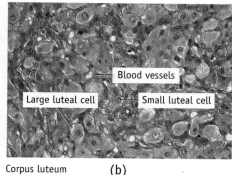

Pre-ovulatory follicle　(a)　　　　Corpus luteum　　　(b)

FIGURE 12-5

Photomicrographs of a pe-ovulatory tertiary follicle (a) and corpus luteum (b) from an ovary of a cow. Note that cells of the follicle are arranged in distinct layers. The granulosa cells are separated from the theca cells by a basement membrane (a). Following ovulation and luteinization, the granulosa and theca cells mix and are transformed into large and small luteal cells, respectively (b). The distinct basement membrane of the follicle degenerates and connective tissue is re-arranged to form a matrix that supports the luteal cells and numerous blood vessels that infiltrate the corpus luteum.

The mechanisms by which LH induces luteinization are not well understood. However, it is clear that LH induces the expression of various genes that give rise to several regulatory factors that induce tissue re-organization and growth via paracrine mechanisms.

Biochemical Changes

The major biochemical change associated with the transformation of a ruptured follicle into a corpus luteum results in a change in pattern of steroidogenesis. As described earlier, the preovulatory follicle produces estradiol at a very high rate, and this process requires cooperation between granulosa and theca interna cells. In the follicle, thecal cells express the enzymes necessary for converting cholesterol to testosterone, but not those mediating the conversion of testosterone into estradiol. In contrast, granulosa cells express the enzymes necessary to produce progesterone and convert testosterone into estradiol. In the follicle, steroidogenesis is controlled by LH and follicle-stimulating hormone (FSH). During luteinization the steroidogenic pathways of these follicular cells are altered such that they produce only progesterone. This involves an increase in expression of enzymes necessary to convert cholesterol into progesterone. In addition, luteal cells are not responsive to FSH.

REGULATION OF THE CORPUS LUTEUM

We now turn to a consideration of the mechanisms that regulate the activity of the corpus luteum. Luteinizing hormone appears to play critical roles in both the development and maintenance of the corpus luteum, but other hormones also contribute to these processes. The regression of the corpus luteum cannot be attributed simply to a removal LH. Rather, there are hormonal mechanisms that actively induce the demise of the corpus luteum. The next three sections will be devoted to detailed accounts of the mechanisms regulating the development, maintenance, and regression of the CL.

Regulation of CL Development

Insight into the endocrine regulation of the CL can be gained from the extensive research done with sheep. In the ewe, disruption of LH secretion via hypophysectomy (surgical removal of the pituitary gland) on day 5 of the estrous cycle prevents growth of the CL as well as the rise in progesterone that marks the early luteal phase. The underdevelopment of corpora lutea in these animals is attributed to a decrease in numbers of luteal cells and fibroblasts as well as a reduction in size of both small and large luteal cells, compared to normal. In addition, the capacity to synthesize progesterone is compromised in these cells.

The effects of hypophysectomy on luteal development are not entirely due to a deficiency of LH. When hypophysectomized ewes are given LH replacement therapy, progesterone production is restored to normal levels, but the size of the corpus luteum remains smaller than normal. The combination of growth hormone, another pituitary hormone, and LH will restore both the size and function of the CL.

Maintenance of the Corpus Luteum

In addition to playing an important role in development of the CL, LH is crucial to the maintenance of normal CL function. Removal of the pituitary gland or selective disruption of LH secretion results in regression of the CL in sheep, cattle, swine, and monkeys. Although LH appears to be the primary requirement for maintaining the CL in these species, the combined effects of LH and growth hormone (GH) may be necessary for normal CL function.

The mechanisms regulating the growth and maintenance of corpora lutea in rodents and rabbits are quite different from those in domestic ruminants. In rodents and rabbits, estradiol appears to be the major luteotrophic hormone. In rabbits, the role of LH is to sustain follicular production of estradiol, which then acts on the luteal cells to stimulate progesterone production. In the rat, prolactin is required for luteal cells to express receptors for estradiol and LH. The role of LH is to stimulate luteal production of estradiol, which then acts to stimulate progesterone synthesis.

Regulation of Progesterone Synthesis/Secretion

As discussed in the previous section, LH is necessary, but perhaps not sufficient for maintaining the corpus luteum. A comprehensive understanding of how progesterone synthesis is maintained in this tissue requires an understanding of how the small and large luteal cells function. In most mammals the basal (hormone-independent) production of progesterone in large luteal cells is 2 to 40 times greater than that in small luteal cells. Moreover, progesterone synthesis in small luteal cells requires LH. Regulation of progesterone production in large luteal cells appears to be LH-independent, but is regulated by other hormones including GH, insulin-like growth factor-1 (IGF-1), and prostaglandin E_2.

Although the activities of small and large luteal cells are regulated by different hormones, the basic biosynthetic pathway for progesterone synthesis is the same in each cell type (Figure 12-6). Cholesterol is the precursor for all steroid hormones. Steroid-producing cells derive most of their cholesterol from **low-density lipoproteins (LDL),** which are produced by the liver and serve as a means of cholesterol transport in blood. The structure of an LDL is analogous to a cell consisting only of its membrane; that is, a shell containing several types of lipids including cholesterol into which proteins are embedded. One of these proteins (apoprotein B) serves as a ligand for a specific receptor located on the membranes of most cells. Binding of the LDL to its receptor is a prerequisite for cholesterol uptake. The major steps in this process include:

- Internalization of the LDL-receptor complex.
- Liberation of cholesterol from the LDL.
- Uptake and esterification of cholesterol by lipid droplets.

Cholesterol is stored in lipid droplets in the form of cholesteryl esters. Cells that produce steroid hormones have numerous lipid droplets. Enzymes known as esterases break down the esterified form of cholesterol and free cholesterol is then liberated from the lipid droplets. Since cholesterol is a lipid it is insoluble in the cytoplasm. A protein known as **sterol carrier protein-2 (SCP-2)** interacts with the cytoskeleton to facilitate transport of cholesterol across the cytoplasm. A protein called **steroidogenic acute regulatory protein (StAR)** seems to play a role in transporting the cholesterol molecule from the outer mitochondrial membrane to the inner mitochondrial membrane where it encounters the side-chain cleavage enzyme. This enzyme

12

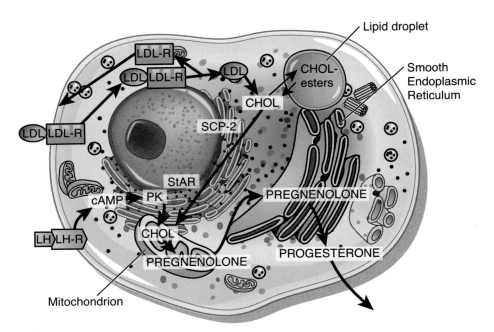

FIGURE 12-6

Schematic illustration depicting the highlights of progesterone synthesis in small luteal cells. LH binds to its receptor and stimulates production of cyclic AMP (cAMP). This second messenger activates protein kinase (PK), which promotes the transport of cholesterol (CHOL) from lipid droplets to the mitochondria. This involves two proteins: SCP-2, a carrier protein, and StAR, a protein that facilitates transport of cholesterol across the mitochondrial membrane. Once inside the mitochondria, cholesterol is converted to pregnenolone. Because pregnenolone is the rate-limiting intermediate in steroidogenesis, the rate of progesterone synthesis is directly dependent on the rate of conversion of cholesterol to pregnenolone. The major source of cholesterol in steroid-producing cells is cholesterol esters that are stored in lipid droplets. These cells derive cholesterol from low-density lipoprotein (LDL), which binds to specific membrane receptors, internalized and broken down to liberate cholesterol. The biosynthesis of progesterone in large luteal cells is identical to that of the small luteal cells with the exception that hormones other than LH regulate the process in the former case.

converts cholesterol to pregnenolone. The rate at which this reaction proceeds is directly proportional to the availability of cholesterol, and is the rate-limiting step in progesterone synthesis. Once pregnenolone is formed it is translocated out of the mitochondria and transported to the smooth endoplasmic reticulum where it is converted to progesterone by the enzyme 3β-hydroxysteroid dehydrogenase. Most of the newly synthesized progesterone leaves the cell and enters the extracellular fluid.

Having described the role of cholesterol in steroidogenesis it is now possible to discuss how LH and other hormones regulate progesterone synthesis in the corpus luteum. LH governs the acute regulation of progesterone synthesis in small luteal cells (Figure 12-6). In this case, the interaction of LH with its receptor results in generation of cyclic AMP (cAMP), which activates protein kinase A, an enzyme that phosphorylates regulatory proteins including StAR. Once phosphorylated, StAR binds cholesterol and facilitates its transport across mitochondrial membranes. Thus LH stimulates progesterone synthesis by increasing the availability of cholesterol, the rate-limiting substrate in this biosynthetic pathway. The more chronic effects of LH on progesterone synthesis may include increasing internalization of the LDL-receptor complex and increasing liberation of cholesterol from lipid droplets.

Less is known about the molecular mechanisms regulating progesterone synthesis in large luteal cells. As mentioned earlier, several hormones seem to play a role in this process. These cells have membrane receptors for GH, IGF-1, and prostaglandin E_2. In addition, each of these hormones has been shown to enhance progesterone secretion in large luteal cells. GH and IGF-1 most likely interact to promote progesterone synthesis. In other words, GH may act directly on these cells as well as stimulate secretion of IGF-1, which acts in an autocrine/paracrine manner to further stimulate progesterone synthesis. The effects of GH, IGF-1, and prostaglandin E_2 in large luteal cells are presumably similar to those of LH on small luteal cells; that is, they enhance the rate-limiting step in progesterone synthesis.

REGRESSION OF THE CORPUS LUTEUM

The regression of the corpus luteum which marks the end of the luteal phase is commonly referred to as **luteolysis** (i.e., degradation of luteal tissue). This process appears to involve two steps. During the initial step, luteal cells lose their ability to synthesize and release progesterone. This is followed by the destruction of cells that make up the corpus luteum. The decrease in progesterone production is likely attributed to reduced blood flow to the corpus luteum as well as a compromised ability to synthesize progesterone.

Endocrine Regulation

In nonprimate species, luteolysis is dependent on the uterus. In other words, hysterectomy during the mid-luteal phase causes a delay in luteolysis. The so-called luteolysin in these species appears to be prostaglandin $F_{2\alpha}$ ($PGF_{2\alpha}$), a hormone produced by the mucosa cells of the endometrium. In species such as the

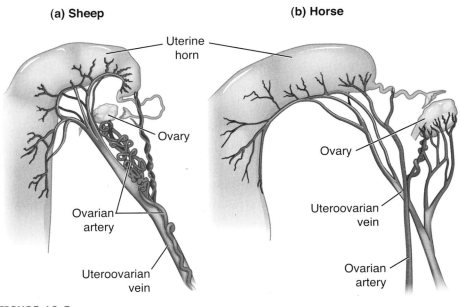

(a) Sheep   **(b) Horse**

Uterine horn

Ovary

Ovarian artery

Uteroovarian vein

Ovary

Uteroovarian vein

Ovarian artery

FIGURE 12-7

Vasculature of the ovaries and uterus in the sheep (a) and horse (b). Note the intimate relationship between the ovarian artery and utero-ovarian vein in the sheep. Numerous contacts between these blood vessels allows a counter-current exchange mechanism whereby $PGF_{2\alpha}$ produced by the uterus and present in high concentrations in the utero-ovarian vein, can diffuse into the ovarian artery and travel directly to the ovary. This type of transport mechanism is not possible in the horse because there are few contact points between the utero-ovarian vein and ovarian artery.

12

guinea pig, sheep, cow, and pig $PGF_{2\alpha}$ is released into the capillaries of the submucosa and leaves the uterus via the utero-ovarian vein. Collateral channels and venules extending from this vein intertwine with and form numerous contacts with the tortuous ovarian artery (Figure 12-7A). These two blood vessels have thin walls at sites of contact which might favor diffusion of $PGF_{2\alpha}$ directly from the utero-ovarian vein into the ovarian artery. This counter-current exchange mechanism ensures that high levels of $PGF_{2\alpha}$ flow directly from the uterus to the ovary without entering the pulmonary circulation where most of it is degraded. In species such as the horse and rabbit, luteolysis is induced by $PGF_{2\alpha}$ produced by the uterus, but the hormone does not travel directly to the uterus via a counter-current exchange system. Anatomic studies reveal that the utero-ovarian vein and ovarian artery are not as intimately related as in species such as the sheep (Figure 12-7B). In species such as the horse and rabbit, $PGF_{2\alpha}$ travels from the uterus through the general circulation before reaching the ovaries.

TABLE 12-2 Results of two studies showing the effects of hysterectomy on lifespan of the corpus luteum (CL) in sheep

Treatment	Lifespan of CL (days)
None	15–17
Hysterectomy [a]	148
Unilateral hysterectomy (contralateral to ovary with CL) [b]	15–17
Unilateral hysterectomy (ipsilateral to ovary with CL) [b]	From 24 to >36

[a] Both uterine horns removed (Wiltbank and Hansel, 1956).
[b] One uterine horn removed (Inskeep and Butcher, 1966).

Aside from anatomic studies that characterized the vasculature of the female reproductive tract, the most compelling evidence that $PGF_{2\alpha}$ can be transported directly from the uterus to the ovary comes from several important studies (Table 12-2). First, the fact that the lifespan of the CL is extended after removal of the uterine horn on the same side as the ovary with the CL, but not after removal of the horn on the side opposite to the CL demonstrates that the luteolytic activity of the uterus involves localized effects. These results, together with evidence showing that when $PGF_{2\alpha}$ is injected into the uterus it is preferentially transferred to the ovarian artery (Figure 12-7) support the hypothesis that the hormone travels from the uterus to the ovary via a counter-current mechanism.

Luteolysis appears to be a tightly regulated event and appears to depend on the pattern of $PGF_{2\alpha}$ secretion. Release of $PGF_{2\alpha}$ is low and nonpulsatile during most of the luteal phase. However, during the late luteal phase (e.g., day 14 in cows) release of the hormone increases dramatically and circulating concentrations take on a pulsatile pattern (Figure 12-9). The pulses last several hours and occur once every 6 to 8 hours. The onset of this pulsatile release of $PGF_{2\alpha}$ appears to be the physiologic trigger for the onset of luteolysis.

At this point you might be wondering what causes the increase in $PGF_{2\alpha}$ release. There is consensus that the trigger is estradiol produced by the preovulatory follicle that is growing rapidly at this time in the ovarian cycle. The increase in estradiol appears to initiate a positive feedback loop between oxytocin and $PGF_{2\alpha}$ (Figures 12-10). According to this hypothesis estradiol stimulates release of oxytocin from the posterior pituitary gland. At about the time estradiol concentrations are increasing, the endometrium begins to express receptors for oxytocin. Thus oxytocin can act on the uterus to stimulate release of $PGF_{2\alpha}$ which then stimulates release of additional oxytocin from the corpus luteum. Thus a positive feedback loop is established between uterine $PGF_{2\alpha}$

12

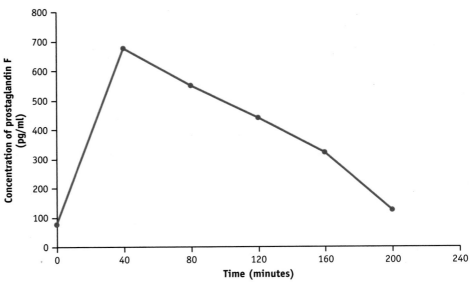

FIGURE 12-8

Concentrations of $PGF_{2\alpha}$ in the plasma of blood collected from the ovarian artery between 0 and 24 min after injection of the hormone into the uterus. Concentrations of the hormone increased in the carotid artery and jugular vein only within the first 5 minutes following injection, suggesting that the increase in the ovarian artery was due to counter current transfer rather than an induced release of endogenous $PGF_{2\alpha}$. Data taken from Hixon and Hansel (1974).

FIGURE 12-9

Circulating patterns of progesterone and $PGF_{2\alpha}$ during the estrous cycle of the ewe. Luteal regression appears to be caused by the appearance of the pulsatile release of $PGF_{2\alpha}$, which is the result of an increase in expression of oxytocin receptors in the uterus.

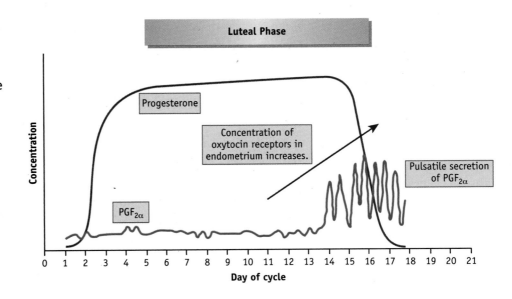

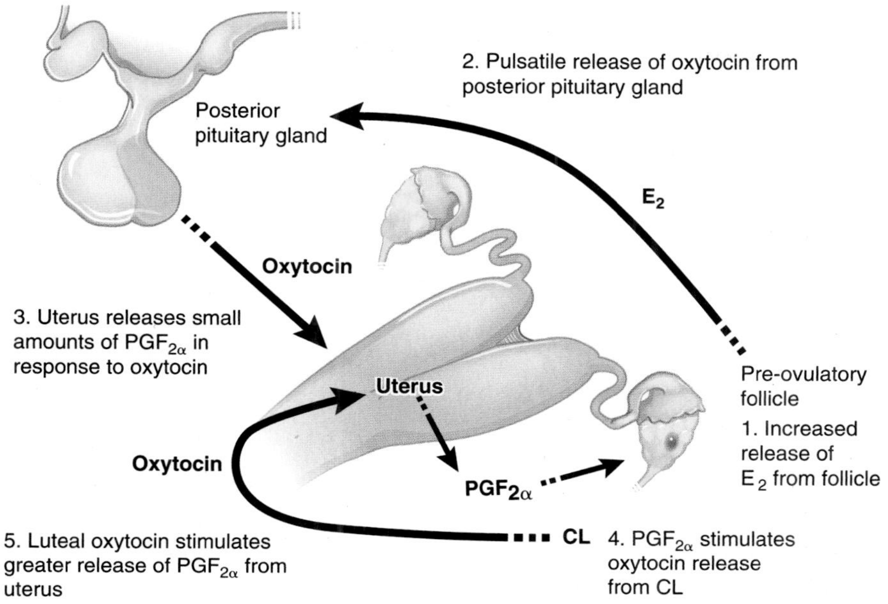

FIGURE 12-10

Summary of the prevailing hypothesis explaining the regulation of luteolysis in the sheep. The high concentrations of estradiol produced by the dominant follicle stimulates release of oxytocin by the posterior pituitary gland. This initiates an oxytocin-PGF$_{2\alpha}$ positive feedback system between the uterus and corpus luteum, which results in the pulsatile release of PGF$_{2\alpha}$ and the demise of the corpus luteum.

and oxytocin produced by the CL. This relationship causes the two hormones to be released in a pulsatile manner. Each pulse of PGF$_{2\alpha}$ is preceded by a pulse of oxytocin. The reason these hormones are not released continuously is that the uterus becomes refractory to PGF$_{2\alpha}$ following each pulse of PGF$_{2\alpha}$ and the luteal tissue becomes refractory to oxytocin following each pulse of oxytocin.

Regulation of luteolysis in other domestic ungulates appears to be similar to that described for the ewe. In contrast, the mechanism in primates appears to be quite different. Removal of the uterus does not extend the life of the CL in humans and other primates. Although the precise mechanism controlling luteolysis in these species has not been elucidated, the prevailing theory to explain this process is summarized in Figure 12-11. Briefly, luteal regression is brought about by enhanced production of PGF$_{2\alpha}$ by the ovary. The resulting decline in production of estradiol and progesterone induces an increase in uterine production of PGF$_{2\alpha}$, which causes constriction of the spiral arteries that supply blood to the endometrium. The resulting necrosis of tissue leads to menstrual bleeding and sloughing of the endometrial lining.

Inhibition of Progesterone Synthesis/Secretion

The decline in progesterone concentrations that characterizes the end of the luteal phase is the result of two PGF$_{2\alpha}$–induced events (Figure 12-12): 1) reduced blood flow to the corpus luteum and 2) inhibition of progesterone synthesis.

PGF$_{2\alpha}$ is well known as a vasoconstrictor. During the early stages of luteolysis PGF$_{2\alpha}$ induces constriction of the arterioles that bring blood into the luteal

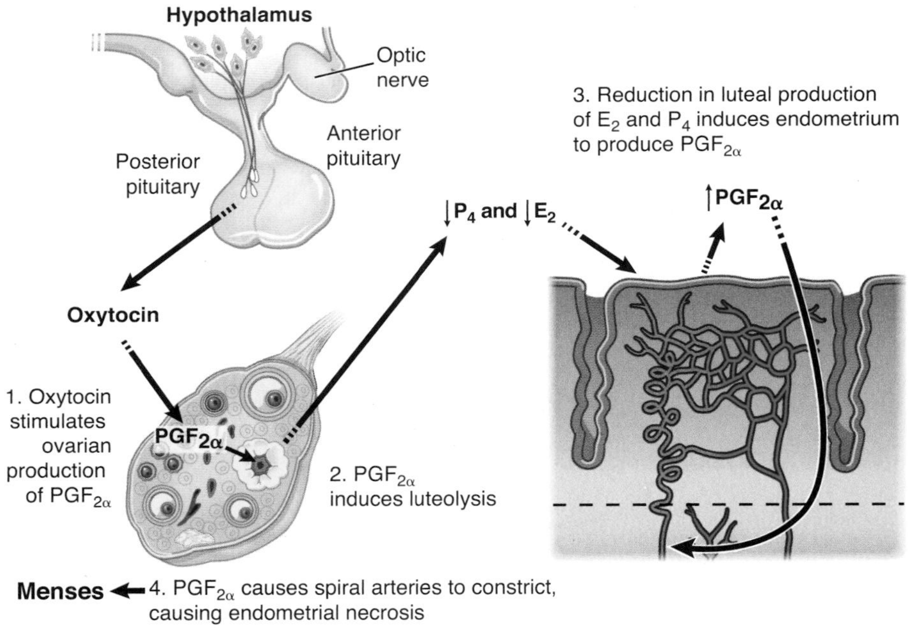

Hypothalamus

Optic nerve

Anterior pituitary

Posterior pituitary

Oxytocin

1. Oxytocin stimulates ovarian production of $PGF_{2\alpha}$

$PGF_{2\alpha}$

2. $PGF_{2\alpha}$ induces luteolysis

$\downarrow P_4$ and $\downarrow E_2$

3. Reduction in luteal production of E_2 and P_4 induces endometrium to produce $PGF_{2\alpha}$

$\uparrow PGF_{2\alpha}$

Menses ← 4. $PGF_{2\alpha}$ causes spiral arteries to constrict, causing endometrial necrosis

FIGURE 12-11

Summary of the current hypothesis explaining regulation of luteolysis and menses in primates. Oxytocin stimulates ovarian production of $PGF_{2\alpha}$, which induces luteal regression. The resulting drop in progesterone and estradiol causes the uterine endometrial cells to release $PGF_{2\alpha}$, which acts locally on the uterine submucosa to cause constriction of spiral arteries. Reduced blood flow causes tissue necrosis, which results in menses.

FIGURE 12-12

12

Schematic illustration of how $PGF_{2\alpha}$ is believed to induce regression of the corpus luteum. The hormone acts via membrane receptors and second messengers to induce structural (lysis of luteal tissue) and functional (decreased progesterone production) luteolysis.

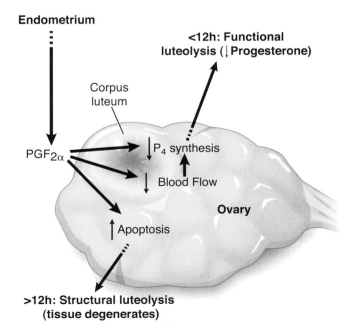

Endometrium

<12h: Functional luteolysis ($\downarrow$ Progesterone)

Corpus luteum

$PGF_{2\alpha}$

$\downarrow P_4$ synthesis

$\downarrow$ Blood Flow

Ovary

$\uparrow$ Apoptosis

>12h: Structural luteolysis (tissue degenerates)

tissue. This action may be mediated by endothelin-1. In addition, receptors for $PGF_{2\alpha}$ have been detected on endothelial cells indicating that this hormone might have direct effects on luteal blood vessels. It has been proposed that $PGF_{2\alpha}$ induces degeneration of capillaries in the corpus luteum. The combination of these effects on luteal vasculature causes a marked reduction in blood flow to the gland and deprives it of vital nutrients (energy substrates and oxygen), substrates for progesterone synthesis (LDL), and luteotrophic support (LH).

In addition to depriving luteal cells of the support necessary to produce progesterone $PGF_{2\alpha}$ also acts within luteal cells to disrupt progesterone synthesis. The primary mode of action appears to be an inhibition of intracellular cholesterol transport. $PGF_{2\alpha}$ has been shown to decrease concentrations of the cholesterol transporter (SCP-2) as well as proteins that make up the cytoskeleton. Each of these components is necessary for movement of cholesterol from lipid droplets to the mitochondria.

It should be noted that the aforementioned effects of $PGF_{2\alpha}$ on progesterone production have been documented in the large luteal cells. We know substantially less about the mechanisms by which $PGF_{2\alpha}$ inhibits progesterone synthesis in small luteal cells. It has been postulated that oxytocin, or some other luteal factor disrupts cholesterol transport to bring about a decrease in progesterone synthesis.

Morphologic Changes in Luteal Tissue

Morphologic changes can be observed in luteal tissue soon after progesterone secretion diminishes. These structural changes are attributed to the previously described effects of $PGF_{2\alpha}$ on blood flow to the tissue, as well as direct effects of the hormone on the various cell types found in the tissue. Within 24 hours following administration of $PGF_{2\alpha}$, the size of the corpus luteum decreases. Endothelial cells are the first cells to display morphologic changes. As noted in the previous section, these cells begin to degenerate thus destroying the extensive capillary network that supplies the corpus luteum with blood. Soon after there is a marked reduction in the numbers of large and small luteal cells. Other noteworthy changes include infiltration by leukocytes and degradation of the extracellular matrix, the connective tissue that provides support for the luteal tissue.

Apoptosis

The disappearance of cells from the corpus luteum during luteolysis is the result of a process known as apoptosis. This is the mechanism by which cells

12

self-destruct. It is commonly observed when support of endocrine cells is removed. For example, when the nondominant follicles undergo atresia due to a decline in FSH, apoptosis is initiated in granulosa cells. Characteristic changes that occur in cells undergoing apoptosis include 1) fragmentation of the nucleus, 2) fragmentation of DNA, and 3) formation of membrane-bound vesicles that contain cytoplasmic materials. The extent to which $PGF_{2\alpha}$ plays a role in inducing apoptosis is unclear, but it has been implicated in domestic ungulates, rats and humans.

Oxidative Stress

Oxidizing agents such as superoxide anion radicals, hydroxyl radicals, and hydrogen peroxide accumulate in luteal tissue during its regression. These so-called free radicals are toxic to cells and may play important roles in luteolysis. The source of most of these compounds is likely to be macrophages that infiltrate the corpus luteum during luteolysis to degrade the extracellular matrix and phagocytize the byproducts of tissue degeneration.

SUMMARY OF MAIN CONCEPTS

- The luteal phase of the ovarian cycle encompasses the period between ovulation and regression of the corpus luteum and is divided into three main phases: 1) an early phase during which the CL forms and gains the ability to produce progesterone, 2) a middle phase during which the corpus luteum is maintained and progesterone secretion is sustained, and 3) a late phase during which the corpus luteum regresses.

- Progesterone serves two main functions: 1) providing the major negative feedback signal regulating LH release in the adult and 2) preparing the female reproductive tract for pregnancy.

- Leutinizing hormone induces luteinization of the ruptured follicle and is the primary luteotrophic hormone in most species of mammals.

- $PGF_{2\alpha}$ induces luteolysis by reducing blood flow to luteal tissue and by disrupting the biosynthetic pathway for progesterone.

DISCUSSION

1. The corpus luteum of the cow is responsive to $PGF_{2\alpha}$ by day 4 of the estrous cycle. In addition, the corpus luteum begins to regress by day 18 of the cycle. Using this information, calculate the percentage of a group

of nonpregnant cows (selected at random) that would be expected to ovulate following an injection of $PGF_{2\alpha}$.

2. Suppose on day 1 of the estrous cycle you provided a cow with an implant that produced physiologic amounts of progesterone. If you leave the implant in the cow for 30 days, what would you expect to happen to the corpus luteum that formed between days 0 and 5? In other words, would you expect it to regress as normal, or continue functioning until the implant is removed? Explain your answer.

3. Based on your understanding of the relationship between progesterone and LH, explain the physiologic basis for using progesterone treatments as a means to prevent pregnancy.

4. What effect would a massive injection of LH (to mimic an LH surge) have on the dominant follicle that emerges early in the luteal phase?

REFERENCES

Ginther, O.J. 1974. Internal regulation of physiological processes through venoarterial pathways: a review. *Journal of Animal Science* 39:550–564.

Hixon, J.E. and W. Hansel. 1974. Evidence for preferential transfer of prostaglandin $F_{2\alpha}$ to the ovarian artery following intrauterine administration in cattle. *Biology of Reproduction* 11:543–552.

Inskeep, E.K. and R.L. Butcher. 1966. Local component of utero-ovarian relationships in the ewe. *Journal of Animal Science* 25:1164–1168.

Johnson, S.K. 2005. Possibilities with today's reproductive technologies. *Theriogenology* 64:639–656.

Niswender, G.D., Juengel, J.L., Silva, P.J., Rollyson, M.K., and McIntush, E.W. Mechanisms Controlling the Function and Life Span of the Corpus Luteum. *Physiological Reviews* 80:1–29.

Patterson, D.J. and M.E. Smith. 2007. Progesterone-based estrus synchronization for beef replacement heifers and cows. In R.E. Youngquist and W. R. Threlfall, *Current Therapy in Large Animal Theriogenology 2.* St. Louis: Saunders, pp. 496–508.

Silvia, W.J., Lewis, G.S., McCraken, J.A., Thatcher, W.W., and Wilson, Jr., L. 1991. Hormonal regulation of uterine secretion of prostaglandin F2α during luteolysis in ruminants. *Biology of Reproduction* 45:655–663.

12

CHAPTER
13 | Sexual Behavior

CHAPTER OBJECTIVES

- Define sexual behavior.

- Discuss sexual behavior from the perspective of behavioral ecology.

- Describe general patterns of sexual behavior in male and female mammals.

- Discuss theories describing the regulation of sexual behaviors.

BACKGROUND

Previous chapters emphasized the physiologic mechanisms regulating production of gametes in males and females. Although the production of spermatozoa and oocytes is necessary for sexual reproduction, it is not sufficient on its own. In order for sexual reproduction to occur, a spermatozoon and oocyte must fuse and form a zygote. This aspect of reproduction depends largely on the behaviors of males and females. Any behavior that facilitates conception is commonly referred to as sexual behavior. It should be emphasized from the outset that the sexual behaviors of males and females are interdependent. In other words, behaviors expressed by one member of the mating pair are influenced by and influence the behaviors expressed by the other member. This makes it difficult, if not impossible, to provide completely independent accounts of the sexual behaviors of males and females. With this in mind, it is useful to differentiate between behavioral and ethologic approaches to studying behaviors. The former deals with the behaviors of individual animals, often studied under experimental conditions. The latter field of study deals with expression of behaviors under the social and environmental conditions in which the behavior is normally expressed. In the case of sexual behavior, an ethologic approach focuses on the behaviors expressed when two or more individuals interact sexually. A derivative of ethology is the field of behavioral ecology; that is, the study of the ecologic and evolutionary bases for behavior. The goal of behavioral ecology is to provide an account of how a behavior contributes to an animal's ability to adapt to its environment. The information provided in this chapter will emphasize an ethologic perspective. However, in order to appreciate why animals express various types of sexual behaviors, it is important to have an appreciation for the insights derived from behavioral ecology.

Definition and Causes of Behavior

A behavior is commonly defined as a response to a stimulus. This general definition may seem insufficient because various stimuli induce responses that are not generally regarded as behaviors. For example, stimulation of the cervix of a queen will induce ovulation, but ovulation is typically regarded as a physiologic response, not as a behavior. We generally regard behaviors as the responses of whole organisms to stimuli. Behaviors can be understood from the level of individual animals, groups of animals, or entire species of animals. For example, howling is a behavior expressed by individual wolves as well as packs of wolves. It is also trait that helps characterize the species of *Canis lupus.* From ecologic and evolutionary perspectives a behavior can be understood to mean an animal's responses to its environment. Sexual behaviors are responses to various stimuli in the context of mating.

What causes behavior? This may seem like a silly question. An obvious, but superficial, answer would be that a stimulus causes a behavior. This question is much more complex than it might first appear to be. Consider howling in canids. What causes these animals to howl? Behavioral ecologists address this issue of causation by differentiating between ultimate and proximate causes. Proximate causes of behavior include so-called ontogenetic variables and mechanistic variables. Ontogenetic variables refer to the development of behavior; that is, how an animal learns to perform a behavior. These variables include the animal's life experiences; for example, interactions with other dogs that howl. Mechanistic variables include the physiologic mechanisms regulating behavior; for example, the physiologic ability to produce this type of vocalization.

Behaviors are also attributed to ultimate causes. These include phylogenetic and adaptive causes. Phylogenetic causes can be viewed as constraints related to the animal's lineage. For example, dogs can howl but they can't vocalize like a cow. Adaptive causes deal with selection of a particular behavior; that is, the extent to which a behavior enables an animal to adapt to its environment. To put this in simpler terms, howling is caused by genes that allow dogs to do so as well as selection pressure that favored this type of vocalization.

The question, "What causes dogs to howl?" is the same as, "Why do dogs howl?" Based on the aforementioned analysis, an appropriate answer would be as follows. Dogs howl because they express a particular set of genes (an ultimate cause) that give rise to physiologic mechanisms (a proximate cause) that allows dogs to vocalize in this way. Moreover, they howl because this trait was selected for (an ultimate cause) and they live in an environment in which

13

they learn how to do it (a proximate cause). The same philosophical framework can be applied to any behavior, including sexual behaviors. It may be useful to keep this discussion in mind as we explore the various sexual behaviors of mammals.

Behavioral Ecology

Behavioral ecologists are primarily concerned with the ultimate causes of behaviors. Because the focus of this book is on physiology, our main concern will be with the proximate causes of sexual behavior; that is, how animals develop and express sexual behaviors. Nevertheless, it is valuable to consider some of the fundamental principles of behavioral ecology. The general goal of this field is to understand behaviors in the context of the relationships between an animal and its environment (both living and nonliving components). From a Darwinian perspective, an animal expresses a particular behavior because that behavior allowed it to adapt successfully to its environment. Consider the homosexual mounting that characterizes estrus in domestic cows (*Bos taurus*) that evolved in Europe. Not all members of the *Bovidae* family express this behavior. For example, in addition to *Bos taurus*, yaks (*Bos grunniens*), a domestic bovine that inhabits central Asia and the American bison (*Bison bison*) express this behavior as well whereas water buffalo (*Bison bubalis*) do not. It is important to keep in mind that the environment in which an animal evolves need not be natural. For example, in the case of domestic animals, human influence can have profound effects on what types of behaviors have been perpetuated. It appears that the selection pressure from humans might have enhanced expression of mounting behavior in *Bos taruus*. The exact reason for this is unclear. Some people have suggested that this trait was perpetuated by humans because it proved to be useful in identifying when a cow is reproductively active, especially in cases where farmers were sharing the use of a bull for breeding.

When evaluating a sexual behavior from an evolutionary perspective, one should consider its cost. In other words, what impact does the expression of a particular behavior have on the overall fitness of the animal? Consider the behavior of the feral Soay sheep (mentioned in the discussion section of Chapter 8). These animals exist in an extremely harsh environment and the breeding season occurs during the autumn, a time when it is important to feed and build up body fat to help survive the brutal winters. The rams of this breed engage in a violent and exhausting rut at a time when it is important for animals to build up a reserve of metabolic fuels. These circumstances make the energetic costs of this rutting behavior tremendous. Not only does the behavior itself require an expenditure of energy, it reduces the amount of time these animals have to feed. Such behavior severely reduces the fitness of rams. The winter mortality rate of Soay rams under natural conditions often exceeds 90 percent. The reason such behavior persists is that these animals attain puberty at an extremely early age, which allows them to mate before the onset of winter. This example illustrates that concept that an understanding of why a certain sexual behavior exists requires consideration of the environment in which the animal must live as well as knowledge of its other reproductive traits.

PATTERNS OF SEXUAL BEHAVIOR IN MAMMALS

It is difficult, if not impossible, to discuss the physiologic control of sexual behavior without first describing some general features of these behaviors in representative species. This section includes an overview of the patterns of sexual behavior in rodents, humans, domestic carnivores, and domestic ungulates. It is appropriate to discuss the sexual behavior of rodents because these species have been studied more extensively and intensively than any other type of mammal. It is also appropriate to consider the sexual behavior of humans. In addition to being a topic with which most people are fascinated, conflicting ideas regarding what constitutes normal sexual behavior underlie several extremely contentious social issues (e.g., should certain sexual practices, such as homosexuality, be condoned?). A biological understanding of such behaviors provides a useful context for analyzing such issues. Finally, an understanding of sexual behavior in domestic animals is justified because such information is often used to manage the reproductive activity of these animals. Moreover, such understanding helps us to better understand the animals with which we interact and upon which we depend.

The following descriptions of sexual behaviors reflect general commonalities for various species. However, behavioral ecologists remind us that there is tremendous variation in behavior among individual members of a particular species. It is this variation in behavior that allows a species to adapt to its environment.

Rodents

Most research concerning the sexual behavior of rodents has been done with rats and has relied on methods consistent with a behaviorist perspective; that is, experiments involving an isolated mating pair in laboratory conditions, usually confined to a cage or arena. More recently investigators have embraced an ethologic approach where the rats are studied in the company of other males

13

and females, and where animals can retreat to other locations to avoid interactions with each other. These conditions correspond to conditions that prevail in settings outside the laboratory. The following description reflects an ethologic perspective of the sexual behaviors of rodents such as rats and hamsters.

Rats and hamsters are opportunistic copulators. In other words, they will engage in sexual activity under a variety of circumstances. These might involve a dyad (one male and one female), a triad (one individual of one sex and two of the other sex), or even groups. Unlike humans and domestic animals, where an uninterrupted period of copulation is followed by orgasm, rodents engage in multiple bouts of copulations during which the male undergoes a series of ejaculations.

When a female becomes sexually receptive (i.e., expresses estrus or heat) it begins to mark its surroundings with its scent in order to attract males to its domain. Once the male encounters a female, the pair engages in mutual anogenital investigation. When a female is in heat, she will face the male and then run away in order to make the male chase after her. If the male chases the female, she will stop abruptly and assume what is known as presentation posture, or "pre-lordosis crouch" (Figure 13-1A). If the male mounts the female and palpates her flanks with his front legs, she will assume the full lordosis position (Figure 13-1B) and the male will attempt intromission. The female will then dart a short distance away from the male and then crouch again to solicit the male. This cycle of solicitation, runaway, and lordosis may

(a)   (b)

FIGURE 13-1

Mating behaviors expressed by male and female hamsters. The female has stopped and expressed the presentation or pre-lordosis position, whereas the male investigates the female's anal-genital area (a). The male then mounts the female and palpates her flank region causing her to express the mating or lordosis posture (b). Intromission soon follows mounting, but ejaculation occurs only after several bouts of mounting. Photographs provided by Dr. Cheryl Sisk and Dr. Heather Figueira, Michigan State University.

be repeated several times and represents a "bout of copulation" from the male's perspective. The interval between intromissions is determined solely by the female. Only after several bouts does the male ejaculate. After ejaculation the male enters a "postejaculatory refractory period," an interval of approximately 5 minutes during which the male is not receptive to a soliciting female. The female also enters a postejaculatory refractory (approximately 1 minute in duration) during which she will not solicit a male.

As estrus terminates, the female solicits males less often and the inter-intromission interval increases. Females may still exhibit lordosis, but they begin to fight with the males rather than copulate with them. The males usually withdraw from their pursuit at this time and seek other receptive females.

There appears to be a limit to the number of ejaculations a male can experience during mating. After several ejaculations, the male becomes sexually exhausted and cannot initiate or complete a bout of copulation with the same or different female. This usually lasts an average of 4 days. During group matings, female rats will compete for males. In other words, one female might attempt to intercept a male that is chasing after another female. Males, on the other hand, typically express cooperative behavior; that is, a male will allow other males to mate with the female after the original male has engaged in several copulations.

Domestic Carnivores (Dogs and Cats)

When male dogs and cats first encounter females, they sniff and lick the area surrounding the **perineum** (area around the anus and vulva) of the female. Males of both species determine if the female is in heat by **pheromones** that are present in the urine and secretions of the anal glands and vulva. A pheromone is a chemical substance secreted by one individual that will induce a behavioral response in another. In this case the pheromones produced by the female trigger male sexual behavior, which, in turn, evokes female sexual behaviors.

When the bitch is in heat, she is likely to accept investigations by the male and may engage in soliciting behavior. This includes licking the male's genitalia, presenting her perineal region to the male and then running away (teasing). If the female expresses this behavior, rather than hostile behavior, the male will attempt to mount. The female will permit the male to mount only when she is in heat. The mounting behavior of the male consists of clasping his forelimbs just anterior to the bitch's hind limbs. As this occurs, the male's penis becomes erect and he attempts intromission. If successful, the male will thrust vigorously for several minutes. During copulation, the penis continues to enlarge and cannot be withdrawn. This is known as the copulatory lock or tie (Figures 13-2 and 13-3). The male typically dismounts the female by

13

FIGURE 13-2

Sequence of mating postures expressed by the dog and bitch. If the bitch is sexually receptive she will allow the male to mount and permit intromission (a). During intromission the dog's penis becomes fully erect and cannot be withdrawn. At this point the dog dismounts and turns away from the bitch (b). The mating pair assumes the so-called "tie" posture until the dog's penis becomes flaccid (c).

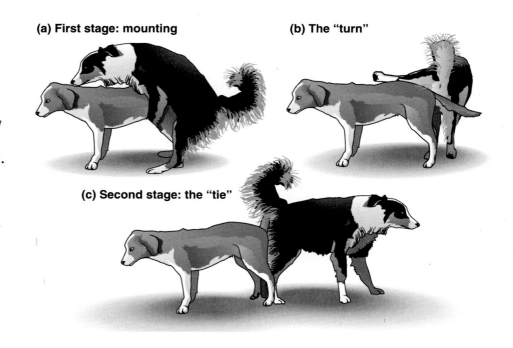

(a) First stage: mounting

(b) The "turn"

(c) Second stage: the "tie"

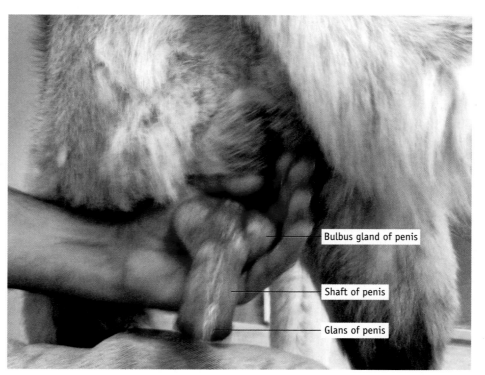

Bulbus gland of penis

Shaft of penis

Glans of penis

FIGURE 13-3

Photograph of a dog with a fully erect penis held in the position it would assume during the copulatory tie. Note the enlarged bulbus gland that prevents the penis from being withdrawn from the female's vagina. Photograph provided by Ms. Remi Tagawa, undergraduate student at the University of Kentucky.

FIGURE 13-4

Mating postures of the tom and queen. The queen has assumed the mating posture (lordosis) and allows the tom to mount and engage in intromission. When the tom ejaculates the queen will cry out and bat at the male to drive him away. Photograph provided by Mr. Michael Meyer, undergraduate student at the University of Kentucky.

stepping off of the female, lifting one of its hind legs over her back such that the two animals are standing "rear to rear." The pair of dogs usually stands quietly for approximately 15 minutes until the penis becomes flaccid. During the copulatory lock, the female's vagina contracts around the penis while it releases pulses of prostatic fluid. This behavior appears to facilitate movement of ejaculatory fluid into the vagina. Disunion normally occurs in a quiet fashion, but the bitch might express aggressive behavior when the tie is broken. In many cases, mating occurs in the presence of several males. If there are more than three males present, the duration between male-female contact and copulation increases and the amount of time spent copulating decreases. This may be due to increased aggression among males, as well as expression of mate preference by the bitch. Male dogs rarely attempt mating more than once per day, whereas females will copulate with different males daily. Therefore, it is possible for a bitch to give birth to pups from different males.

Sexual contact between a queen and a tom is typically due to the female entering the male's territory. Male cats are extremely territorial and will not usually breed away from their homes. When a queen is in heat, she will express soliciting behavior. This includes assuming lordosis, treading their hind limbs, and presenting their perineum to the male (Figure 13-4). In response to an estrual female, toms might caterwauler (make a harsh cry). When the tom encounters a queen expressing soliciting behavior, he will grip the scruff of the queen's neck with his teeth and then mount. Intromission and ejaculation proceed quickly once the male mounts the female. In most cases, these behaviors last less than 60 seconds. As the male ejaculates, the female will cry out, roll over, and bat at the male with her from paws. This causes the male to disengage and jump away. Once the male separates the female will roll

13

frantically and lick at her vulva, a response known as the "after reaction." Following the after reaction, which lasts from 1 to 7 minutes, the female will allow the male to approach and the process is repeated. Domestic cats will mate at a rate of approximately twice per hour, but rarely mate more than 15 times each day. Under feral conditions, a male will share its territory with a harem of females and can service as many as 20 queens.

Domestic Ungulates

Detailed accounts of the sexual behaviors of domestic animals can be found in the scientific literature and will not be presented herein. This chapter will focus only on the general principles concerning expression of sexual behavior in these animals, specifically in cattle, sheep, swine, and horses. Like rodents, the females of these species express estrous cycles. Therefore, the focal point of the sexual behaviors of males and females is expression of sexual receptivity of females.

Precopulatory Behavior

Table 13-1 summarizes the general characteristics of behaviors that precede copulation in farm animals. In general these behaviors are more variable (both within and between species) than those expressed during copulation.

TABLE 13-1 Common Behaviors Expressed by Males and Female Livestock Prior to Mating

Sex	Searching	Courtship
Male	• Approaches group of sexually active females (bull) • Approaches and moves among females and investigates (stallions, bulls, rams, and boars) • Exhibits flehmen (bulls, stallions, and rams)	• Investigates, licks, and nuzzles anogenital region (bull and boar) • Grinds teeth and salivates (boar) • Rests chin on females (bull) • Becomes excited and restless (stallion) • Stretches neck and holds head horizontally (ram)
Female	• Increases locomotion (cows and mares) • Becomes restless (ewes and sows) • Increases vocalization (cows) • Twitches (cows) or elevates tail (cows and mares)	• Grooms other individuals (cows) • Attempts to mount other females (cows) • Urinates in presence of male (mares and ewes) • Assumes immobile stance (sow)

FIGURE 13-5

An example of homo-sexual mounting in Holstein cows. Domes-tic cows will permit other cows to mount them when they are in estrus, and will also aggressively mount other females that are in heat. It is common for sexually receptive cows to form "sexually active groups" consist-ing of cows that will take turns mounting each other, often sepa-rated from the rest of the herd. Photograph provided by Dr. Peter J. Hansen, University of Florida.

Precopulatory Behaviors

Precopulatory behaviors include those that begin with searching for a mate and end with mounting. Some of the behaviors leading up to copulation appear to be necessary for copulation. In other words, such behaviors may be necessary for sexual arousal. For example, some wild mammals undergo extensive and elabo-rate courtship rituals consisting of complex sequences of motor patterns and multisensory stimulation lasting several hours. In general, the precopulatory be-haviors of domestic animals consist of only a brief courtship period. For exam-ple, when placed in a small herd of cows (<30), a bull will move among the fe-males and sniff the perineal region of each individual. Once the bull encounters an estrual female, the pair may engage in mutual anogenital investigation and move in a circular fashion. Within a few minutes the bull will rest its head on the rump of the female. Once the bull makes contact, he will attempt to mount.

In addition to investigatory behavior, males and females express other types of behaviors including vocalizing, urinating, and licking. It is not un-common for the female to initiate sexual activity. For example, in cattle, es-trual females may approach bulls and push against them or even mount them to arouse sexual interest.

One of the more peculiar sexual behaviors is the homosexual mounting ex-pressed by domestic cattle and some other bovids (Figure 13-5). A cow that is in heat will attempt to mount other cows, and will also allow other cows to mount

13

FIGURE 13-6

A stallion exhibiting the Flehmen response. This particular stallion was sexually excited and had investigated the perineal region of an estrual mare before displaying this behavior.

her. In a herd setting, groups of estrual cows form sexually active groups and engage in mutual mounting. It is difficult to classify this behavior in terms of copulation. Nevertheless it seems plausible to suggest that homosexual mounting in cows is a precopulatory behavior because it may play a role in attracting males from a distance. Once the male makes contact with this sexually active group he then engages in the precopulatory behaviors described previously.

A behavior that is frequently associated with mating in ruminants and horses is the flehmen response (Figure 13-6). Although males and females may express this behavior at any time, it is typically expressed by males during the precopulatory period. During the flehmen response an animal's head is raised and its upper lip is curled as though it were grimacing. The behavior is usually triggered when the male is exposed to urine or when he makes direct contact with the female's perineal region during anogenital investigations. Such behavior restricts air flow through the nasal passages and exposes the paired nasopalatine ducts that are continuous with the vomeronasal organs, chemoreceptors located in the basal region of the nasal cavity. These paired organs are enclosed by a bony or cartilaginous capsule and are divided by the nasal septum. The epithelial lining of the vomeronasal organs is lined with pseudostratified cells, including sensory neurons. Microvilli protrude from these nerve cells into the lumen of the organ. Axons emanating from these cells merge to form vomeronasal nerves that enter the accessory olfactory bulbs, which are located at the posterior-dorsal portion of the main olfactory bulbs. Presumably these neurons interact with other nerve cells that project to other parts of the central nervous system, including those that regulate sexual behavior. The vomeronasal chemoreceptors are stimulated by pheromones that are be present in bodily fluids aspirated into the nasopalatine ducts.

Copulatory Behaviors

As noted earlier, copulatory behaviors are very similar across species. The only copulatory behavior typically described for females is the mating posture. The willingness of a female to stand firm for mounting is the tell-tale sign that she is in heat.

The copulatory behaviors of males include erection, mounting, intromission, and ejaculation. The former three behaviors can be easily observed, whereas ejaculation is difficult to confirm.

Once the male becomes sexually aroused, its penis will become erect and mounting soon follows. Mounting behavior is quite similar among the various species of four-legged mammals. In all cases the male rises on its hind legs with its chest resting on the rump of the female and its front legs clasping each side of the female's hips. Almost anyone who has contact with nonhuman mammals knows that mounting behavior is not necessarily expressed in the context of mating. For example, young, sexually immature animals frequently mount each other regardless of sex. In addition, homosexual mounting has been observed in all wide variety of both domestic and wild mammals (see Box 13-1).

As the male mounts, its penis enters the female's vagina (intromission). In each of the species of domestic ungulates, the penis protrudes just before mounting. In bulls, seminal fluid dribbles from the penis before intromission. This behavior has not been observed in rams, stallions, or boars. Once intromission has been achieved, the males exhibit pelvic thrusting. Bulls and rams typically exhibit only one deep thrust during ejaculation. In contrast, stallions will thrust multiple times. Boars will exhibit multiple shallow thrusts that serve to position the penis in the sow's cervix. Ejaculation begins once the penis is positioned appropriately. The duration of mounting lasts only several seconds in bulls and rams, 20 to 60 seconds in stallions, and between 5 and 20 minutes in boars. The physiologic regulation of penile erection and ejaculation will be described in detail in a subsequent section of this chapter.

13

BOX 13-1 | Focus on Fertility: The Kinsey Scale

No discussion of human sexual behavior is complete without mentioning the work of famed sexologist Alfred Kinsey (1894–1956). By 1937, Kinsey was a renowned zoologist who made important scientific contributions to taxonomy and evolution theory. Some of his work dealt with the mating patterns of the gall wasp. This work might have provided the incentive for Kinsey to initiate studies of human sexuality. Before Kinsey's work, there were no detailed accounts of how human beings expressed their sexualities. In 1939, Kinsey assumed responsibility for teaching a marriage course at Indiana University and began collecting case histories of human sexual behavior. He and his colleagues conducted over 18,000 interviews, which served as the basis for two important books: *Sexual Behavior in the Human Male,* published in 1948, and *Sexual Behavior in the Human Female,* published in 1953. Perhaps the most noteworthy, and controversial, aspect of Kinsey's work is the idea that although men and women engage in a variety of sexual activities that can be classified as homosexual or heterosexual, individuals themselves should not be strictly classified in this dualistic manner. According to Kinsey, "Males do not represent two discrete populations, heterosexual and homosexual. The world is not to be divided into sheep and goats. It is a fundamental of taxonomy that nature rarely deals with discrete categories. . . The living world is a continuum in each and every one of its aspects." Kinsey developed a seven-point scale to characterize human sexual behavior (Table 13-2) and reported distributions of scale ratings for populations of males and females during a particular period of time. For example, within a given year approximately 75 percent of males were rated as completely heterosexual, whereas 7 percent could be classified as exclusively homosexual. Around 15 percent of men had a rating of 3.

Are Kinsey's ideas applicable to other mammals? It is clear that homosexual behavior occurs in a large number of animal species including mammals. For example, such behavior has been documented in all of the domestic livestock species. The most detailed studies have been done with rams. According to recent research, an average of 8 percent of rams are male-oriented, meaning that when given a choice between males and females they choose to engage in sexual contact with males. Are there degrees of homosexuality in rams? Some evidence suggests that this might be the case. For example, within all-male groups of Bighorn sheep, the dominant ram typically mounts subordinate rams, but also copulates with females during the breeding season.

TABLE 13-2 Descriptions of Kinsey scale ratings for human sexual behaviors and distribution of scale ratings for single and married men in any single year[1]

Kinsey Scale	Description	% men
0	Exclusively heterosexual with no homosexual.	75
1	Predominantly heterosexual, only incidentally homosexual.	22
2	Predominantly heterosexual, but more than incidentally homosexual.	20
3	Equally heterosexual and homosexual.	15
4	Predominantly homosexual, but more than incidentally heterosexual.	10
5	Predominantly homosexual, but only incidentally heterosexual.	8
6	Exclusively homosexual.	6

[1]From Kinsey, *et al.* (1948)

Postcopulatory Behaviors

Following copulation both the male and female enter a so-called refractory period during which no sexual contact occurs. The length of this refractory period varies among species, but usually lasts no longer than several minutes. The period of time between copulations is often referred to as the postejaculatory interval. Once the individuals have recovered they will resume mating behaviors. Females will engage in sexual activity so long as they are in estrus. However, it is not uncommon for females to terminate sexual contact with a particular male following several copulations. On the other hand, males may attempt to mate repeatedly with the same female and ignore other sexually active individuals. Complete cessation of sexual activity in males occurs after repeated ejaculations. This phenomenon is known as sexual exhaustion. Boars, stallions, rams, and bulls will exhibit exhaustion after approximately 8, 20, 70, and 35 ejaculations, respectively.

Humans

Research on human sexual behavior under laboratory conditions has been limited due to ethical constraints and social taboos. Nevertheless, sex researchers including A. Kinsey, W.H. Masters, V.E. Johnson, and more recently S. Hite, have compiled a massive amount of data based on observations, interviews, and surveys. Such studies reveal that humans, like rodents, are opportunistic; that is, they will engage in sex in a myriad of circumstances and in an almost limitless combination of partners and sexual orientations. This realization lead Alfred Kinsey (see Box 13-1) to remark, "the only unnatural sex act [in humans] is the one that cannot be performed."

In spite of the difficulties studying human sexual behavior, there have been objective measurements of sexual arousal and response in both men and women. Masters and Johnson conducted landmark studies on the human sexual response and identified several distinct phases: 1) increasing levels of sexual excitement and arousal, 2) plateau, 3) orgasm, and 4) resolution (the aftermath of a dramatic release of tension). The basic response is similar for men and women, but the occurrence of multiple orgasms is more frequent in women.

Compared to the rat, human copulatory behavior is less stereotypic (more likely to be learned), less sexually dimorphic (aside from the male's pelvic thrusting the behaviors of men and women are quite similar), and more likely to be continuous (continuous genital simulation until orgasm rather than bouts of copulation leading to refractoriness).

It is important to emphasize that in most primates, including humans, the female does not express estrus. Estrus refers to the behavioral state during which

13

a female is sexually receptive to a male; that is, will accept a male. This state is caused by high circulating levels of estradiol. It would be incorrect to claim that female primates are always sexually receptive. It is more appropriate to say that their sexual activities are not confined to a particular time during the menstrual cycle. Interestingly, the rate of sexual activity in women is highest during the middle of the menstrual cycle. However, this does not appear to be due to a direct effect of estradiol on the brain, as is the case in animals that express a true estrus. The increase in sexual activity at this time during the menstrual cycle seems to be attributed to an increase in the sexual interests of males, not an increase in female receptivity per se. This response may be due to changes in aromatic compounds produced by microorganisms in the female's reproductive tract, which act as pheromones to stimulate the sexual interest of males.

THEORETICAL FRAMEWORK FOR UNDERSTANDING SEXUAL BEHAVIOR

One approach to learning the sexual behaviors of mammals is to become familiar with the specific behaviors expressed by males and females of each species during sexual encounters. The obvious disadvantage of this approach is that it would quickly become overwhelming. The patterns of sexual behavior described in the previous section illustrate that there is considerable variation among species. A more manageable approach is to determine if there are some general guidelines or principles that apply to all mammals. Decades of studies involving laboratory rodents have provided a wealth of information about the sexual behaviors of these animals. These data have been used to construct a theoretical framework for understanding the sexual behavior of other species including humans. In the next several sections, we shall consider this framework and then use it to provide an account of the sexual behaviors of several representative species.

Sex as Incentive

Most ethologists regard sex as a consummatory act. Consummatory acts are those that are associated with the termination of some goal-directed behavior; that is, **consummation.** Sex is a consummatory act in the sense that it terminates behaviors directed toward the goal of finding and having sex with another individual. Sex is also an intrinsically rewarding act. In other words, it is rewarding independent of whether or not it reduces a biological need. Thus it is as incorrect to say that an animal seeks sex to reduce some deprivation of sex as it is to say that a person seeks a piece of chocolate in order to alleviate some deprivation of chocolate. In both cases the things that are sought are the incentives for seeking them. These examples are quite different from the case where an animal searches for food in order to alleviate a deficit in nutrients. In the latter case, food has instrumental value; that is, it is used for a particular purpose. The main point to remember is that sex is the incentive that motivates an animal to express behaviors directed at seeking and having sex with a potential mate.

The best evidence to support the claim that sex is a powerful incentive for certain behaviors comes from experiments where rats have to traverse an electric grid to in order to copulate. Male rats will cross the grid and risk being shocked only when estrual females are used as stimuli. Likewise, female rats will cross the grid when intact males are present, but not when castrated males are used.

Sexual Behavior as Adaptive Behavior

Having established that the act of sex itself provides the incentive that motivates animals to seek sexual encounters, it is now possible to consider how this incentive activates sexual behaviors. The theory of adaptive behavior provides a useful framework for understanding this issue. The central assumption of this theory is that the physiologic mechanisms regulating an animal's behavior can adjust (adapt) to a particular environment such that the animal can better cope with that environment. The behavior an animal expresses in a particular environment is known as a "state." For example, a ram that is grazing on pasture is said to be in the grazing state. If the ram encounters a sexually receptive ewe, the ram will enter the reproductive state, and its behavior will change accordingly. The extent to which the ram will complete the sexual act depends on two variables: 1) how motivated it is and 2) previous experiences with receptive ewes (learned behaviors). These ideas are central to "the sexual incentive motivation model," an hypothesis proposed by Dalbir Bindra during the 1970s (Figure 13-7) to explain how motivation and learning activate the transition from one behavioral state to another.

According to Bindra's theory, transitions in behavioral states involve two regions in the brain: 1) the "central motive state" and 2) the "central representation of the incentive." The name of the former region may be confusing because the term "state" usually is used to refer to an animal's behavioral condition, not a physical location in the brain. It seems more appropriate to think of this region as a neuronal system that becomes more active during a particular behavioral state. The central representation of the incentive should be understood as a neuronal system that integrates and provides a physiologic

13

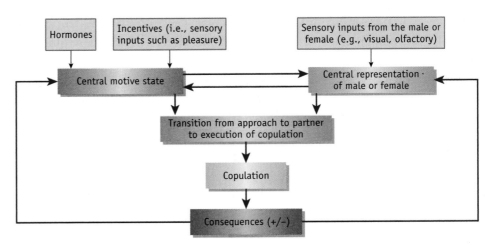

FIGURE 13-7

Theoretical model for explaining the regulation of sexual behavior in mammals. Whether or not an animal will copulate with another depends on the animal's motivation to engage in sexual activity as well as being able to recognize an appropriate sexual stimulus. Sexual motivation is thought to be regulated within some region of the brain (central motive state) and is influenced by various reproductive hormones and sensory inputs that serve as rewards. The ability to recognize a sexual stimulus is also regulated by some brain centers (central representation), which receive and process sensory information about the stimulus. Interaction between these two brain centers is required for an animal to engage in copulation with a sexual partner. However, the consequences of such activity can feed back on the brain to enhance or suppress sexual activity.

representation of the inherent features of the goal object; for example, the visual, olfactory, gustatory, auditory, and tactile signals generated by the potential mate. These two regions are mutually excitatory. At some point both systems reach some threshold level of excitation and interact to promote a transition from observing to acting on its motivation. The behaviors involved with acting are divided into those that bring the animal into contact with the goal object (**appetitive** behaviors) and those that are performed once the animal has made contact with the goal object (consummatory behaviors). Bindra's model also addresses the fact that the execution of the goal can feed back on the aforementioned regions of the brain to affect subsequent sexual activity. This explains how learning can affect behavior.

The Central Motive State

Bindra and other behavioral scientists view this region as a hypothetical entity consisting of neuronal circuits that promote goal-directed actions in response to particular stimuli. The so-called state is generated by hormonal inputs (sex hormones such as testosterone, estradiol, and progesterone) as well as sensory inputs. The main physiologic role of this area is to regulate the degree of motivation generated by a stimulus. For example, the region will generate appetitive behaviors under the influence of appropriate sex hormones and positive sensory inputs (e.g., pleasure). In contrast the region will generate aversive behaviors when there is little to no stimulation from sex hormones and/or the sensory inputs are negative (e.g., pain).

In the case of sexual behavior, the medial preoptic area (MPOA) of the hypothalamus appears to play an important role in generating the central motive state. Receptors for estrogens and androgens are found in the MPOA and when this region is destroyed, the motivation of male and female rats to seek mating partners is diminished.

Bindra suggested that sexual motivation was regulated in part by gonadal hormones, but did not propose specific roles for these hormones. A later section of this chapter provides a discussion of the role of gonadal steroid hormones in generating the central motive state. At this point it is only necessary to point out that these hormones exert two types of action. First, these hormones play a role in the sexual differentiation of this region, much like the actions of these hormones on differentiation of the genitalia. Some scientists refer to this action as **organizational** or morphogenic. Such effects are permanent and irreversible. The embryonic brain appears to start out as inherently female or indifferent. During the fetal and early prenatal stages of development portions of the brain that regulate sexual behavior either remain feminine in the case of females or become **defeminized** (suppression of a behavior that is characteristically female) and/or **masculinized** (acquisition of a behavior that is characteristically male) in the case of males. A second role of these hormones becomes important once the animal becomes pubertal. In sexually mature animals, gonadal steroids are necessary for expression of sexual behavior. A familiar example is the induction of estrus behavior by estradiol in females. These types of actions are often referred to as **activational**. These effects are reversible; that is, they are expressed only when the particular steroid hormone is present. It is important to remember that both the organizational and activational actions of steroid hormones are necessary for an adult to express sexual behaviors. In other words, the adult must have the appropriate neuronal architecture to generate the behaviors as well as a stimulus (steroid hormones) to induce the behavior.

The Central Representation of the Incentive

Although the central motive state is involved with activating sexual motivation, it does not act alone. Sexual motivation also requires an incentive stimulus, which is represented somewhere in the brain. This region, like the central motive state, should be viewed as a hypothetical system of neuronal circuits. It is likely located somewhere in areas of the brain that deal with integration of sensory information as well as memory. The physiologic role of this region is to provide the animal with an accurate image of the stimulus such that a particular behavior is expressed in the appropriate context. For example, if a male lacked or had an erroneous central representation of a receptive female, he might not express sexual behavior in the presence of a receptive female and/or might express such behavior in the presence of inappropriate stimuli (e.g., presence of nonreceptive females, or other species).

Transition from Approach to Consummatory Behaviors

As noted earlier in this section, the central motive state and the central representation of the incentive act together to induce approach (appetitive behavior). In other words, when there are positive inputs to the motive state and the incentive stimulus is clearly represented, an animal will initiate behaviors that bring it into contact with the incentive. Once contact is made, the animal engages in consummatory behaviors. Details of these types of behaviors will be considered in the next section.

Incentive Sequence Model

Bindra's model of sexual behavior provided a foundation for research that is aimed at characterizing the sequence of behavioral events that characterize the transition from the nonsexual to the sexual state. As noted earlier, this sequence of events involves appetitive and consummatory behaviors. This transition period can be thought of as a cascade of behavioral events that begins with initiation of approach behavior and ends with completion of copulation. Figure 13-8 illustrates a common sequence of events that is associated with copulation in mammals. This model divides sexual behavior into several phases including: 1) courtship, 2) precopulation, 3) mounting, 4) intromission, 5) ejaculation, and 6) postejaculatory. The courtship and precopulatory behaviors are generally regarded as appetitive behaviors, whereas mounting, intromission, and ejaculation are usually classified as consummatory behaviors. The major problem with this account of sexual behavior is that it is overly simplistic and fails to capture the interactive nature of sex. Moreover, it fails to account for the fact that appetitive and consummatory behaviors are not necessarily distinct categories. The transition from one behavioral phase to the other is more subtle and involves behaviors that are truly transitional in nature; that is, they fit into both categories. For example, once the male rat

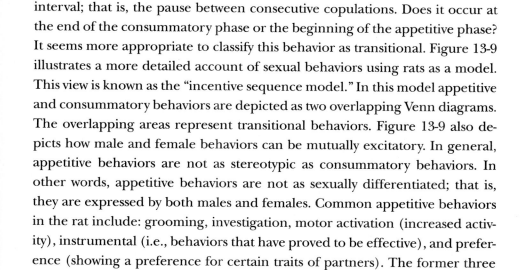

FIGURE 13-8

Sequential model of sexual interactions between males and females before, during, and after copulation.

comes into close contact with a female rat, the female begins solicitation and copulation quickly follows. It is difficult to classify solicitation as either appetitive or consummatory. On the other extreme is the so-called postejaculatory interval; that is, the pause between consecutive copulations. Does it occur at the end of the consummatory phase or the beginning of the appetitive phase? It seems more appropriate to classify this behavior as transitional. Figure 13-9 illustrates a more detailed account of sexual behaviors using rats as a model. This view is known as the "incentive sequence model." In this model appetitive and consummatory behaviors are depicted as two overlapping Venn diagrams. The overlapping areas represent transitional behaviors. Figure 13-9 also depicts how male and female behaviors can be mutually excitatory. In general, appetitive behaviors are not as stereotypic as consummatory behaviors. In other words, appetitive behaviors are not as sexually differentiated; that is, they are expressed by both males and females. Common appetitive behaviors in the rat include: grooming, investigation, motor activation (increased activity), instrumental (i.e., behaviors that have proved to be effective), and preference (showing a preference for certain traits of partners). The former three behaviors are sometimes classified as excitement or anticipatory behaviors, whereas the latter two are sometimes called preparatory behaviors. Consummatory behaviors expressed by rats are sexually differentiated, but are common to most mammals. Transitional behaviors include those that can occur during the transition from the appetitive and consummatory phases, or between the transition from consummatory to appetitive phases.

Human sexual behavior can also be understood in terms of this incentive sequence model (Figure 13-10). As noted earlier, there are few if any

FIGURE 13-9

The "incentive sequence model" of sexual behavior as applied to rats. This model classifies sexual behaviors as appetitive and consummatory and acknowledges that some sexual behaviors are transitional in the sense that they are expressed between periods of copulatory activity. Redrawn from Pfaus (1996).

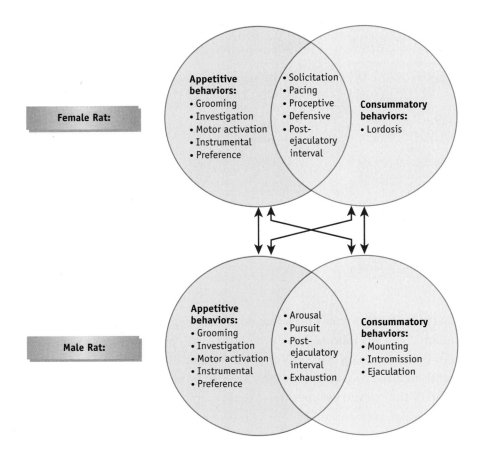

differences in sexual behaviors expressed by men and women. Nevertheless, human sexual behaviors can be classified into appetitive, transitional, and consummatory. There have been no attempts to classify the sexual behaviors of livestock in terms of this model. However, it seems reasonable that this model could be easily applied to these species.

PHYSIOLOGY OF SEXUAL BEHAVIOR

The ability to execute behaviors directed toward copulation in response to appropriate sexual stimuli is dependent on the presence of a particular neuronal circuitry as well as the ability of these neuronal circuits to become activated. For example, the behaviors required for a bull to approach and copulate with an estrual cow are dependent on neuronal pathways that convey information about the cow as well as those that control erection, mounting, intromission, and ejaculation. Comparable neuronal pathways are necessary for the cow to interact sexually with the bull. A comprehensive understanding of these physiologic aspects of behavior requires an appreciation of how the brain develops

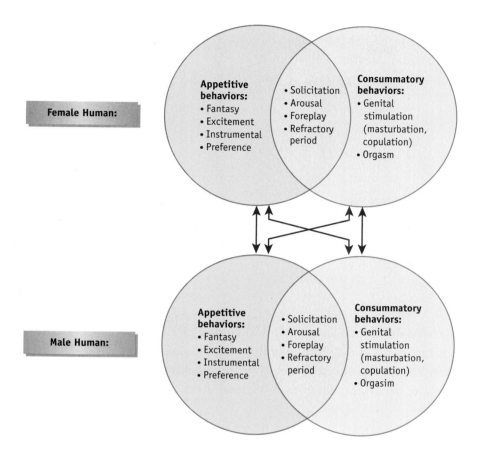

FIGURE 13-10

The "incentive sequence model" of sexual behavior as applied to humans. Note that the appetitive, transitional, and consummatory behaviors of men and women are the same. Redrawn from Pfaus (1996).

the neural structures necessary for sexual behavior as well as how these structures function in adults.

Sexual Differentiation of the Brain

The fact that males and females express different types and patterns of sexual behavior suggests that the neuronal structures regulating these behaviors are also different. An extensive amount of research done in rodents suggests that at some time during development, the nervous system becomes sexually differentiated via mechanisms that are similar to those controlling sexual differentiation of the genital organs. In other words, sexual differentiation of the nervous system depends on the actions of gonadal hormones (see Chapter 3). The sexual differentiation of the brain is expressed both behaviorally and physiologically. Behavioral differences were described in earlier sections of this chapter. Physiologic differences include mechanisms controlling release of reproductive hormones, particularly LH. Some of the best-documented examples of

13

TABLE 13-3 Effects of Neonatal Castration and Steroid Replacement on Sexual Behavior and Patterns of LH Release in Laboratory Animals[1]

Type of Animal	Sexual Behavior as Adult[2]	Adult Pattern of LH Release[2]
Non-treated Male	Masculine	Tonic
Non-treated Female	Feminine	Tonic + Surge
Castrated Female	Feminine	Tonic + Surge
Female + Testosterone	Masculine	Tonic
Castrated Male	Feminine	Tonic + Surge
Castrated Male + Testosterone	Masculine	Tonic

[1]Treatments imposed within 7 days after birth.
[2]In the presence of appropriate gonadal steroids.

sex-related differences in LH release include 1) ability of estradiol to induce an LH surge in rats and 2) timing of the prepubertal increase in pulsatile LH release in sheep. Our investigation of the sexual differentiation of the brain will emphasize behavioral differences between males and females.

Hormonal Control

Early studies of the sexual differentiation of the genital organs provided the basis for investigating the sexual differentiation of the brain. An important underlying assumption of these studies is that gonadal steroids play a central role in regulating the sexual differentiation of the brain. Table 13-3 provides a summary of results from these early studies which were done in laboratory animals (rats and guinea pigs primarily). It is clear from these results that gonadal steroids influence development of sexual behavior. Removal of the testes in males during the early postnatal periods leads to expression of feminine behaviors and patterns of gonadotropin release (tonic or pulsatile as well as surge modes) during adulthood. In contrast, exposure of females to testosterone during the same periods leads to expression of masculine behaviors and gonadotropin patterns (tonic only) during adulthood. The hypothesis that emerged from these studies is that the early brain is indifferent or inherently female and that exposure to testosterone at a critical time in development induces structural changes that are necessary for expression of male behavior. Testosterone can bring about these effects by masculinization (acquisition of behaviors characteristic of males) and defeminization (suppression of behaviors characteristic of females). According to this view, the absence of androgens causes the brain to feminize. The extent to which feminization requires estrogen is unclear. Male rats that are castrated neonatally (during a critical period of sexual differentiation) will express lordosis as adults. Complete development of the female behavior also requires exposure to estrogens early in life.

Although it is well-accepted that gonadal steroids play important roles in the sexual differentiation of the brain, other nonendocrine mechanisms play a role as well. These mechanisms involve neural expression of genes located on the X and Y chromosomes. For example, sexually differentiated patterns of cell migration within the central nervous system occur early in embryonic development before these cells express receptors for gonadal steroids.

The results presented in Table 13-2 reflect the organizational and activational actions of gonadal steroids that were described previously. The neonatal effects of testosterone in males are organizational in the sense that they influence the organization of neuronal structures that regulate sexual behavior in the adult. In contrast, the sexual behaviors and LH patterns expressed in adult rats require exposure to androgens and estrogen, respectively.

Compared to the rat, our knowledge of the sexual differentiation of the brain in other mammalian species is incomplete. In all cases, there is anatomic, physiologic, and behavior evidence to suggest that the central nervous system is sexually differentiated. Major differences are associated with which traits are different as well as the degree of difference between males and females. For example, regulation of gonadotropin secretion does not appear to be sexually differentiated in primates; that is, both male and female rhesus monkeys will express cyclic gonadotropin patterns characteristic of the ovarian cycle when provided with ovarian transplants or given injections of estradiol and progesterone in patterns that mimic the ovarian cycle. The extent to which sexual behavior is differentiated in primates is unclear. Most of the data concerning this issue has been derived from individuals that express disorders that affect the production and/or response to gonadal steroids. Moreover, the fact that the sexual behaviors of some primate species (e.g., humans) are not as stereotypic as those observed in the rat makes it difficult to clearly differentiate masculine and feminine behaviors. Nevertheless some clinical information from human patients has been interpreted to support the hypothesis that testosterone has a masculinizing effect on sexual behavior. For example, men who are androgen insensitive develop female gender identities, whereas women exposed to high levels of androgen during development have a tendency towards a masculine identity. Such results should be viewed with caution since sexual identity in humans is a complex trait and is likely influenced by social as well as developmental factors.

13

Our understanding of the sexual differentiation of the brain in domestic livestock is superficial compared to laboratory rodents. The major difficulty associated with studying sexual differentiation in livestock is that a significant portion of sexual differentiation occurs prenatally (Recall that the critical period of sexual differentiation in the rat and guinea pig occurs during the first few days after birth). Therefore, it is extremely difficult, if not impossible, to study the effects of gonadectomy on development of sexual behavior in larger animals. However, it is possible to treat pregnant females with testosterone and study its effects on the behavioral development of their offspring. Much of our understanding of the hormonal control of sexual differentiation of the brain in these species is based on this experimental approach.

Normal adult ewes appear to retain the potential to express either male or female behaviors. In addition to expressing estrous behavior in response to estrogen, they will show strong masculine behaviors when treated chronically with testosterone. In contrast rams express only masculine behaviors and will not express estrous behavior when treated with estradiol. Ewes that were exposed to androgens prenatally do not express estrus as adults. These results support the hypothesis that testosterone promotes development of male behavior by defeminizing as well as by masculinizing neural structures that control sexual behavior.

Sexual differentiation of the brain in cattle and swine appears to involve defeminization more than masculinization. Cows and sows will express male mounting behavior when treated chronically with testosterone. Moreover, cows routinely display mounting behavior during estrus. In contrast neither bulls nor boars express estrous-like behavior when treated with estradiol. One plausible explanation is that the females of these species retain the neural structures necessary for both male and female behaviors, whereas males lose the ability to express female behaviors due to defeminization. Presumably the trigger for defeminization is prenatal exposure to testosterone.

Critical Periods

As noted earlier, gonadal steroids exert their organizational effects at certain critical periods during development. In rodents, the critical period for testosterone-induced masculinization of the brain is between embryonic day 18 (E18) and up to 30 days after birth, depending on the trait studied. Testosterone-induced defeminization of behavior can occur from E18 up to 7 days postpartum. With respect to masculinization, there may be two critical periods for sensitivity to testosterone, one spanning the neonatal and prepubertal periods and another coincident with onset of puberty. Each of these critical periods coincides with surges of testosterone production. The male rat experiences surges in circulating concentrations of testosterone at E18 and on the day of birth. A third surge of testosterone occurs at puberty. Recent evidence suggests that testosterone produces organizational effects on the brain of male rats throughout adolescence. Castration of male rats after the neonatal period, but before onset of puberty permanently impairs expression of sexual behavior.

A critical period for sexual differentiation of the human brain has not been precisely identified. Regions of the brain that show sexual dimorphisms do not become fully differentiated after birth (in some cases not until adulthood). Gender identity is believed to become established within the first year of life, but it is unclear if this is related to particular structural changes in the brain.

It is generally agreed that the critical period for sexual differentiation of the brain in mammals with long gestation lengths occurs during the prenatal phase of development. This appears to be the case for cattle, and sheep, but not for swine. The critical period for masculinization and/or defeminization in cattle may be between days 80 and 100 of the 284-day gestation period. This is a time when androgen concentrations increase in male fetuses. Ewes that were exposed to androgens between days 50 and 80 of pregnancy (gestation length = 180 days) do not show sexual receptivity suggesting that a critical period of sexual differentiation occurs during this time window. In swine, there is little evidence to support the idea of a prenatal critical period. Sows exposed to androgens from days 29 to 35 or days 39 to 45 of the 113-day gestation period express normal estrous cycles as adults. Moreover, boars that are castrated before 2 months of age express sexual behaviors similar to those expressed by females. Taken together these data support the idea that a critical period for de-feminization of the brain occurs postnatally in male pigs.

Hormonal Control

Although surges of testosterone production coincide with critical periods of masculinization and defeminization of the brain, testosterone is not the principal hormone responsible for these changes. Estradiol, a metabolite of testosterone, is responsible for most of the organizational changes necessary for masculine behavior in most mammals. Regions of the brain that are involved with sexual behavior express the aromatase enzyme, which converts testosterone to estradiol. Estradiol then acts via its receptors to induce structural changes in these areas. These effects are mediated by two types of estrogen receptors. The alpha form of the receptor is involved with masculinization, whereas the beta form is involved primarily with defeminization. In addition to these effects, there are a number of regions in the rat brain that become masculinized by testosterone or dihydrotestosterone acting on androgen receptors.

13

Does estrogen have any effect on sexual differentiation of the brain in females? Although much of feminization occurs independent of estrogen, there is evidence to suggest that complete feminization of the rat brain requires estrogens. When an anti-estrogen is administered to female rats during the neonatal period, these animals fail to express normal estrous cycles and sexual receptivity. This raises an interesting question; how can estrogen both masculinize and feminize the brain? The answer may lie in the dosages required to produce these changes. It seems likely that low amounts of estrogen cause the brain to feminize, whereas higher doses cause it to masculinize. Thus sexual differentiation of the brain may be more of a continuum than a dichotomous process.

In species where estradiol serves as the primary masculinizing agent, there are mechanisms that protect the female brain from exposure to high levels of estrogen. For example, throughout the prenatal and prepubertal periods rats produce **α-fetoprotein,** a serum protein which circulates in high concentrations and has a higher affinity for estradiol than for testosterone (Figure 13-11). The protein binds estradiol thereby preventing the hormone from entering the brain. Testosterone, on the other hand, does not bind the protein and freely enters the brain.

In some species (e.g., guinea pig, rhesus monkey, and human) testosterone acts directly to masculinize the brain (i.e., it doesn't have to be converted to estradiol to be effective). In these cases, there are no mechanisms protecting the female brain from estrogens (females produce insignificant amounts of androgen under normal circumstances).

Neuroanatomic Changes Associated with Sexual Differentiation of the Brain

Although the hormonal mechanisms controlling sexual differentiation of the brain have been well documented, we are only beginning to understand where these hormones act as well as what effects these hormones have on neural structures. The search for where steroid hormones act can be limited to those regions of the brain that express high densities of steroid receptors. Of these areas, those that are sexually dimorphic are of particular interest. Although there are as many as 18 brain structures that meet these criteria, the sexually dimorphic nucleus of the pre-optic area (SDN-POA) appears to be particularly important in regulating sexual behavior, at least in rodents. The volume of this area is six times larger in males than in females. This larger size is attributed largely to a difference in number of neurons. The prevailing hypothesis linking this structural difference to the physiologic

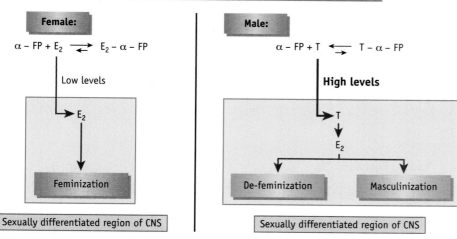

FIGURE 13-11

Mechanism whereby α-fetoprotein (α-FP) protects the female brain from exposure to high levels of the estradiol (E_2) in neonatal female rodents. The affinity of this protein for estradiol is much greater than that for testosterone (T). Therefore, most of the estradiol present is bound and unable to exert effects on development of neuronal systems that regulate sexual behavior. In contrast, most of the testosterone exists in the free form and is available to exert organizational actions on neurons that regulate sexual behavior.

effects of steroid hormones is illustrated in Figure 13-12. Briefly, it is proposed that the masculinizing effects of steroid hormones (e.g., estradiol) prevent apoptosis in SDN-POA neurons thereby causing the volume of this area to be larger in males than females. Exactly how this structural feature is related to expression of behavior is unclear. One idea is that estrogen promotes development of a particular neuronal circuitry that is necessary for expression of male behaviors.

Behavioral Reflexes

In this final section, we return our attention to the actual behaviors expressed before and during mating. However, our focus now will be on the physiologic regulation of such behaviors rather than descriptions of them. It is useful to view sexual behaviors as component of behavioral reflexes which are analogous to the neural reflex arc described earlier in Chapter 6. In other words, various stimuli are detected by sensors that activate afferent

FIGURE 13-12

Proposed mechanism whereby testosterone (T) masculinizes/defeminizes the brain. Testosterone is converted to estradiol (E$_2$), which then interacts with its receptor (ER) to promote expression of genes that regulate neuronal growth. The pattern of neuronal growth resulting from this action is somehow related to expression of masculine sexual behaviors.

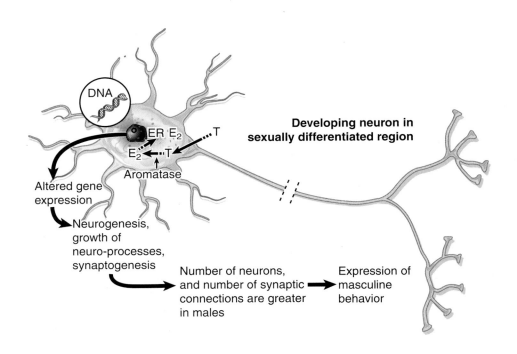

neuronal pathways that impinge upon efferent neurons that induce responses in one or more effector organs.

We have a detailed physiologic understanding of only a few of the behavioral reflexes associated with reproduction. Penile erection and ejaculation have been studies more extensively than any other sexual behaviors. Our most complete understanding comes from studies with men.

Physiologic Control of Penile Erection

Your understanding of erection will be facilitated by reviewing the anatomy of the penis described in Chapter 5. Of primary importance are the penile erectile tissue (corpus cavernosum and corpus spongiosum) and the smooth muscle layers of the arteriolar and arterial walls. During the flaccid state, the smooth muscle cells are tonically contracted, thus restricting blood flow to the erectile tissue. Although blood flow is reduced during the flaccid state it is sufficient to supply nutrients to the penile tissues. During sexual arousal, neurotransmitters are released from efferent nerves that innervate the smooth muscle and induce relaxation of this tissue. This allows blood flow to the penis to increase, leading to the following events that cause erection:

- Due to increased blood flow, blood becomes trapped in the expanding sinusoids of the erectile tissue.

- The venous plexuses that allow blood to flow from the penis become compressed, thus reducing venous outflow.

- The increase in blood volume causes an increase in pressure within the cavernous tissues, causing the penis to elevate.

- Contraction of the ischiocavernosus muscle leads to a further increase in intracavernous pressure causing a rigid erection.

In recent years our understanding of the molecular mechanisms controlling penile erection has been advanced. Figure 13-13 illustrates how the smooth muscle cells of penile smooth muscle are regulated. The tonic contraction of smooth muscle cells during the flaccid state (Figure 13-13A)

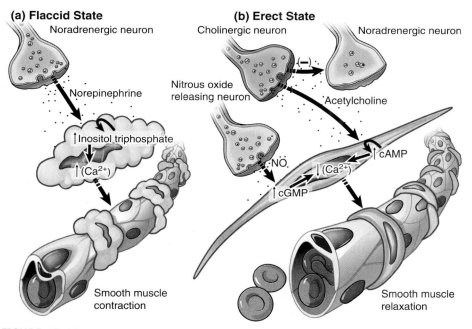

FIGURE 13-13

Schematic illustration showing how penile erection is regulated. Smooth muscle of vasculature supplying blood to the penis is tonically constricted during the flaccid state due to the actions of norepinephrine, a neurotransmitter released by noradrenergic neurons of the sympathetic nervous system (a). In contrast, erection of the penis results from dilation of these blood vessels which is due to activation of cholinergic (parasympathetic) and nitrous oxide (NO)–producing neurons (b). The cholinergic neurons produce acetylcholine, which acts on noradrenergic neurons and smooth muscle cells to counteract the actions of norepinephrine. Nitrous oxide acts on smooth muscle cells via a cyclic GMP (cGMP)-mediated pathway to inhibit contraction.

13

is maintained by norepinephrine, a neurotransmitter released by sympathetic neurons that innervate this tissue. Norepinephrine binds to membrane receptors to enhance production of inositol triphosphate, which acts as an intracellular messenger to increase intracellular concentrations of calcium. Calcium interacts with another protein (calmodulin) which promotes the interaction between actin and myosin that is necessary for contraction. During erection, the mechanisms that promote muscle contraction are suppressed (Figure 13-13B). Of primary importance is the activation of parasympathetic neurons, which release acetylcholine, a neurotransmitter that inhibits the sympathetic neurons that release norepinephrine. The reduction in norepinephrine release removes the stimulus that is necessary to sustain smooth muscle contraction. It has recently become apparent that other neurons play important roles in erection. These neurons innervate smooth muscle cells and release nitrous oxide, which enters smooth muscle cells and promotes the synthesis of cyclic GMP. This second messenger acts on a protein which promotes the uptake of calcium from the cytosol. This decrease in calcium promotes relaxation of the smooth muscle cells.

Physiology of Ejaculation

Although erection and ejaculation are usually thought of as separate processes, the two behaviors are closely related. In fact, it may be useful to view erection as part of the ejaculatory process because it is a necessary condition for ejaculation under normal circumstances. Ejaculation should be viewed as a complex behavior consisting of two components: emission and expulsion. Emission refers to the entry of seminal fluid into the urethra. In contrast, expulsion refers to release of the seminal fluid from the urethra. Ejaculation is typically portrayed as a spinal reflex, and this may be the case in some mammals. However, it is clear that ejaculation includes cerebral components as well, especially in humans. Interestingly it appears that ejaculation in human males can be regulated by both spinal and cerebral reflexes. In normal men, ejaculation appears to be primarily a cerebral process which inhibits the spinal reflex. However, men with spinal cord injuries (i.e., lack descending input from the brain) retain the ability to ejaculate suggesting that they retain spinal ejaculatory centers. Figure 13-14 illustrates the concepts of spinal and cerebral ejaculatory reflexes. In each case stimulation of the penis leads to activation of somatic sensory neurons which enter the central nervous system. In ways that are not fully elucidated, these neurons interact with other nerve cells, leading to activation of autonomic efferents. Sympathetic neurons innervate smooth muscles cells in the accessory sex glands, the ductus deferens, and neck of the

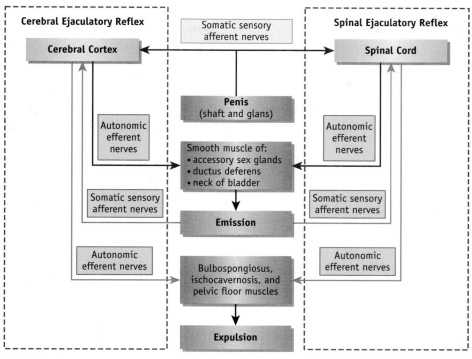

FIGURE 13-14

Schematic illustration of neural reflexes regulating ejaculation in men. Similar mechanisms may be operative in other species of mammals, such as the dog. Ejaculatory behavior appears to be part of two separate neuronal reflexes: one involving the spinal cord and the other involving the brain. In each case, the ejaculatory response can be divided into emission (red lines) and expulsion (blue lines). The reflex begins with stimulation of the penis and induction of afferent nerve impulses. These afferent neurons enter the central nervous system and activate efferent neurons which then innervate the accessory sex glands, ductus deferens, and bladder and promote movement of seminal fluid into the genital ducts (emission). Emission then induces the second phase of the response, which involves contraction of muscles that induce expulsion of semen from the ducts (ejaculation).

bladder and cause these tissues to contract. This produces a two-fold response: 1) entry of seminal fluids into the urethra (emission) and 2) closure of the bladder to prevent retrograde flow of ejaculate into the bladder. The entry of fluids into the urethra stimulates other sensory neurons which enter the central nervous system and ultimately activate parasympathetic efferents that innervate the bulbospongiosus muscle, ischiocavernosus muscle, and muscles of the pelvic floor. These nerves cause rhythmic contractions of these skeletal muscles to expel semen from the urethra (expulsion).

SUMMARY OF MAJOR CONCEPTS

- A sexual behavior is a response of an animal to a stimulus in the context of mating and can be classified on the basis of when the behavior occurs relative to copulation.

- The causes of sexual behavior are ultimate (how the traits of a species allow it to adapt to its environment) and proximate (the physiologic and psychologic mechanisms regulating behavior).

- All mammals express a particular sequence of behaviors leading up to and following copulation and these can be understood as incentive-based behaviors.

- According to the prevailing theory, sexual behavior is the result of an animal having an incentive to engage in sex, as well as a clear depiction of the sexual stimulus. Various regions of the brain regulate the motivation and perception of sex and these areas are sexually differentiated.

DISCUSSION

1. The interaction between mares and stallions can be violent. During estrus, mares will allow a male to approach, but often show defensive behavior toward stallions when they are not in heat. Typically a mare will kick at a stallion when she is not sexually receptive. Based on this observation and your understanding of the causes of sexual behavior, explain why experienced stallions approach mares from the side during the appetitive phase of copulation.

2. Using the incentive sequence model of sexual behavior, categorize the sexual behavior of males and females of your favorite species of mammal.

3. What type of sexual behavior would you expect to observe in an adult female rat that was treated with extremely high doses of estradiol during the first 3 days after birth? Explain your answer.

4. Homosexual behavior is not uncommon in males of many species of mammals. There is intense disagreement over the cause of such behavior. Some argue that it is the result of how an individual brain is structured, whereas others assert that it is a learned behavior. Using your understanding of sexual behavior develop hypotheses to support each point of view.

REFERENCES

Agmo, A. 1999. Sexual motivation—an inquiry into events determining the occurrence of sexual behavior. *Behavioural Brain Research* 105:129–150.

Bindra, d. 1974. A motivational view of learning, performance and behavior modification. *Psychological Review* 81:199–213.

Dean, R.C., M.D. and T.F.Lue. 2005. Physiology of penile erection and pathophysiology of erectile dysfunction. *Urologic Clinics of North America* 32:379–395.

Ford, J.J. and M.J. D'Occhio. 1989. Differentiation of sexual behavior in cattle, sheep and swine. *Journal of Animal Science* 67:1816–1823.

Gorski, R. A. and C.D. Johnson. 1982. Sexual differentiation of the brain. *Frontiers in Hormone Research* 10:1–14.

Katz, L.S. and T.J. McDonald. 1992. Sexual behavior of farm animals. *Theriogenology* 38:239–253.

Kinsey, A.C., W.R. Pomeroy, and C.E. Martin. 1948. *Sexual Behavior in the Human Male*. Philadelphia: W.B. Saunders, pp. 636–659.

McDonnell, S.M. 2000. Reproductive behavior of stallions and mares: comparison of free-running and domestic in-hand breeding. *Animal Reproduction Science* 60–61:211–219.

Morris, J.A., C. L. Jordan and S.M. Breedlove. 2004. Sexual differentiation of the vertebrate nervous system. *Nature Neuroscience* 7:1034–1039.

Motofei, I.G. and D. L. Rowland. 2005. Neurophysiology of the ejaculatory process: developing perspectives. *BJU International* 96:1333–1338.

Owens, I.P.F. 2006. Where is behavioural ecology going? *Trends in Ecology and Evolution* 21:356–361.

Pfaus, J.G. 1996. Homologies of animal and human sexual behaviors. *Hormones and Behavior* 30:187200.

Resko, J.A., A. Perkins, C.E. Roselli, J.A. Fitzgerald, J.V.A. Choate, and F. Stormshak. 1996. Endocrine correlates of partner preference behavior in rams. *Biology of Reproduction* 55:120–126.

Root-Kustritz, M.V. 2005. Reproductive behavior of small animals. *Theriogenology* 64:734–746.

Wilson, C.A. and D.C. Davies. 2007. The control of sexual differentiation of the reproductive system and brain. *Reproduction* 133:331–359.

Wood, R.I. and D.L. Foster. 1998. Sexual differentiation of reproductive neuroendocrine function in sheep. *Reviews of Reproduction* 3:130–140.

13

OVERVIEW OF PREGNANCY

Prior to this point our consideration of mammalian reproduction has been limited to the production gametes and the physiologic and behavioral strategies that ensure mating occurs when a viable oocyte is available. Our major concern in the next three chapters will be with the mechanisms that are directly involved with the creation of new individuals. In other words, we will consider how the male and female gametes come to interact with each other to result in fertilization. In mammals, fertilization occurs within the oviduct. In all but the monotreme mammals, procreation requires development of offspring within the female's reproductive tract; that is, pregnancy. The pregnant state begins with fertilization and ends with parturition (Figure 14-1). For pedagogical reasons it is useful to divide discussions of pregnancy into three major areas of concern. The first area includes the transport of gametes and embryo as well as fertilization. The second includes events that precede implantation: the development of the embryo (embryogenesis), formation of the fetal membranes that will develop into the placenta, and maternal recognition of pregnancy. The third area of concern deals with how the pregnancy is maintained and finally terminated (parturition). This chapter will deal with the first area of concern.

TRANSPORT OF GAMETES AND ZYGOTES

Fertilization is predicated on the production of viable gametes as well as the successful delivery of spermatozoa and oocytes to the oviduct, the site of fertilization. The journey of spermatozoa to the oviduct is long and involves translocation from the testes, through the male genital ducts, and finally through most of the female's reproductive tract. The journey of the oocyte is much shorter; that is, from the ovary into the oviduct. Once fertilization occurs, the embryo is translocated from the oviduct into the uterus where it eventually implants and establishes pregnancy.

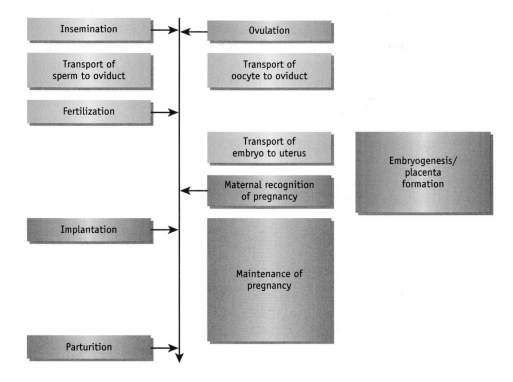

FIGURE 14-1

Sequence of major events during pregnancy grouped according to major aspects of the process: 1) transport of gametes, fertilization, and transport of the embryo (blue), 2) development of the embryo, extraembryonic membranes, and maternal recognition of pregnancy (pink), and 3) maintenance and termination of pregnancy (yellow).

Transport of Spermatozoa Through the Male Reproductive System

As spermatids undergo morphogenesis to form spermatozoa, they remain in pockets of the Sertoli cells and move within the nurse cells toward the lumen of the seminiferous tubule. Such movement is facilitated by tracks formed by microtubules. Once sperm detach from Sertoli cells (spermiation) they enter the lumen of the tubule and are passively transported (at this stage spermatozoa are nonmotile) toward the rete testis. The precise mechanism of tubular transport has not been elucidated, but it is unidirectional. The force that moves sperm through the tubules is the hydrostatic pressure arising from fluid produced by the Sertoli cells.

Once the spermatozoa enter the rete testis, they pass through the efferent ducts into the caput epididymis. This too is a passive process and is largely attributed to the positive pressure of the testes, which results from production of fluid by the seminiferous tubules. Once inside the epididymis, the spermatozoa continue to move through the lumen, due in part to ciliary action of luminal epithelial cells as well as regular peristaltic contractions of the muscularis. Movement along the epididymis is slow, but constant. On average it takes 10 to 12 days for spermatozoa to move from the caput to the cauda epididymis. While spermatozoa move through the epididymis, they undergo maturational changes, including the acquisition of motility.

14

There is a steady trickle of spermatozoa from the cauda epididymis into the vas deferens. These cells enter the urethra and are usually excreted in urine. The release of spermatozoa into the urethra is much more dramatic during ejaculation. As noted in Chapter 13, ejaculation consists of emission and expulsion. During emission, pulsatile contractions of the epididymal muscularis move spermatozoa into the vas deferens and into the pelvic urethra. During ejaculation, movement of spermatozoa along the vas deferens is caused by rhythmic contractions of the ischiocavernosus and bulbospongiosus muscles. These contractions force seminal emissions into the pelvic urethra. As these fluids enter the urethra, secretions from the accessory sex glands are added to the ejaculate. These fluids make up 90 percent of the volume of semen. The total volume as well as the contribution of each gland to semen varies among species. In general, the vesicular glands contribute the most volume followed by the prostate and bulbourethral glands. Because these glands line the pelvic urethra in series, ejaculate consists of several fractions. For example, in humans the first fraction contains the highest concentration of sperm cells and primarily fluid secreted by the prostate gland. This portion of semen is rich in citric acid. The second fraction is primarily from the vesicular glands and contains high concentrations of fructose. Overall, the major difference between semen and other body fluids (e.g., blood plasma) is that semen contains high concentrations of citric acid, fructose, and numerous other compounds including some proteins unique to the prostate gland. The functions of these compounds are poorly understood. However, it is likely that they create an environment that supports sperm cells once they are released into the female's reproductive tract.

Transport of Spermatozoa Through the Female Reproductive System

During mating, the ejaculated semen is normally deposited into the anterior vagina (e.g., humans, cattle, sheep, and rabbits) or uterus (e.g., horses, pigs, and rodents), depending on the species.

In either case, spermatozoa face a series of barriers that must be successfully negotiated in order for fertilization to occur. Table 14-1 shows the number of spermatozoa in ejaculates and the site of semen deposition for several species of mammals. Several million sperm are normally deposited in the female reproductive tract. Most of these are lost at various locations in the female genital ducts, leaving only several hundred sperm at the site of fertilization in the oviduct (Figure 14-2). Only the most viable sperm complete the journey to the oviduct. Many die due to the hostile conditions of the

TABLE 14-1 Number of sperm ejaculated, site of sperm deposition, and number of sperm reaching the oviduct in several species of mammals[1]

Species	Number of Sperm in Ejaculate (millions)	Site of Deposition (Natural Mating)	Number of Sperm in Oviduct
Rat	58	Uterus	500
Rabbit	280	Vagina	250–500
Cattle	3000	Vagina	<100
Sheep	1000	Vagina	600–700
Swine	8000	Uterus	1000
Human	280	Vagina	200

[1]From Austin and Short (1982)

FIGURE 14-2

Schematic illustration summarizing the fate of sperm after entering the female reproductive tract. Sperm are propelled in an anterograde direction due to contractions of the female reproductive tract. However, many are lost due to retrograde flow and absorption.

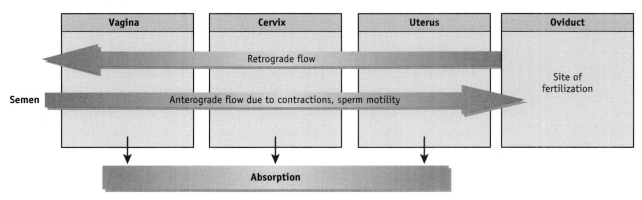

female's reproductive tract. These nonviable sperm are removed from the female reproductive tract via retrograde (outward) flow as well absorption following degradation by leukocytes.

Sperm in the Vagina

The vaginal environment is extremely hostile to sperm. This is due to its low pH (less than 5.0) that is produced by the metabolic activity of a lactic acid-forming bacterium. Although seminal plasma may provide some buffering capacity, this effect is only temporary. Thus survival of sperm requires that they move rapidly into the cervix. A large percentage of ejaculated sperm are

14

lost from the female reproductive tract via retrograde flow. Although much of this seems to be to outflow from the vagina, such loss can also occur following insemination into the uterus.

Sperm Transport through the Cervix

Contractions of the vagina and uterus appear to play important roles in transporting sperm through the female genital ducts. Spermatozoa have been found in the oviducts within 5 minutes after insemination; a response that can not be attributed to sperm motility alone. Such movement can be attributed to a negative vaginal pressure as well as an increase in vaginal and uterine contractions which occur following coitus. Together these responses propel sperm from the vagina into the cervix and through the uterus. This so-called rapid phase of sperm transport may not be too important with respect to fertility. The vast majority of sperm that reach the oviduct do so via a slow phase, which requires several hours. The cause of the postcoital contractions is unclear, but some evidence suggests that they might be induced by constituents of the seminal plasma (e.g., prostaglandins).

The cervix acts is a major barrier to sperm transport in cases of intravaginal insemination. In rabbits, less than 2 percent of the sperm deposited in the vagina during mating could be recovered in the uterus within 12 hours. With respect to sperm transport, this organ is best understood as a filter. Recall that the lumen of the cervix is highly convoluted and consists of many crypts. In addition, the cervical epithelium produces copious amounts of mucus which flows in a retrograde direction. Progressively motile sperm enter the cervix and become lodged in the crypts as they make their way through the lumen. In this way, viable sperm are physically protected from attack by marauding leukocytes. In contrast, less viable and dead sperm are carried out of the cervix via the mucus and are more prone to phagocytosis by leukocytes.

Whereas contractions of the vagina and uterus play a major role in the rapid transport of sperm through the cervix, progressive motility appears to be a major factor in the sustained or slow transport phase. The ability of spermatozoa to move through the cervix is directly dependent on the nature of the cervical mucus. This cervical gel consists of two major elements: a glycoprotein-rich fraction called mucin (40 percent) and an aqueous phase containing various soluble components consisting of inorganic salts, proteins, and low-molecular weight organic compounds (e.g., simple sugars, amino acids, and lipids). The consistency of the mucus gel changes with the stage of the ovarian cycle in response to changing concentrations of ovarian steroid hormones. During the peri-ovulatory period, when estradiol concentrations are high, the gel consists of 95 percent water and has a low viscosity. In this form, the mucin molecules line up in parallel chains (micelles). The aqueous spaces between these fibrils allow sperm cells to pass. However, during the luteal phase, when progesterone concentrations are high, the water content drops to 90 percent causing the fibrillar structure to disappear and raise viscosity of the gel. This latter structure is not compatible with sperm migration.

Spermatozoa are not evenly distributed within the lumen of the cervix. Most sperm are found near the mucosa. Those that migrate through the cervix toward the uterus occupy the crypts. Apparently the mucus produced in the base of the crypts is of lower viscosity than that produced in the apical regions. Thus sperm tend to migrate within these low-resistance pathways. The gradual emergence of sperm from the crypts into these "privileged pathways" is believed to account for the sustained release of spermatozoa from the cervix. The extent to which the cervix plays a role of a reservoir for sperm is unclear.

Sperm Transport through the Uterus

The morphology of the uterus doesn't impede migration of sperm in the way the cervix does. Nevertheless, the uterus is a barrier to sperm transport in the sense that these cells must cross the lumen of this organ in order to be transported to the oviducts. As noted earlier, uterine contractions play a much more important role in movement of spermatozoa through the uterus than sperm motility. This conclusion is based on the fact that 1) there is no known mechanism that ensures that sperm will move only toward the uterotubal junction, 2) uterine motility increases around the time of mating, and 3) inert particles are transported across the uterus when its motility is high (as it is around the time of mating). Once sperm enter the uterus, they are suspended in fluid secreted by the uterine mucosal cells. In addition to serving as a transport medium, this fluid provides support for the sperm as well as stimulates sperm activity.

Spermatozoa are found in the uterus up to 24 hours after insemination, but are not typically present thereafter. Most of the sperm move to the oviduct within a few hours after insemination (e.g., 6 to 8 hours in sheep and cattle). Those that remain in the uterus after this time are likely phagocytized by leukocytes that infiltrate the uterus within 4 hours after insemination. Interestingly, dead sperm and other debris are removed from the uterus via transport to the oviducts.

Sperm Transport Through the Uterotubal Junction

The final barrier to sperm transport is the uterotubal junction. This is the major barrier for sperm migration when insemination occurs within the

14

uterus (e.g., pig, rodents). This structure is also a significant barrier for species in which intravaginal insemination occurs. The mechanism whereby the uterotubal junction impedes sperm migration is virtually unknown. Its ability to impair sperm transport does not appear to be related to its structure. Although the lumen of this region is tortuous, the junction does not contain a sphincter or valve. Research in some species supports the idea that sperm motility plays an important role in transport through this region.

Sperm Transport Through the Oviduct

Sperm migrate along the length of the oviduct and those that are not lost enter the peritoneal cavity via the infundibulum. Oviductal transport of sperm occurs in two ways. First, those that enter the oviduct shortly after mating move rapidly to the ampullary region. The second phase of transport is much slower. As spermatozoa pass through the uterotubal junction they accumulate in the proximal isthmus of the oviduct forming a reservoir. As sperm enter the isthmus, they become immotile and adhere to the epithelium, possibly due to the effects of mucus-rich fluid that is produced in this area. After a period of several hours, the sperm gradually regain their motility, break away from the epithelial cells, and resume their journey. As they leave the isthmus, spermatozoa become hyperactive and move quickly into the ampullary region. Whatever the cause of this delay in oviductal transport of sperm, it seems that this phenomenon is an important prerequisite for fertilization. For example, in sheep and cows, an accumulation period of 6 to 8 hours is necessary to achieve good fertilization rates.

Hyperactivity of sperm may be necessary for them to break away from the isthmus and resume their migration. However, this alone is not sufficient for transport into the ampulla. The oviduct itself appears to play an important role. In order to understand how the oviduct affects sperm transport it is necessary to review its morphology. The isthmus is a narrow, thick-walled segment containing a well-developed muscularis. The mucosal cells are not ciliated, but contain many secretory cells, which produce a thick, mucus-containing fluid. The anterior region of this segment opens into the larger ampullary region. This is the longest section of the oviduct and is thin walled with a thin muscularis. In addition, the mucosal epithelium consists of both secretory and ciliated cells. Secretions in the ampulla contain more water and are therefore less viscous than those produced in the isthmus. The most anterior segment is the infundibulum, which, like the ampulla, has a thin wall. The infundibulum is a funnel-shaped structure. The edges of the widest portion have a lacey appearance and the walls are densely folded (fimbriae). The mucosal cells are heavily ciliated throughout the infundibulum.

The oviduct effects anterior movement of sperm cells in the following way. Fluids produced by the anterior oviduct are pushed toward the uterus due to the beating of cilia. However, this fluid does not enter the uterus because the thicker oviductal fluid and the narrower lumen in the isthmus create resistance, which reverses the flow of fluids. Thus sperm transport is facilitated by a current, which moves fluid toward the infundibulum. The muscularis of the oviduct also plays a role in sperm transport. The oviduct exhibits peristaltic contractions at most times during the ovarian cycle. However, during the ovulatory period when estradiol concentrations are elevated, these contractions become more regular and are directed toward the ovary. The major consequence of this action is a mixing of oviductal contents, which might enhance the probability that sperm cells will encounter the oocyte.

Transport of the Oocyte

There are two main features of oocyte transport. The first deals with the "pick up" of the oocyte by the oviduct following ovulation. The second deals with migration of the oocyte to the site of fertilization.

The ovulated oocyte is embedded in a mass of granulosa cells (Figure 14-3). Cells of the corona radiata are attached to the zona pellucida. This layer of follicular cells is surrounded by a gelatinous matrix containing

FIGURE 14-3

Schematic illustration of the ovulated oocyte encased by a gelatinous matrix and granulosa cells of the cumulus oophorus.

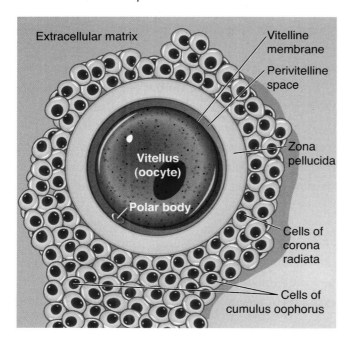

14

numerous cells of the cumulus oophorus. This "oocyte-cumulus complex" appears to be necessary for pick up and transport by the oviduct. Once ovulated, the complex of cells is translocated across the fimbriae and into the oviduct through the **ostium** (opening into the infundibular oviduct). This involves four mechanisms. First, negative pressure resulting from contractions of the muscularis of the oviduct may create a suction that draws the oocyte into the oviduct. Second, rhythmic contractions of smooth muscle of the ligament supporting the oviduct and ovary may cause the fimbriae to make contact with the ovary. Third, the cilia of the fimbriae make contact with the oocyte-cumulus mass and push it toward the lumen of the oviduct. Fourth, strands of oviductal mucus appear to make contact with the ovary during ovulation and may help guide the oocyte toward the oviduct.

Once the oocyte enters the oviduct it migrates toward the uterotubal junction. The time required to complete the journey from the fimbriae to the uterus averages 3 to 4 days in most species. Fertilization normally occurs in the ampullary-isthmic junction and does not appear to disrupt transport in most species. Transport of the oocyte or zygote through the oviduct is a complex process that is poorly understood. The extent to which oviductal cilia play a role in this process is unclear. It appears likely that contraction of the circular smooth muscle of the oviduct together with the flow of oviductal secretions interact to regulate movement of the oocyte and zygote through the duct towards the uterus. Movement of the oocyte or zygote appears to be almost random, but contractions of the oviduct seem to promote a directional flow towards the uterus. However, as the oocyte approaches the isthmus, its movement is impeded by the reverse flow of fluids described previously. The oocyte reaches this site within 24 hours after ovulation, but is retained there before resuming its migration through the isthmus. Resumption of transport may be related to the increase in diameter of the isthmus and uterotubal junction which occurs as estradiol levels fall and progesterone concentrations increase following ovulation.

The oocyte loses its complex of follicular cells by the time it enters the ampullary-isthmic junction. The timing of this process varies greatly among species. In most mammalian species the cumulus oophorus is present at the time of fertilization (e.g., primates). In others (e.g., cattle and sheep), the oocyte is encased only by the zona pellucida at the time of fertilization.

GAMETE MATURATION

The aforementioned mechanisms governing gamete transport permit sperm to make contact with an oocyte. However, this alone is not a sufficient condition for successful fertilization. It is also necessary for the gametes to be fully mature. This means that they are capable of interacting in a way that results in syngamy. Maturation of the oocyte typically refers to meiotic maturation; that is, the conversion of a full-grown primary oocyte into a secondary oocyte (unfertilized ovum). This occurs in response to the pre-ovulatory surge of LH and involves both the completion of meiosis 1 and metabolic changes that allow the oocyte to become fertilized and develop further. Unlike the oocyte, spermatozoa are not fully mature upon leaving the gonad. As noted earlier, sperm mature to the point where they are motile and have the potential to fuse with an oocyte. However, the capacity to fertilize isn't gained until the sperm resides in the female reproductive tract for several hours.

Capacitation of Spermatozoa

The physiologic changes that make spermatozoa capable of fertilizing an oocyte are referred to collectively as **capacitation.** This phenomenon is associated with two major changes in the physiology of spermatozoa (Figure 14-4). First, capacitation permits induction of the **acrosome reaction,** which is necessary for a sperm to penetrate the zona pellucida and ultimately fuse with the plasma membrane of the oocyte. Second, capacitation causes hyperactivation of the sperm. When this occurs the sperm becomes highly motile due to a frantic whipping motion of its tail. This may provide the thrust necessary to penetrate the zona pellucida.

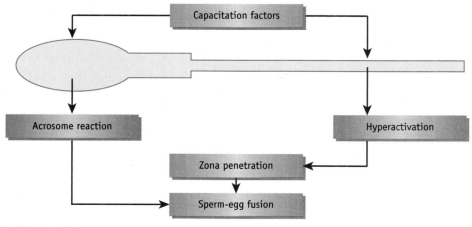

14

FIGURE 14-4

Schematic illustration summarizing major consequences of sperm capacitation. Various chemical factors secreted by the female reproductive tract activate the sperm rendering it capable of recognizing and binding to an oocyte as well as inducing hyperactivity. Both responses are prerequisites for fertilization.

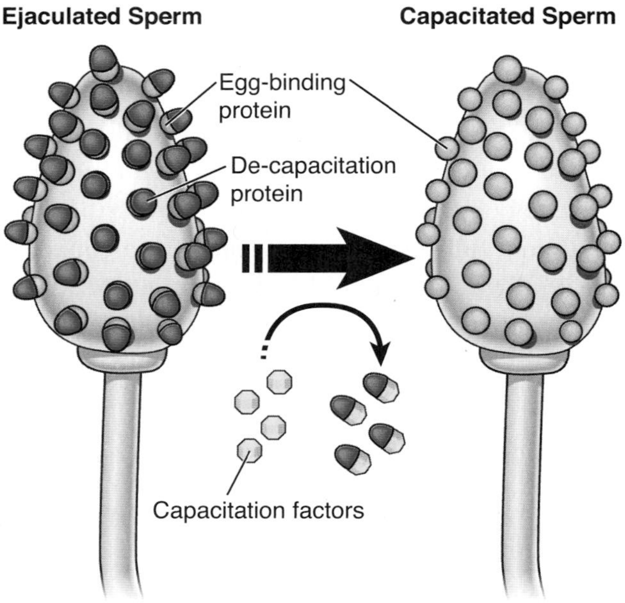

FIGURE 14-5

Schematic illustration of the currently accepted hypothesis for sperm capacitation. The surface of the head of ejaculated sperm are coated by de-capacitation proteins which mask egg-binding protein. Capacitation factors produced by the female reproductive tract interact with the sperm membrane and strip it of the de-capacitation proteins. This exposes regions of egg-binding protein that interact with proteins protruding from the surface of the zona pellucida.

The molecular basis for capacitation is poorly understood. Nevertheless, there is general agreement that capacitation involves changes in the plasma membrane of the sperm. One change that seems to be especially important is the removal or alteration of protein molecules that stabilize or mask membrane proteins that are necessary for the sperm cell to interact with an oocyte. Likewise, structural changes in the membrane in the tail region may be responsible for hyperactivation. Figure 14-5 illustrates the prevailing hypothesis of capacitation. Certain egg-binding proteins located on the surface of the plasma membrane of sperm have been shown to interact with proteins located on the zona pellucida. Prior to capacitation, these surface proteins are covered by so-called decapacitation factors, or protein constituents of seminal plasma. Solutes in fluids produced by the female reproductive tract strip the de-capacitation factors away from the sperm membrane, allowing them to interact with zona pellucida proteins, which in turn induce the acrosome reaction. Another effect of the capacitation-inducing compounds may be to alter the lipid composition of the plasma membrane, thereby altering the disposition of proteins involved with capacitation.

The identity(ies) of the chemical(s) that induce capacitation remain(s) unknown. However, they are presumed to be constituents of fluids secreted by the cervix and/or oviduct. In species where sperm are deposited in the uterus, the oviduct is clearly the site of capacitation. Recall that sperm accumulate in the caudal isthmus before moving on. In species where insemination occurs in the vagina, capacitation may occur in the cervix as well as in the oviduct. Due to variations in sperm characteristics as well as migration rate through the female tract, spermatozoa are capacitated at different times. This helps maintain a steady supply of capacitated sperm over several hours. Oviductal fluids are effective in inducing capacitation of sperm in vitro, as are artificial media designed to mimic the fluid produced by the oviduct. However, no single constituent appears to be *the* capacitation factor. Interestingly, production of capacitation factors doesn't appear to be organ specific.

FERTILIZATION

The site of fertilization is the ampullary-isthmic junction of the oviduct. Only a minute fraction of the spermatozoa ejaculated into a female's reproductive tract reach the oviduct, and only a small percentage of these find their way to the site of fertilization. According to some estimates as few as 100 spermatozoa occupy the ampullary-isthmic junction at the time of fertilization. It is unclear whether the meeting of the oocyte with sperm is due to chance, or if this is due to the oocyte producing a chemical agent that attracts spermatozoa (chemotaxis).

Fertilization can be divided into five major steps: 1) passage of the sperm through the cumulus oophorus (in species where this structure is present at the time of fertilization), 2) binding between the spermatozoon and oocyte, 3) induction of the spermatozoon's acrosome reaction, 4) fusion of the spermatozoon and oocyte, and 5) activation of the oocyte. Each of these steps will be considered in the following three sections.

14

Interaction of Spermatozoon with the Cumulus Oophorus

In most mammals, the oocyte is ovulated with its accompanying corona radiata and cumulus oophorus. However, in some ungulates such as the sheep and cattle, these cells are shed at the time of or shortly after ovulation. In the former cases, successful fertilization requires sperm to penetrate the cumulus before interacting with the zona pellucida. Penetrating the cumulus requires capacitation as well as an intact acrosome. Details regarding how sperm pass through this layer of cells is unclear, but the plasma membrane of sperm contain surface enzymes such as hyaluronidase, which lyses hyaluronic acid, a major structural component of the matrix supporting the cumulus cells.

Binding Between Spermatozoon and Oocyte

The next step in fertilization is the binding of the spermatozoon to the zona pellucida of the oocyte. Cell-to-cell adhesions are not uncommon in nature. Some familiar examples include the binding of bacteria and viruses to host cells. The mechanisms that mediate these types of interactions also mediate the interaction between a sperm and the zona pellucida. In all cases, it seems that binding is the result of interactions between complementary proteins located on the membranes of the adhering cells. As described in the previous section, capacitation appears to involve the unmasking of an **egg-binding protein** located on the plasma membrane of the sperm. This protein is a ligand for a sperm receptor located on the surface of the zona pellucida. The zona pellucida is composed entirely of three glycoproteins: ZP1, ZP2, and ZP3. ZP2 and ZP3 are bound by covalent bonds to form long filaments (Figure 14-6). These filaments are cross-linked by ZP1 molecules. It appears that

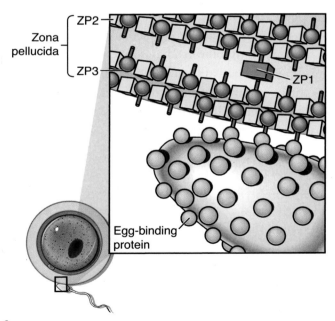

FIGURE 14-6

Schematic illustration depicting the molecular basis of binding between the membrane of the sperm and the zona pellucida surrounding the oocyte. Zona protein (ZP3) acts as a ligand for egg-binding protein, whereas ZP1 and ZP2 act to stabilize the binding of sperm to oocyte.

ZP3 is the ligand for the sperm's egg binding protein, which is analogous to a receptor. This conclusion is based on the fact that disruption of expression of the ZP3 gene causes infertility in female mice.

The identity of the egg-binding protein on the sperm membrane remains elusive. However, the existence of such a protein is supported by the observation that purified ZP3 binds only to the plasma membrane covering the sperm head.

The Acrosome Reaction

You may recall that the acrosome is analogous to a large secretory vesicle that covers the nucleus in the apical region of the sperm's head (Figure 14-7). The outer acrosomal membrane lies just beneath the plasma membrane, whereas the inner acrosomal membrane overlies the nucleus. The acrosome reaction refers to structural changes in this arrangement that occur in

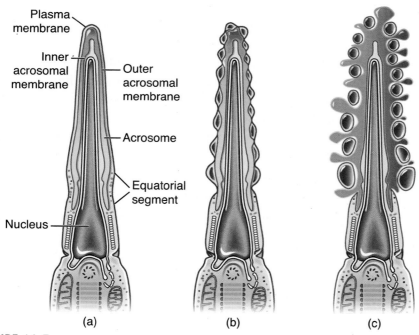

(a)   (b)   (c)

14

FIGURE 14-7

Schematic illustration showing changes in the ultrastructure of the head of a sperm during the acrosome reaction. The plasma membrane of the sperm covers the entire cell and the acrosome resides between the plasma and nuclear membranes (a). Upon interacting with ZP3 of the zona pellucida, the plasma membrane fuses with the outer acrosomal membrane at various sites (b). Soon after fusion of these two membranes, vesicles form and create small channels through which acrosomal materials can escape (c).

response to the sperm binding the zona pellucida. During the reaction, the outer acrosomal membrane fuses with the plasma membrane in multiple locations along the head of the sperm, leading to the formation of numerous vesicles separated by channels through which acrosomal contents can escape. The vesicles are eventually dispersed, leaving only the inner acrosomal membrane overlying the nucleus. The contents of the acrosome include proteolytic enzymes such **acrosin.** These acrosomal lysins dissolve the zona pellucida creating a small hole through which the sperm can pass. At this point the hyperactivity of the sperm cell becomes important. The thrust created by the vigorous beating of the tail propels the sperm through the breach in the zona pellucida, leaving a small slit. The time required for a sperm to penetrate the zona averages between 7 and 30 minutes, depending on the species. Once the sperm enters, the membrane is repaired. In addition, once a sperm binds to the zona pellucida, ZP3 molecules are altered in a way that renders them unable to interact with egg-binding proteins.

At this point you might be wondering how binding of the sperm to the zona pellucida induces the morphologic changes associated with the acrosomal reaction. The best way to understand this process is to view the interaction between the egg-binding protein and ZP3 as analogous to the interaction between a receptor and its hormone. In other words, ZP3 binds to its receptor (egg binding protein) on the sperm membrane and induces intracellular responses that trigger vesicularization of the acrosome (Figure 14-8). Although the ZP3 receptor has not been identified, we

FIGURE 14-8

Summary of the cascade of intracellular events triggered by ZP3 binding to egg-binding protein. Interaction between these two proteins induces production of a second messenger that causes an increase in Ca^{2+}, which induces the acrosome reaction.

14

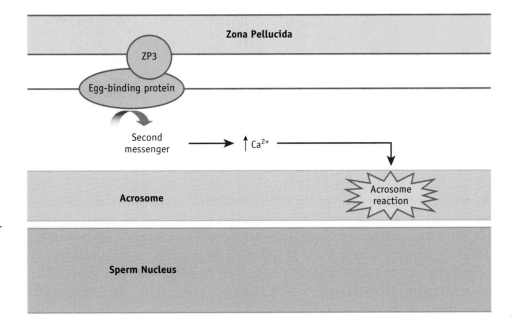

know that its activation leads to an increase in intracellular calcium, which acts as a second messenger to trigger the acrosome reaction. In fact, an increase in calcium appears to be both a necessary and sufficient condition for this process.

Fusion of Sperm and Oocyte

Once the spermatozoon passes through the zona pellucida, it becomes lodged in the perivitelline space, makes contact with the membrane of the oocyte (vitelline membrane), and is eventually engulfed by the oocyte (Figure 14-9). The point of contact on the sperm is in the region of the equatorial segment. The inner acrosomal membrane and the plasma membrane of the sperm adhere to some of the numerous microvilli of the vitelline membrane. This is possible because the acrosome reaction renders the sperm capable of

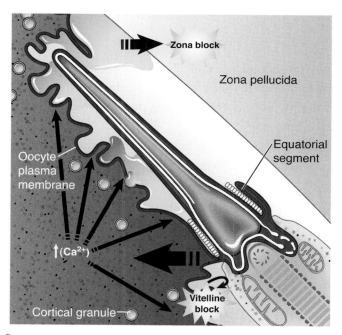

FIGURE 14-9

Schematic drawing of a sperm docking to the vitelline membrane and becoming engulfed by the cytoplasm of the oocyte. Interaction between the membranes of the two gametes induces the cortical reaction; that is, movement of secretory vesicles towards the cortex of the oocyte. The contents of these vesicles acts on both the zona pellucida and vitelline (oocyte) membrane to induce the zona and vitelline blocks, reactions that prohibit additional sperm from penetrating the zona and vitelline membranes.

14

FIGURE 14-10

Schematic drawing of a sperm completely engulfed by the oocyte. As this occurs, the head is severed from the tail, the plasma and nuclear membranes degenerate, and the nuclear contents disperse (decondense).

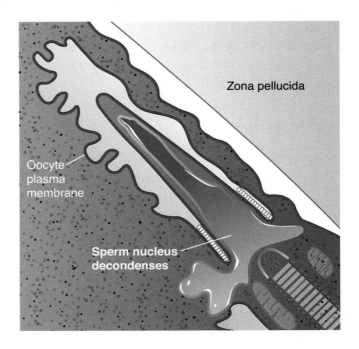

binding to the vitelline membrane. The exact mechanism is unknown, but some scientists hypothesize that dissolution of the acrosome induces changes in the remaining membrane that exposes proteins that allow it to dock with the membrane of the oocyte.

One of the hallmarks of fusion between sperm and egg is the cessation of movement by the sperm tail. Eventually the entire sperm, including the tail, is engulfed by the oocyte (Figure 14-10). The tail, including its mitochondria, degenerates quickly. While this is occurring the membranes in the head of the sperm degenerate and the nuclear contents decondense (disperse).

Activation of the Oocyte

As the sperm fuses with the oocyte, it induces a series of events known as activation. These include 1) the **cortical reaction,** 2) completion of meiosis 2 with extrusion of the second polar body, and 3) formation of pronucleus and syngamy. The timing of these events in the hamster is illustrated in Figure 14-11.

Soon after a sperm cell makes contact with the oocyte membrane, the so-called cortical granules develop along the outer edge (cortex) of the oocyte. These are actually secretory vesicles which release their contents via exocytosis soon after fusion begins (Figure 14-9). Exocytosis is preceded by an increase in intracellular Ca^{2+}, which occurs in response to the sperm binding to the oocyte. The compounds released by the cortical granules act on the zona pellucida and

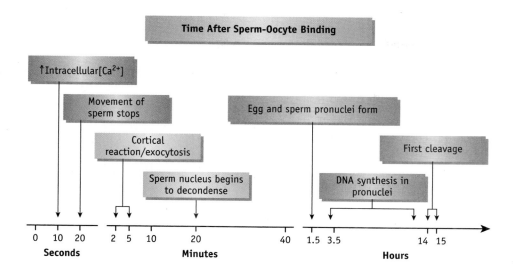

FIGURE 14-11

Timeline showing sequence of events associated with activation of the oocyte following fusion with the sperm cell in hamsters.

the vitelline membrane altering their structure in ways that prevent them from binding other sperm. These changes are known as the **zona reaction** and the **vitelline block**, respectively, and serve the function of preventing multiple fertilizations (i.e., polyspermy). The precise molecular mechanisms responsible for these responses remain unclear, but they appear to involve some sort of masking or chemical alteration of the proteins that interact with sperm proteins. For example, in the acrosome reaction, ZP3 loses its ability to bind to sperm cells.

Once the nucleus of the sperm cell enters the oocyte, it decondenses. In other words, its nuclear membrane disperses and the dense chromatin swells. As sperm cells mature in the epididymis, disulfide cross-links form among various nuclear proteins causing the nuclear material to become insoluble and condense. Once inside the oocyte, these disulfide bonds are chemically reduced, thus dispersing the proteins and liberating the chromosomes. As this occurs a new nuclear envelope organizes to form the male pronucleus. Little is known about the formation of the female pronucleus.

Once the two pronuclei have formed, they merge to form a single nucleus. This process is known as syngamy. Syngamy begins with the apposition of the two pronuclei. This is followed by the disintegration of the apposed membranes, reorganization of chromosomes, and formation of a single nuclear envelope encompassing the male and female chromosomes.

TIMING OF COPULATION, OVULATION, AND FERTILITY

We have encountered two major strategies whereby ovulation occurs harmoniously with delivery of spermatozoa. In the first case, sexual receptivity of females (estrus) coincides with ovulation (most mammals). In the second case,

14

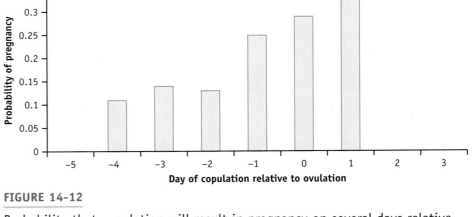

FIGURE 14-12

Probability that copulation will result in pregnancy on several days relative to ovulation in women. The observation that pregnancy is possible within the 5 days preceding ovulation is explained by the fact that sperm remain viable in the female tract for several days. The observation that pregnancy is possible only within 1 day following ovulation is attributed to the fact that the oocyte does not remain viable for more than 24 hours. Data taken from Wilcox et al., (2004).

females are receptive continuously (e.g., most primates). However even with these adaptations there is a large chance that copulation will not result in fertile mating. Some of the variation in fertility can be attributed to the timing of copulation relative to ovulation. The key to successful fertilization is to have an oocyte arrive at the site of fertilization at a time that overlaps with the presence of viable spermatozoa. Sperm that arrive too early are likely to die before the oocyte arrives. On the other hand, an oocyte can lose its viability before the arrival of viable sperm cells. In general, sperm will live in the female tract for only a few days. The life span of oocytes is usually less than 24 hours. Therefore conception is possible only for a short period of time preceding and including the day of ovulation. Figure 14-12 illustrates this concept for humans. You may be surprised by the fact that the chance of pregnancy during this fertile period averages 20 percent. Because this period represents only 21 percent of the menstrual cycle, the average probability of conception for a particular copulation is only about 4 to 5 percent. This might seem very low in light of the exponential growth in global human population discussed in Chapter 1. However, when one considers the fact that the global population of humans includes billions of women of reproductive age, it becomes clear how such a low fertilization rate can result in tremendous

14

increases in population. Consider the following example. There are approximately 80 million females of reproductive age (15 to 55 years of age) living in the United States. Assuming that this pool remains stable and that women engage in regular sexual activity without using birth control, a 5 percent conception rate would mean that 4 million new pregnancies are expected each year. This number is extremely close to the annual birth rate in the United States, which is surprising because an estimated 80 percent of women use some form of birth control in this country. It may be that the actual chance of pregnancy in the absence of birth control is higher than 5 percent because sexual intercourse is not evenly distributed across all days of the menstrual cycle. According to one recent study, women show the highest degree of sexual activity during the middle of the ovarian cycle, a time coinciding with the aforementioned window of fertility. Such a phenomenon might also account for the higher birth rates reported in many developing countries where contraception is not widely available to women.

BOX 14-1   Focus on Fertility: Postcopulatory Sexual Selection

One of the foundational principles of evolution theory is sexual selection; that is, the idea that natural selection promotes an increase in frequency of genes that confer reproductive advantage. Darwin understood this process to be confined to the precopulatory period; that is, selection of a mate with favorable features. This view is based on the assumption that females are predominantly monogamous. During the 1980s the notion of female monogamy was challenged, and now there is wide consensus that females of most animal species copulate with multiple partners **(polyandry).** The idea of female promiscuity raises the possibility that sexual selection can occur following copulation. In other words, it seems plausible that there is "competition" among sperm from different males. Evolutionary biologists have identified two types of postcopulatory selection: 1) competition among sperm of different males to fertilize the oocyte, and 2) the ability of the female to favor the sperm of some males over others (cryptic female choice). The

former idea includes the notion that sperm from one male can disrupt the fertilizing capacity of sperm from another male ("kamikaze sperm"). The latter idea appears to be the most plausible in mammals.

Evidence of postcopulatory selection can be demonstrated by so-called heterospermic insemination experiments. In these studies equal numbers of sperm from two or more males are mixed and inseminated in equal proportions. This approach eliminates sire-related variations in fertilization rate that can be attributed to differences in sperm number and insemination time relative to ovulation. Results of heterospermic insemination studies demonstrate that the sperm of some sires display a fertilization advantage over other sires. Recent research has been devoted to understanding the physiologic basis for such an advantage. The most widely accepted idea is that the female is somehow selecting "desirable" sperm based on certain phenotypic traits that are the products of "fertilizing efficiency genes." Traits that appear to be most important are motility

14

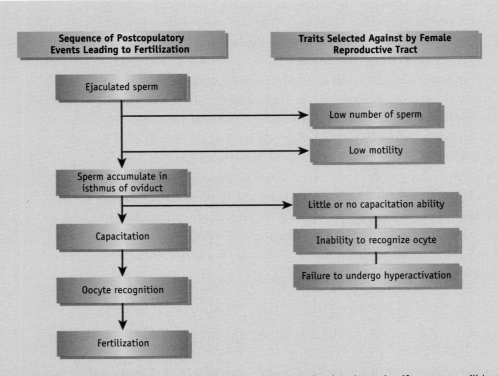

Figure 14-13 Summary of key postcopulatory events (pink) and sperm traits that determine if an oocyte will be fertilized (yellow). Characteristics of the female reproductive tract acts as barriers to sperm therefore creating selection pressure. Under these circumstances males that produce large numbers of motile sperm that readily become capacitated will have a higher reproductive fitness than those that produce fewer sperm with lower motility and low capacitation rates.

and the ability to undergo capacitation (Figure 14-13). The idea of the female "selecting" sperm requires elaboration. As noted earlier in this chapter, the female reproductive tact can be portrayed as hostile to sperm, meaning that it provides various barriers to sperm. Sperm that overcome these barriers are more likely to fertilize oocytes and therefore pass on genes responsible for this fertilization fitness. In addition to favoring genes that regulate sperm physiology, the selection pressure imposed by the female reproductive tract might also favor anatomic, physiologic, and behavioral traits that impact copulation. For example, large testicles can be understood to be adaptations that ensure delivery of large numbers of sperm cells. One of the more unusual hypotheses is that the shape of the penis of some animals is the result of selection pressure that favors efficient delivery of sperm to the female reproductive tract as well as to remove seminal fluid from "competitors." Finally the deep pelvic thrusting of males during copulation has been portrayed as a behavioral adaptation that serves to remove "rival" sperm. Whether or not such ideas have scientific merit remains to be determined.

The theory of postcopulatory selection has been used to explain why there is such tremendous variation in the size, shape, and activity of sperm cells within the animal kingdom. One of the more practical implications of postcopulatory selection theory is the possibility of developing new methods for evaluating the fertility of males for livestock breeding programs. If such tests were developed, the fertility of a sire would be understood in terms of the fertilizing ability of his sperm relative to those of other sires.

SUMMARY OF MAIN CONCEPTS

- The maturation and transport of gametes to the ampullary isthmic junction of the oviduct are necessary conditions for fertilization in mammals.

- Less than 1 percent of the sperm cells deposited in the female reproductive tract appear at the site of fertilization. The majority of sperm cells are lost due to retrograde flow and death.

- Transport of sperm in the oviduct involves movement of oviductal fluids by ciliary movement, peristaltic contractions of the muscularis, and motility of the sperm cell.

- Sperm gain the ability to fertilize an oocyte via the process of capacitation, which unmasks oocyte binding proteins and induces hypermotility of sperm.

- Fertilization is a multi-step process that includes binding of the sperm to the zona pellucida, induction of the acrosome reaction and penetration of the zona pellucida by the sperm, fusion of the sperm with the oocyte, and activation of the oocyte.

DISCUSSION

1. When spermatozoa collected from an ejaculate are added to a Petri dish containing an oocyte suspended in a common cell culture medium there is no fertilization. What is a reasonable explanation for this?

2. Injecting rabbits with certain drugs that induce contraction of smooth muscle, increases the number of sperm cells that are recovered from the oviduct following insemination. Explain how these drugs bring about this effect. Would you expect such treatments to increase fertilization rate (number of embryos per rabbit doe)? Why or why not?

3. Suppose a flock of 100 ewes is inseminated artificially with a mixture of semen from two different rams. The semen samples have equal concentrations of semen from each ram. You expect that the number of lambs from each sire would be equal (about 50 percent). However, you discover that 70 percent of the lambs are the offspring from one ram and 30 percent are from the other ram. Based on your knowledge of sperm transport and fertilization, develop a hypothesis to explain this response.

4. Some researchers have successfully decapacitated sperm cells. What does this mean? How might you go about accomplishing this?

14

REFERENCES

Bedford, J.M. 1982. Fertilization. In: C.R. Austin and R.V. Short, *Reproduction in Mammals, Book 1: Germ Cells and Fertilization,* Second Edition. Cambridge: Cambridge University Press, pp. 128–163.

Birkhead, T.R. and T. Pizzari. 2002. Postcopulatory sexual selection. *Nature Reviews: Genetics* 3:262–273.

Harper, M.J.K. 1982. Sperm and egg transport. In: C.R. Austin and R.V. Short, *Reproduction in Mammals, Book 1: Germ Cells and Fertilization, Second Edition.* Cambridge: Cambridge University Press:102–127.

Harper, M.J.K. 1994. Gamete and Zygote Transport. In: E. Knobil and J.D. Neill, *The Physiology of Reproduction Vol. 2.,* Second Edition. New York: Raven Press, pp. 123–188.

Holt, W.V. and K.J.W. Van Look. 2004. Concepts in sperm heterogeneity, sperm selection and sperm competition as biological foundations for laboratory tests of semen quality. *Reproduction* 127:527–535.

Luke, M.C. and D.S. Coffey. 1994. The Male Accessory Sex Tissues: Structure, Androgen Action, and Physiology. In: E. Knobil and J.D. Neill, *The Physiology of Reproduction Vol. 2.,* Second Edition. New York: Raven Press, pp. 1435–1488.

Wilcox, A.J., D.D. Baird, D.B. Dunson, D.R. McConnaughey, J.S. Kesner, and C.R. Weinberg. 2004. On the frequency of intercourse around ovulation: evidence for biological influences. *Human Reproduction* 19:1539–1543.

Wasserman, P.M. 1999. Mammalian fertilization: molecular aspects of gamete adhesion, exocytosis, and fusion. *Cell* 96:175–183.

Dean, J. 1992. Biology of mammalian fertilization: role of the zona pellucida. *The Journal of Clinical Investigation, Inc.* 89:1055–1059.

Wilcox, A.J., C.R. Weinberg and D.D. Baird. 1995. Timing of sexual intercourse in relation to ovulation: effects on the probability of conception, survival of the pregnancy and sex of the baby. *New England Journal of Medicine* 333:1517–1521.

Yanagimachi, R. 1994. Mammalian Fertilization. In: E. Knobil and J.D. Neill, *The Physiology of Reproduction Vol. 2. Second Edition.* New York: Raven Press:189–318.

14

CHAPTER 15

Establishing Pregnancy: Embryogenesis, Maternal Recognition of Pregnancy, Implantation, and Placentation

CHAPTER OBJECTIVES:

- Describe early embryonic development.

- Describe the development of the extra-embryonic membranes.

- Describe the implantation and formation of the placenta.

- Describe different ways of classifying placentas.

- Describe how the maternal system recognizes pregnancy.

OVERVIEW OF EARLY PREGNANCY

Although pregnancy can be said to begin with fertilization, this process is only one of several necessary conditions for successful production of offspring. In order for pregnancy to continue the zygote must develop into an embryo (embryogenesis), and establish with the mother a physiologic relationship that permits it to develop to the point where it can live outside the mother's body. In order to establish such a relationship, the embryo must achieve two goals: 1) provide a pregnancy recognition signal that causes the uterus to remain in a state that supports pregnancy and 2) develop a placenta that interacts with the uterine endometrium to facilitate exchange of nutrients and wastes with the mother. A comprehensive understanding of early pregnancy requires an appreciation of these important events in the context of both space and time. In other words, it is important to keep in mind *when* these events occur relative to ovulation or fertilization as well as *where* these events take place within the female reproductive tract (Figure 15-1 and Table 15-1).

EARLY EMBRYOGENESIS

We begin our discussion with the presence of a zygote within the isthmus of the oviduct. Our current focus is on how the zygote develops into an embryo that has the ability to generate a pregnancy-recognition signal and then establish a functional interaction with the uterus (Figure 15-2). In most species of mammals, the zygote undergoes a mitotic division within 24 hours after ovulation, creating a 2-cell embryo. A series of mitotic divisions then ensues to create 4-, 8- and finally 16-cell embryos. These early cell divisions are often called cleavage divisions because the

339

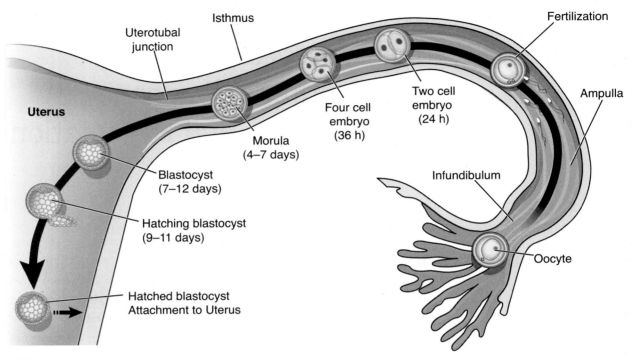

FIGURE 15-1

Overview of major events occurring within the reproductive tract of the cow between ovulation and attachment of the embryo. Note that fertilization occurs near the ampullary isthmic junction of the oviduct and that the embryo undergoes numerous cell divisions as it moves toward the uterotubal junction. By the time the embryo enters the uterus it has developed to the blastocyst stage and hatches from the zona pellucida before attaching to the endometrium.

TABLE 15-1 Timing of several key events during early pregnancy in several species of mammals

Species	Blastocyst stage (days after ovulation)	Arrival in uterus (days after ovulation)	Implantation (days after ovulation)
Cattle	8-9	3-4	17-20
Sheep	6-7	2-4	15-16
Horses	8-9	4-10	28-40
Swine	5-6	2-2.5	11-14
Dogs	5-6	8-15	17-21
Cats	5-6	4-8	13-14
Rodents	3-4	3-4	5-10
Humans	4-5	4-5	6.5-26

FIGURE 15-2

Schematic illustration of early embryonic development showing morphological changes that occur between formation of the zygote and hatching of the blastocyst.

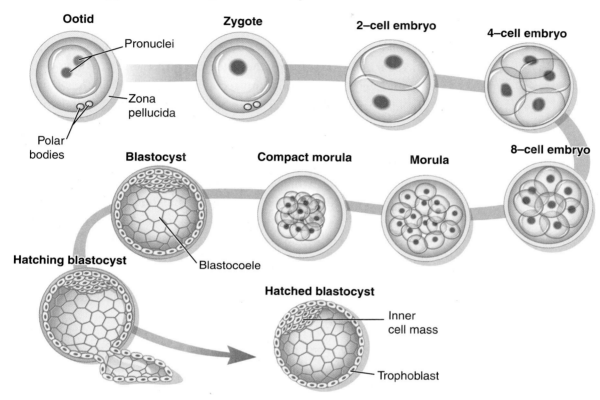

cells divide into two equal halves, each of which has only half the cytoplasm of the parent cell. In other words, the number of cells within the embryo increases with each division, but the overall volume of the structure remains the same as the original zygote. This allows the embryo to develop within the zona pellucida. The cells of the embryo (**blastomeres**) are **totipotent** up to the 16-cell stage. This means that each cell has the ability to give rise to an individual. In fact, it is possible to create identical twins by removing individual blastomeres, inserting each one into a separate zona pellucida and then transferring them to the uteri of host mothers. At the 16-cell stage, the embryo consists of a sphere of densely packed cells known as a **morula** (from the Latin term *morum*, which means mulberry). Formation of the morula is the result of blastomeres undergoing a process called **compaction.** This refers to individual cells merging to form what appears as a single mass of cells. At this point in development, the embryo exhibits the first visible signs of differentiation. The blastomeres on the surface of the mass form tight junctions to create an outer membrane, therefore partitioning the embryo into two major classes of cells: an inner mass of compacted cells

15

known as the inner cell mass and an outer trophectoderm (called the trophoblast later in development) that envelopes the inner cell mass. Formation of the
trophectoderm causes water to be retained within the extracellular space. As fluids accumulate, the inner cell mass is displaced to a polar region of the embryo,
resulting in formation of a fluid-filled cavity called the blastocoele. The cells of
the inner cell mass move as a unit because they are joined together by gap junctions. Once the blastocoele forms, the embryo is referred to as a blastocyst.

Continued cell division along with accumulation of fluids in the blastocoele
causes the blastocyst to expand in size. Much of the increase in cellular mass of
the embryo is attributed to growth of the trophectoderm, which is known as the
trophoblast in later stages of embryogenesis. The cells of this membrane begin
producing and releasing proteolytic enzymes which digest the zona pellucida to
create a small fissure through which the embryo can escape. The emergence of
the embryo from the zona pellucida is known as **hatching.** By the time of hatching, the inner cell mass has differentiated into an outer layer of cells called the
primitive endoderm (or **hypoblast**) and an inner layer of cells called the primitive ectoderm (or **epiblast**). The differentiation of the embryo into embryonic
and extraembryonic tissues is a complex process (Figures 15-3 and 15-4). The
hypoblast grows outward to form a layer of tissue (primitive endoderm) that

FIGURE 15-3

Flow chart depicting
early differentiation
of the blastocyte into
embryonic and extra-
embryonic tissues.

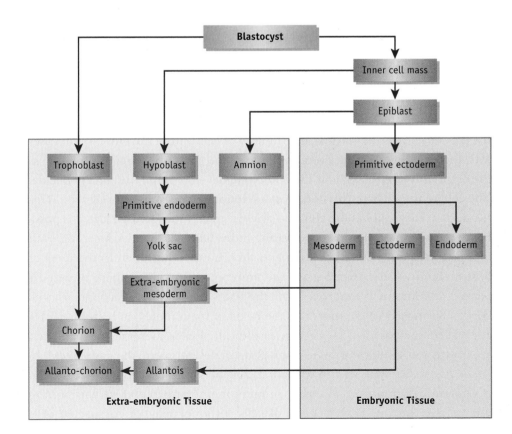

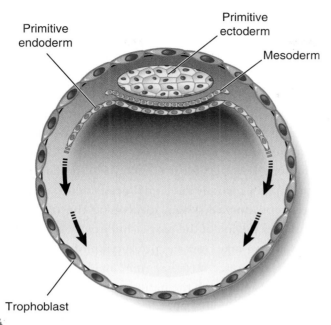

FIGURE 15-4

Schematic depiction of the blastocyst during early stages of development of extra-embryonic membranes. The primitive ectoderm will differentiate to form the body of the embryo, whereas the primitive endoderm expands to form the yolk sac. At this stage, a thin layer of tissue called the mesoderm develops from the primitive ectoderm and occupies the space between the primitive ectoderm and primitive endoderm. This tissue will develop further and join with the trophoblast to form the chorion.

lines the blastocoele and eventually gives rise to the yolk sac. The epiblast will form an extraembryonic membrane known as the amnion as well as the body of the embryo, which differentiates into the embryonic germ layers: mesoderm, endoderm, and ectoderm. The trophoblast remains on the exterior of the embryo and eventually gives rise to the membranes that become the placenta. In humans the trophoblast forms a syncytium (multinucleated mass of cytoplasm formed by merged cells) called the **syncytiotrophoblast,** which can be distinguished from the inner cellular layer known as the **cytotrophoblast.** The cytotrophoblast together with embryonic connective tissue (mesenchyme) form the placental tissue known as the chorion.

PROGESTERONE IS ESSENTIAL FOR PREGNANCY

The uterus must remain under the influence of progesterone in order to sustain pregnancy. Mammals rely on two general strategies to ensure that progesterone levels remain elevated while a conceptus is present in the uterus.

15

Some species accomplish this by having a luteal phase that covers the duration of gestation. Recall that in induced ovulators (e.g., rabbit and cat) copulation induces ovulation and development of corpora lutea, which remain functional whether or not there are embryos present. Some spontaneous ovulators employ similar strategies. For example, the length of the bitch's luteal phase is the same for the pregnant and nonpregnant states. Finally, it is mating, not the presence of embryos, that prevents luteolysis in female rats. A second strategy applies to species where the length of the luteal phase is shorter than the time required for the embryo to implant in the uterus. In these cases a pregnancy-recognition signal generated by the conceptus acts on the maternal reproductive system to prolong the life of the corpus luteum. In some species the life of the corpus luteum is extended only until the developing placenta can take over the production of progesterone. In other cases, the embryo generates a signal that maintains the corpus luteum throughout pregnancy. Although the nature of maternal recognition of pregnancy signals varies considerably among species, embryos usually begin to generate these signals within a few days after hatching.

MATERNAL RECOGNITION OF PREGNANCY

For the purpose of understanding maternal recognition of pregnancy, it may be useful to view pregnancy as a conflict-based relationship between the conceptus and mother. Each is an autopoietic system that requires energy for vital metabolic processes. However, each system competes for metabolic fuels to support their metabolic requirements. From a metabolic perspective, the goal of the conceptus is to derive nutrient support from the mother, whereas the goal of the mother is to consume and conserve fuels. In order to survive in the maternal reproductive tract, the conceptus (essentially a foreign body) must invade the maternal body, intervene with the mother's ovarian cycle, and suppress maternal defensive mechanisms that are intended to resist such physiologic disruptions. According to this perspective, maternal recognition of pregnancy can be viewed as a strategy for disrupting the maternal drive toward luteolysis; that is, toward maintaining the nonpregnant state.

Ruminants

The concept of maternal recognition of pregnancy is nicely demonstrated by an experiment done with sheep. Briefly, injections of extracts from the trophectoderm delay luteolysis in nonpregnant ewes. This type of experiment supports the idea that the embryo has the ability to disrupt the luteolytic process. Additional research has lead to the identification, purification, and synthesis of a small protein (molecular weight of 18,000 to 20,000 daltons) that is responsible for this effect. The peptide, ovine interferon-τ (oIFN-τ), is produced by trophoblasts and can be detected in the uterus between 13 and 21 days after ovulation in pregnant ewes, a time window that coincides with the so-called critical period for pregnancy recognition; that is, a time during which the corpus luteum must be rescued in order for pregnancy to continue. oIFN-τ prevents normal onset of luteolysis by interfering with release of $PGF_{2\alpha}$ by the endometrial epithelium. The mode of action involves blocking expression of receptors for oxytocin on these epithelial cells (Figure 15-5). Recall that oxytocin is responsible for inducing the pulsatile release of $PGF_{2\alpha}$, which is required for luteolysis.

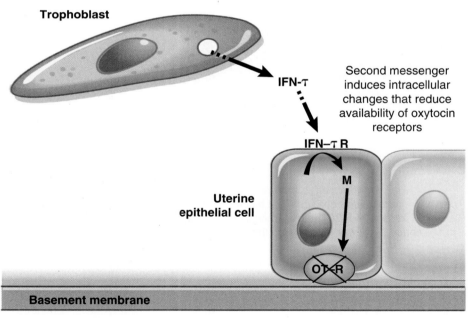

FIGURE 15-5

Schematic diagram showing how the trophoblast of the embryo interacts with endometrial mucosal cells to prevent luteolysis in the ewe and cow during early pregnancy. Interferon-τ is produced by the trophoblast and acts locally on uterine endometrial cells to block expression of oxytocin receptors (OT-R) which are required for oxytocin to induce release of $PGF_{2\alpha}$, the luteolytic hormone.

15

It is generally believed that the mechanism of maternal recognition of pregnancy for cows is very similar to that for ewes. In cows the pregnancy-recognition signal is called bovine interferon-τ (bIFN-τ). It is likely that ruminants rely on additional mechanisms to prevent the regression of the corpus luteum. For example, ovine embryos produce a protein that acts on large luteal cells to protect them from the luteolytic actions of $PGF_{2\alpha}$.

Swine

Maternal recognition of pregnancy in pigs is similar to that in ruminants in the sense that the pregnancy recognition signal disrupts the actions of $PGF_{2\alpha}$. It is different from the ruminant mechanism with respect to the nature of the signal, as well as how it acts (Figure 15-6). The pregnancy recognition signal in pigs appears to be estradiol. The porcine blastocyst produces large amounts of this steroid hormone between the tenth and

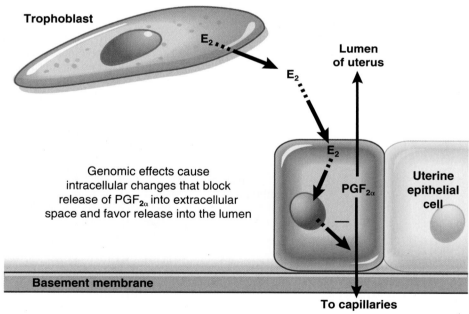

FIGURE 15-6

Schematic diagram showing how the trophoblast of a pig embryo prevents luteolysis in the sow. Estradiol, produced by the trophoblast, acts on endometrial cells to redirect release of $PGF_{2\alpha}$ from the mucosa into the lumen of the uterus, thereby reducing uptake of the luteolysin by blood.

15

twelfth day of pregnancy. Moreover, injections of estradiol on days 11 to 15 of the estrous cycle induce pseudopregnancy; that is, corpora lutea are maintained for the duration of normal pregnancy (113 days). The mode of action of estradiol is remarkable. Its overall effect is to alter the direction of $PGF_{2\alpha}$ secretion by the uterine epithelium. During the estrous cycle, $PGF_{2\alpha}$ is released in an endocrine fashion. In other words, epithelial cells release the hormone into the extracellular space of the submucosa. As described earlier, $PGF_{2\alpha}$ then enters capillaries and travels to veins that drain the uterus. Due to countercurrent exchange between the uterine vein and ovarian artery, the hormone travels to the ovary to act on the corpora lutea. Exposure of endometrial cells to estrogen causes them to re-route the release of $PGF_{2\alpha}$. In other words, the luteolytic hormone is released into the lumen of the uterus; that is, in an exocrine fashion. Although this mechanism appears to be the primary way estradiol acts to lengthen the life of the corpus luteum, other secondary mechanisms might also play a role. For example, estradiol appears to protect luteal cells from the luteolytic actions of $PGF_{2\alpha}$.

Horses

Maternal recognition of pregnancy in mares appears to be more complicated than in other farm animals (Figure 15-7). The critical period for maternal recognition in the mare is between 12 and 14 days after ovulation. Although it is clear that the conceptus generates a signal that prolongs the life of the corpus luteum, the precise mechanism responsible for this action is unclear. We do know that the equine conceptus reduces the concentration of $PGF_{2\alpha}$ found in the uterine vein until 35 days after pregnancy. This effect seems to depend on migration of the conceptus throughout the uterus. Some reproductive physiologists hypothesize that this movement is necessary to distribute a blocker of $PGF_{2\alpha}$ synthesis over the uterine mucosa. Whatever the signal is, it is not sufficient to extend the life of the corpus luteum to the point where the placenta takes over production of progesterone (160 days after ovulation). The bulk of progesterone production between days 35 and 160 of pregnancy comes from accessory corpora lutea. During this time, the chorion of the conceptus produces an LH-like hormone called equine chorionic gonadotropin (eCG). The major action of this hormone is to induce luteinization of follicles. These accessory corpora lutea produce sufficient progesterone to maintain pregnancy and then regress by the time the placenta takes over.

15

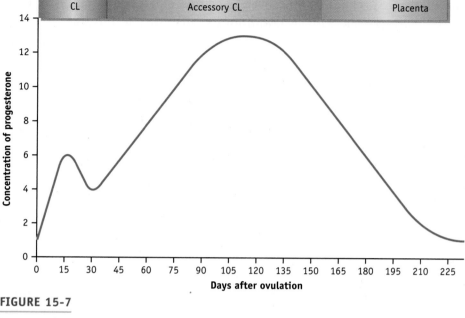

FIGURE 15-7

Pattern of progesterone during the first 230 days of pregnancy in mares. The major sources of this hormone during various stages of pregnancy are indicated by the colored bars. The primary CL is the major source of progesterone for the first 15 days after ovulation. As the primary CL begins to regress, the endometrial cups begin to release large amounts of eCG, which causes follicles to form accessory CL. By 120 days these structures are supplying most of the progesterone. However, these CL begin to regress and the placenta takes over as the major producer of progesterone by 160 days.

Primates

Maternal recognition of pregnancy in humans and other so-called higher primates involves a signal that extends the life of the corpus luteum until the placenta begins producing enough progesterone to maintain pregnancy (Figure 15-8). You may recall that the lifespan of the primate corpus luteum is 14 to 16 days. The human embryo attaches to the uterus between 9 and 12 day after ovulation, but the placenta does not produce appreciable amounts of progesterone until 25 days after ovulation. At approximately 7 days after ovulation the chorion (trophectoderm) of the developing embryo begins producing chorionic gonadotropin. Concentrations of CG increase steadily as the embryo develops, reaching peak concentrations around 30 days after ovulation and then falling throughout the next 30 days. This boost in luteotrophic support counteracts luteolytic signals, resulting in the so-called rescue of the corpus luteum. Progesterone production by

BOX 15-1 | **Focus on Fertility: Is it Necessary to Treat Pregnant Mares with Progestins?**

Each year in North America thousands of pregnant mares are treated with progesterone-like drugs (progestogens) to prevent abortions. This practice has become so widespread that some horse producers are reluctant to not provide supplemental progestogens to pregnant mares. One might expect that such a popular practice is based on sound scientific evidence that proves progestogen therapy to be highly efficacious. One would be wrong in this case. Although no one can be absolutely certain that progestogen treatment cannot prevent abortion in mares, there is little if any scientific evidence to support the idea that it does. So how did such a practice become so popular?

Several facts provide a rational basis for the idea that progestogen supplementation might prevent abortion in mares. First, somewhere between 10 percent and 15 percent of equine pregnancies fail at some time during gestation and most of these occur during the first 40 days when the sole source of progesterone is the primary CL. Such losses can be both economically and emotionally significant to the horse owner. Second, there is no doubt that progesterone is necessary to maintain pregnancy in almost all mammalian species including horses. Destruction of the primary CL has been shown to induce abortion in the mare, and progestogen replacement therapy will prevent pregnancy loss under these circumstances. However, in spite of these well-accepted facts, it would be incorrect to conclude that any or all abortions in mares can be attributed to insufficient progesterone production. To do so is simply "jumping to conclusions." It seems that other pieces of information are necessary to link equine abortions to progesterone insufficiency. Specifically, we must know if 1) abortion is related to low progesterone levels and

2) pregnant mares are likely to have progesterone levels insufficient to support pregnancy.

Based on available research, it appears that low progesterone is not a common problem during most equine pregnancies and when it does occur there is no reason to believe that it is a primary cause of embryo loss. First, the incidence of equine abortions in regions where progestin supplementation is common is no different from those in regions where it is not used. Second, according to a study of 287 mares, only one of the 17 documented abortions was associated with abnormally low concentrations of progesterone in maternal blood. These data should not be interpreted to mean that this one abortion was caused by low progesterone. It is impossible from such studies to determine if the low progesterone caused the abortion or the abortion resulted in low progesterone concentrations. Finally, there seems to be little scientific basis to assume that mares typically express insufficient progesterone concentrations during times when abortions typically occur. In certain rare occasions (e.g., when a mare is afflicted by certain infectious diseases) the primary corpus luteum CL regresses prematurely causing progesterone levels to drop below levels required to maintain pregnancy. However, in most cases of infection-induced embryonic death, the CL actually persists after abortion and must be lysed with $PGF_{2\alpha}$ such that the mare will return to estrus. Finally, secondary CL begin to develop around the time the primary CL begins to regress and these are quite resistant to luteolysis due to the heavy luteotropic supported by LH and eCG. In light of these observations, it appears irrational to believe that progestin supplementation could provide any sort of protection against embryonic loss.

15

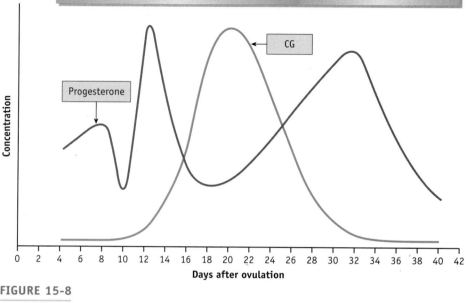

FIGURE 15-8

Patterns of progesterone and chorionic gonadotropin (CG) in female rhesus monkeys during early pregnancy. Production of progesterone by the CL begins to diminish within 8 to 10 days after ovulation. However, it does not regress due to production of CG by the trophoblast of the embryo. This hormone has LH-like activity and provides a luteotropic stimulus that boosts luteal production of progesterone for an additional 2 weeks. By this time the newly formed placenta begins producing progesterone and eventually becomes the major source of this hormone.

the corpus luteum eventually declines, but by this time the production from the placenta is adequate to maintain the uterus in a pregnant state. This change in the source of progesterone production is referred to as the luteal-placental shift.

SUPPRESSION OF THE MATERNAL IMMUNE SYSTEM

Successful pregnancy depends not only on the embryo's ability to sustain maternal production of progesterone, but also its ability to counteract mechanisms that defend the maternal system from invasion by foreign bodies. It is important to remember that because the conceptus contains both maternal and paternal genes it will express proteins that would normally be considered foreign by the maternal immune system. Although the uterus is an immunocompetent organ (i.e., capable of mounting an immunologic attack), and generates antibodies in response to embryo-specific antigens which circulate in maternal blood, the conceptus somehow counteracts these maternal defense mechanisms. It is likely that the conceptus uses diversionary tactics to evade destruction by immune responses. The exact mechanisms employed have not been elucidated, but it has been suggested that the conceptus behaves much like an invading parasite. It is well known that certain parasitic worms produce chemical mediators that mask antigenic surface proteins and/or interfere with other steps in immune responses. Embryos might produce the same types of mediators.

DEVELOPMENT OF THE PLACENTA

The development of a **placenta (placentation)** is a critical prerequisite for development of the embryo past the blastocyst stage. Embryos that fail to accomplish this develop various morphologic defects and die. There are two major steps in the placentation process. First, the trophectoderm of the embryo attaches to the luminal epithelium of the uterus. Second, the embryo develops extraembryonic (or fetal) membranes that become intimately associated with maternal tissue to form the placenta. In species such as humans and rodents, the fetal membranes invade, or implant, into the submucosa of the uterine epithelium. In others (e.g., ruminants, swine, and horses) there is little to no invasion and implantation is synonymous with attachment. Implantation depends on two conditions: 1) development of the embryo to the blastocyst stage and 2) development of the uterus to a state that favors implantation. The latter process is dependent on progesterone, but may also require estradiol in some cases.

Attachment

Attachment of the embryo to the uterus involves the direct interaction of the trophoblast with the uterine epithelium. The process can be broken down into two distinct phases: apposition and adhesion. Apposition refers to development of intimate contact between the embryonic and maternal tissues. This process begins when the embryo becomes immobilized within the uterus. This can occur via two mechanisms. In carnivores, ruminants, horses, and swine, the extraembryonic membranes expand to fill the lumen of the uterus. This causes these membranes to become apposed to the uterine epithelium. In primates and rodents, the embryo is relatively small, so its membranes do not fill the uterus. In humans the uterine epithelial cells "grasp"

15

the trophectoderm and immobilize the embryo. Some investigators describe this phenomenon as analogous to a ball rolling over a surface coated with syrup or some other sticky substance. In this case the "syrup" is a layer of mucin that binds to a protein produced by trophoblasts. Once the trophoblast is stabilized against the uterine epithelium, the process of adhesion begins. This involves the intertwining of microvilli on the maternal and embryonic epithelial cells. At this stage it is difficult to dislodge the embryo from the uterine tissue.

Invasion

In primates and rodents, the trophectoderm penetrates the basal lamina of the uterine mucosa to establish connections with the maternal vasculature. Once the conceptus adheres to the maternal cells, it induces **decidualization** of stromal cells (fibroblast-like cells in the submucosa) and endothelial cells. These cells are transformed into decidual cells, which form the maternal portion of the placenta in these species.

Formation of Extra-Embryonic Membranes

It is important to bear in mind that embryogenesis continues during the process of implantation. By the time the embryo attaches to the uterus, it has rudimentary organ systems and its extra-embryonic membranes have developed (Figure 15-9). As noted earlier, after blastocoele formation the hypoblast forms the yolk sac and the epiblast begins to differentiate into the three major germ cell layers. Part of the mesoderm grows outward and fuses with the trophectoderm to form the chorion. The mesoderm also projects bilaterally on each side of the embryo to form the amniotic folds. These eventually meet to form the amniotic cavity that surrounds the embryo. While this is occurring, the yolk sac regresses and the **allantois,** a vascularized pouch from the hindgut enlarges to fill the cavity between the chorion and amnion. Eventually the allantois fuses with the chorion to form the **allantochorion** (also known as the chorioallantoic membrane), the membrane that becomes the fetal component of the placenta. It should be noted that the chorion is avascular and the fetal blood supply to the placenta comes from the allantois. It is also worth noting that in development of the human placenta, the allantois does not expand and fuse with the chorion. Rather the tissue combines with embryonic mesenchymal tissue to form the umbilical cord which supports vasculature transporting blood between the fetus and placenta.

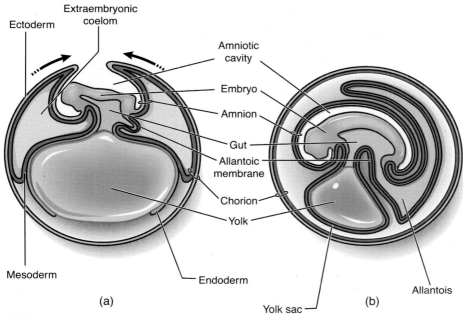

FIGURE 15-9

Schematic illustration depicting the formation of the extraembryonic membranes in the mammalian embryo. During the early stages of this process (a), the yolk sac is the major extra-embryonic membrane and is a reservoir for the yolk, which provides nutrients for the embryo. By this time, the extra-embryonic mesoderm has expanded and formed a sac that envelopes the yolk sac and folds upward to form wing-like structures on either side of the embryo. As development progresses, the mesodermal sac continues to grow until the tips of the wings merge and envelope the embryo in the newly formed amnion. Once the amnion is formed, the yolk sac begins to regress and the allantois forms from an outgrowth of the embryonic hind gut (b). The allantois expands into the extra-embryonic coelom (cavity) and eventually fuses with the chorion to form the allantochorion (not shown). By this time, the yolk sac is a rudimentary structure.

Classification of Placentae

There is remarkable diversity in the types of placentae present in mammals (Table 15-2). Therefore it is useful to have some means by which to classify these organs. One way to classify placentae is based on the origin of fetal blood vessels that vascularize the chorion. In metatherian mammals, it is the yolk sac that serves this function. This tissue fuses with the chorion to form a choriovitelline placenta. In eutherian mammals, the allantois is the source of blood vessels and combines with the chorion to form the chorioallantoic

15

TABLE 15-2 Classification of placentae of several types of mammals based on two criteria

Species	Gross Morphology	Maternal-Fetal Barrier
Cattle	Cotyledonary	Epitheliochorial
Sheep	Cotyledonary	Epitheliochorial
Horses	Diffuse	Epitheliochorial
Swine	Diffuse	Epitheliochorial
Dogs	Zonary	Endotheliochorial
Cats	Zonary	Endotheliochorial
Rodents	Discoidal	Hemochorial
Humans	Discoidal	Hemochorial

placenta. Although this scheme is useful, it does not address the diversity in placenta morphology expressed within eutherian mammals.

Classification Based on Gross Morphology

The most obvious way placentae differ among eutherian mammals is with respect to their gross morphologies. More specifically, there is variation in the distribution of the area within which there is a physical interaction between the maternal and fetal tissues. Contact between these tissues is the result of interaction between the microvilli of the maternal and fetal epithelium. The distribution of the microvilli of the chorion can be diffuse (covering the available surface) or restricted to specialized regions. Examples of the diffuse arrangement are the pig and horse (Figures 15-10 to 15-12). The major difference between these species is that the villi are scattered throughout the surface in the pig, whereas in the horse they are organized in small clusters called **microcotyledons.** The horse placenta develops structures that are unique among mammals. **Endometrial cups** consist of cells that develop from the chorionic girdle, a narrow band of thick trophoblast that develops around the circumference of the fetus at the point where the allantois and yolk sac meet. Between days 36 and 38 of gestation, these girdle cells invade the underlying endometrium and destroy it. They penetrate deep into the endometrial glands and break through the basement membrane. Within 2 to 3 days the cells become spherical and differentiate into clusters of tightly packed cells; that is, the endometrial cups. The major function of these cells is to produce eCG.

15

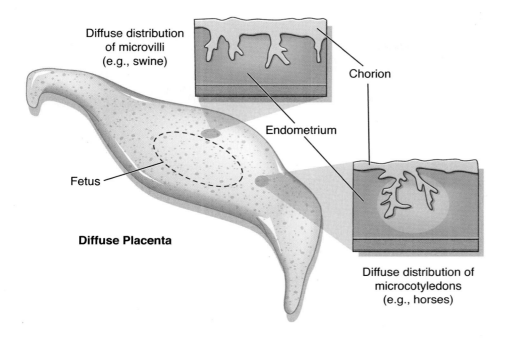

Diffuse distribution of microvilli (e.g., swine)

Chorion

Endometrium

Fetus

Diffuse Placenta

Diffuse distribution of microcotyledons (e.g., horses)

FIGURE 15-10

Schematic illustration of a diffuse placenta. Points of attachment between fetal and maternal tissues are dispersed widely throughout the surface of the allanto-chorion. In the pig, contact between the two tissues occurs via chorionic villi. In the horse, the chorionic villi are organized into microcotyledons.

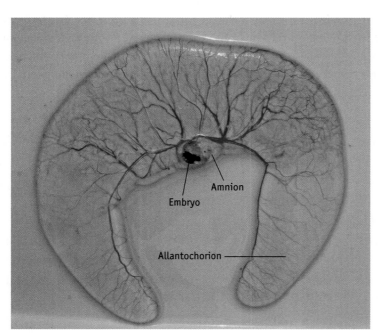

Amnion

Embryo

Allantochorion

FIGURE 15-11

Photographs of a pig embryo and accompanying extra-embryonic membranes removed from a pregnant sow's reproductive tract at an early stage. Note the elongated allantochorion with its extensive vasculature. Also note the embryo that is suspended in the amniotic sac. At this stage of development the chorionic villi are not highly developed, so the allantochorion appears transparent.

15

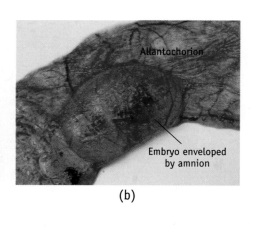

(b)

(a)

FIGURE 15-12

Photographs of a pig embryo and accompanying extra-embryonic membranes removed from a pregnant sow's reproductive tract at an early stage. Note the yellowish-brown regions of the allantochorion (a). As pregnancy progresses, the tips of this membrane become necrotic due to removal of iron by the fetus. Also note the translucent appearance of the allantochorion. This is especially evident in the enlarged view of the membranes covering the embryo (b). This is due to the diffuse distribution of chorionic villi, which serve as sites for attachment to the maternal endometrium.

In the case where the sites of implantation are confined to specialized areas, the distribution of microvilli can be cotyledonary (ruminants), zonary (carnivores), or discoidal (primates and rodents). A cotyledonary placenta consists of discrete zones of microvilli called **cotyledons** (Figures 15-13 to 15-16). The placentae of sheep and cattle contain anywhere between 70 and 120 of these structure. Cotyledons are derived from the trophoblasts and have capillary beds that are supplied by and drained by blood vessels of the fetus. Each cotyledon establishes a physical interaction with a maternal caruncle. Together these structures form the **placentome.**

In zonary placentas (Figure 15-17), a band of microvilli encompasses the equatorial region of the allantochorion. This type of placenta typically fills the lumen of the uterine horn such that the fetal membrane makes contact with the maternal tissue along its entire circumference. This zone of implantation (transfer zone) lies between two zones that are pigmented due to small blood clots which accumulate in these regions. The remaining portion of this type of placenta is an avascular portion of the allantochorion.

15

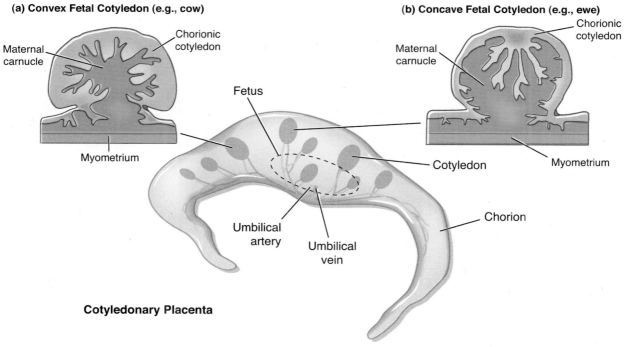

(a) Convex Fetal Cotyledon (e.g., cow)

Chorionic cotyledon

Maternal carnucle

Myometrium

Fetus

(b) Concave Fetal Cotyledon (e.g., ewe)

Chorionic cotyledon

Maternal carnucle

Myometrium

Cotyledon

Chorion

Umbilical artery

Umbilical vein

Cotyledonary Placenta

FIGURE 15-13

Schematic illustration of a cotyledonary placenta. Specialized regions of the chorion (cotyledons) attach to caruncular tissue of the endometrium to form placentomes. In cattle, the cotyledon takes on a convex shape (a), whereas in sheep the cotyledons are convex (b).

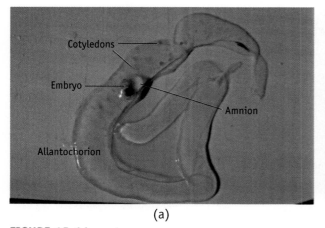

Cotyledons

Embryo

Amnion

Allantochorion

(a)

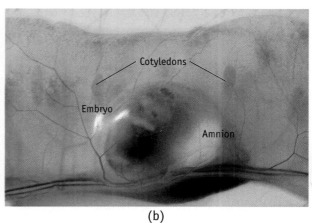

Cotyledons

Embryo

Amnion

(b)

15

FIGURE 15-14

Photographs of a bovine embryo and accompanying extra-embryonic membranes removed from a pregnant cow during an early stage. Note the developing cotyledons and associated blood vessels dispersed on the allantochorion (a). Also note the embryo suspended in the amniotic fluid (b). At this stage the cotyledons are not fully developed and the placenta has not yet attached to the endometrium.

FIGURE 15-15

Photograph of a bovine fetus and accompanying membranes removed from a cow during late pregnancy. The fetal membranes have been excised to expose the fetus and inner surface of the allantochorion. Note the well-developed cotyledons.

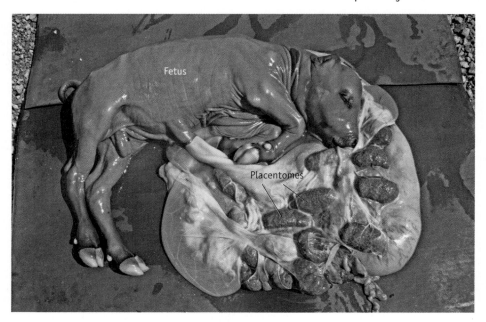

FIGURE 15-16

Photographs showing the structure of the bovine placentome. Note the intimate association between the maternal caruncular tissue and the fetal cotyledon which forms a cap-like structure (a). Attachments between the two tissues form as the chorionic villi interdigitate with folds in the mucosal epithelium of the caruncle (b).

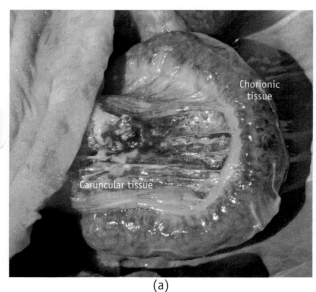

(a)

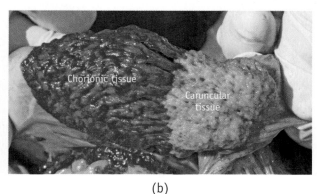

(b)

Zonary Placenta

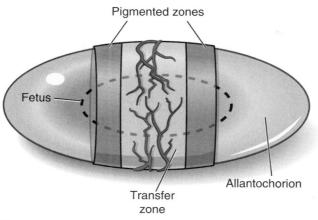

FIGURE 15-17

Schematic drawing of the zonary placenta characteristic of canids. Attachment of the placenta to the uterus occurs along a band which extends around the circumference of the organ. The center portion (transfer zone) is highly vascularized and is the site of transfer of nutrients and wastes between the maternal and fetal systems. This zone is bordered on each side by pigmented regions which develop from maternal hemorrhage and tissue necrosis. The remaining allantochorion does not contain blood vessels.

In humans and rodents, chorionic villi are confined to a disc-shaped region on one pole of the placenta (Figure 15-18).

Classification Based on Maternal-Fetal Barrier

The final method for characterizing the placenta is based on the histology of the maternal-fetal barrier (Figures 15-19 to 15-21). According to this classification scheme, the names for the various types of placentas include the name of the maternal tissue that is in contact with the chorion. As mentioned in an earlier section of this chapter, the fetal membranes do not invade the endometrial mucosa in swine, sheep, cattle, and horses. In these cases, the epithelium of the chorion is apposed to the luminal epithelium of the uterus. These are referred to as epitheliochorial placentas. It might be more appropriate to refer to the ruminant placenta as a syndesmochorial rather than epitheliochorial. Although much of the surface area of these placentas fits the epitheliochorial structure, large binucleate cells from the trophectoderm migrate into the underlying endometrial epithelium and establish direct contact with the basement membrane.

15

FIGURE 15-18

Schematic drawing of the discoid placenta characteristic of primates. A disc-shaped area of chorionic tissue invades the uterine endometrium and forms an intimate exchange zone between maternal and fetal vascular systems.

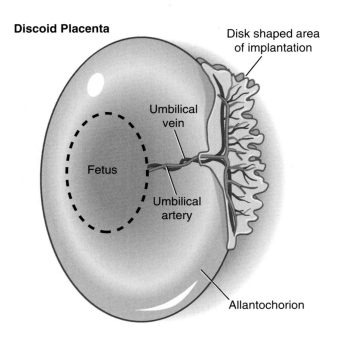

Discoid Placenta

Disk shaped area of implantation

Umbilical vein

Fetus

Umbilical artery

Allantochorion

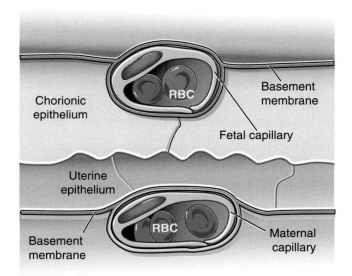

Epitheliochorial Placenta

Chorionic epithelium

RBC

Basement membrane

Fetal capillary

Uterine epithelium

Basement membrane

RBC

Maternal capillary

FIGURE 15-19

Schematic illustration showing the fetal and maternal tissues that make up the epitheliochorial placenta. Note that a total of six tissue layers exists between maternal and fetal blood: 1) maternal endothelial cells, 2) basement membrane of endometrium, 3) epithelium of endometrium, 4) epithelium of chorion, 5) basement membrane of chorion, and 6) fetal endothelial cells.

Endotheliochorial Placenta

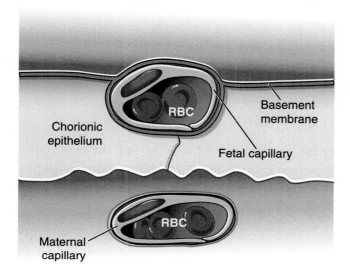

FIGURE 15-20

Schematic illustration showing the fetal and maternal tissues that make up the endotheliochorial placenta. Note that in this type of placenta, the chorionic epithelium has penetrated beyond the epithelium of the uterus leaving only five layers of tissue between the maternal and fetal blood.

Hemochorial Placenta

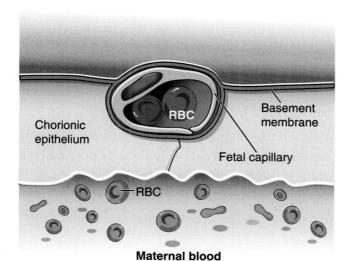

FIGURE 15-21

Schematic illustration showing the fetal and maternal tissue that make up the hemochorial placenta. Note that the chorionic epithelium has penetrated into the submucosa of the endometrium. Chorionic villi protrude into pools of blood that form in the uterine submucosa.

15

In carnivores such as the dog and cat, the chorion invades through the uterine mucosa into the underlying stromal tissue (Figure 15-20). Thus the chorionic epithelial cells are in direct contact with the endothelial cells of capillaries; that is, endotheliochorial placenta. In primates and rodents the chorion penetrates the endothelial cells such that fetal epithelial cells are in direct contact with pools of maternal blood (Figure 15-21).

SUMMARY OF MAJOR CONCEPTS

- In order for pregnancy to advance beyond the blastocyst stage, the conceptus must provide a signal to maintain maternal production of progesterone and form a placenta with the maternal uterus.
- Development of the trophectoderm is a prerequisite for maternal recognition of pregnancy and implantation of the embryo into the uterus.
- Maternal recognition of pregnancy signals prevent/delay luteolysis by interfering with $PGF_{2\alpha}$–induced regression of the CL or by boosting the luteotropic signal.
- Placentas can be categorized based on the distribution of chorionic villi and by the histology of the interface between maternal-fetal tissues.

DISCUSSION

1. Would it make sense to search for a maternal recognition of pregnancy signal in embryos from cats or rabbits? Why or why not?

2. It is common for veterinarians to prescribe progesterone supplementation to pregnant mares in order to prevent early embryonic loss. What is the rationale for this treatment? What would you want to know in order to determine if such a treatment were effective?

3. It is hypothesized that migration of the horse embryo throughout the uterus is necessary for maternal recognition of pregnancy. How would you determine if this hypothesis is correct?

4. How would you distinguish between a maternal recognition of pregnancy signal that is antiluteolytic and one that is luteotrophic?

REFERENCES

Allen, W.R. 2001. Luteal deficiency and embryo mortality in the mare. *Reproduction of Domestic Animals* 36:121–131.

Aplin, J.D. and S. Kimber. 2004. Review: Trophoblast-uterine interactions at implantation. *Reproductive Biology and Endocrinology* 2:48–60.

Fazlaebas, A.T., J.J. Kim, and Z. Strakova. 2004. Implantation: embryonic signals and the modulation of the uterine environment—a review. *Placenta* 25 (Supplement A, Trophoblast Research):S26–S31.

Kaufmann, P. and G.J. Burton. 1994. Anatomy and genesis of the placenta. In: E. Knobil and J.D. Neill, *The Physiology of Reproduction Vol. 2. Second Edition*. New York: Raven Press, pp. 441–484.

Niswender, G.D., J.L. Juengel, P.J. Silva, M.K. Rollyson, and E.W. McIntush. 2000. Mechanisms controlling the function and life span of the corpus luteum. *Physiological Reviews* 80:1–29.

Roberts, R.M., S. Xie, and N. Mathialagan. 1996. Maternal recognition of pregnancy. *Biology of Reproduction* 54:294–302.

Schäfer-Somi, S. 2003. Cytokines during early pregnancy of mammals: a review. *Animal Reproduction Science* 75:73–94.

Spencer, T.E. and F.W. Bazer. 2004. Conceptus signals for establishment and maintenance of pregnancy. *Reproductive Biology and Endocrinology* 2:49–64.

Weitlauf, H.M. 1994. Biology of implantation. In: E. Knobil and J.D. Neill, *The Physiology of Reproduction Vol. 1. Second Edition*. New York: Raven Press, pp. 391–440.

15

CHAPTER

16

Physiology of Pregnancy, Parturition, and Puerperium

OVERVIEW OF MAJOR CONCEPTS

In this chapter we will be concerned with the physiologic events that occur after the placenta is formed and the fetus establishes a physiologic relationship with the mother. The discussion is divided into three main areas: 1) mechanisms that maintain pregnancy, 2) mechanisms that terminate pregnancy, and 3) mechanisms that regulate the recovery of the mother from pregnancy. As the placenta forms and pregnancy is established, the maternal system undergoes numerous changes that allow the mother to cope with the nutritional demands of the fetus. Meanwhile the needs of the fetus change as it develops. The topic of fetal physiology is vast and will not be considered in this chapter. Likewise, pregnancy requires a re-setting of homeostatic mechanisms in the mother and as a consequence the maternal system undergoes numerous physiologic changes. An entire textbook is required to develop a comprehensive understanding of these phenomena. In light of these considerations, our discussion must be restricted to the most significant and best understood adjustments.

CHANGES IN MATERNAL PHYSIOLOGY DURING PREGNANCY

The success of viviparity (live birth) in mammals requires the creation and maintenance of a uterine environment that accommodates and supports fetal growth and development, a means by which the mother provides adequate amounts of the substances upon which the growth and development of the fetus depend, and a means to terminate pregnancy at a time when offspring can survive outside the womb. Each of these requirements is achieved by the actions of hormones produced by the mother's ovaries, the placenta, and the fetus (Figure 16-1). The ovarian hormones act to prepare the reproductive tract for pregnancy,

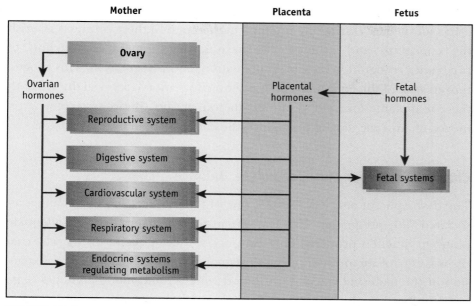

FIGURE 16-1

Endocrine interactions between the fetus, placenta, and mother during pregnancy. Steroid hormones produced by the maternal ovaries alter activities of major organ systems in the mother in order to accommodate the nutritional demands of pregnancy. Hormones produced by the placenta also regulate these organ systems as well as interact with fetal hormones to regulate homeostasis in the fetus.

whereas placental and fetal hormones act to complete the reproductive process. Once birth has occurred, the maternal system must recover from the physiologic changes associated with pregnancy. In many cases this results in a period of infertility following parturition.

Changes in the Maternal Uterine Environment

The most noticeable changes that occur during pregnancy are those of the female reproductive tract. Of these the changes, those associated with the uterine endometrium are most important with respect to maintaining a uterine environment that is compatible with pregnancy. The elevated levels of estradiol that characterize the follicular phase of the ovarian cycle promote proliferation of the uterine mucosa, particularly development of endometrial glands. The subsequent rise in progesterone during the luteal phase promotes secretory activity of these glands. The secretory products of the endometrial glands are known collectively as histotroph and consist of enzymes,

16

growth factors, cytokines, lymphokines, hormones, transport proteins, and other substances. During the preimplantation period these products nourish the conceptus and promote its survival, development, production of pregnancy recognition factors, implantation, and placentation. This progestinized state of the endometrium is essential for embryo survival during both the preimplantation and postimplantation period. If progesterone is removed at any time during pregnancy, the conceptus will be aborted.

Changes in Water and Electrolyte Balance

The development of the placenta is one of the most remarkable changes associated with pregnancy. This vital organ must be supplied with adequate blood to function properly. This must occur without jeopardizing maternal tissues. One of the most important pregnancy-induced changes is an expansion of the maternal vasculature (blood volume) and blood contents (e.g., red blood cells), which accommodates fetal demands for water and oxygen. Expansion of blood volume involves increased retention of sodium, potassium, and calcium by the kidneys. As these ions are reabsorbed by the nephron, water follows to maintain blood osmolality. The bulk of the increased blood volume is retained in large veins. Sometimes blood pools in these areas and can lead to edema.

Changes in Hemodynamics

Profound changes occur in the dynamics of blood flow during pregnancy. These include not only a change in systemic blood flow, but also a change in the distribution of blood flow. In general, blood is shunted toward the uteroplacental unit and mammary gland. One of the major consequences of increased flow to the uterus is facilitating loss of heat generated by fetal metabolism. This is an important aspect of the thermoregulatory abilities of both the mother and fetus. The increased blood flow to the mammary gland is related to the enhanced development of lactogenic tissue in preparation for lactation following parturition.

Cardiac output increases during pregnancy. This is due to an increase in heart rate as well as an increase in stroke volume. In spite of these changes, blood pressure doesn't deviate from the pre-pregnancy state. This is due to an overall decrease in vascular resistance. Major decreases in resistance occur in the vascular beds of the uterus and mammary glands. Blood flow to the skin also increases during pregnancy. This likely helps the mother dissipate heat generated by the fetus.

Changes in Respiration

In most mammals, pregnant females exhibit hyperventilation during pregnancy. This lowers the amount of carbon dioxide in the blood, which causes a slight alkalosis. The slight rise in blood pH does not have an appreciable effect on affinity of hemoglobin for oxygen. As the fetus grows, the expanding uterus will exert pressure on the diaphragm changing the configuration of the mother's thoracic cavity. This lowers the residual capacity of the lungs. The increased respiration rate along with a decrease in resistance of the airway compensate for the lower lung capacity.

Changes in Blood

Pregnancy has been referred to as a state of hypercoagulability. The ability of blood to coagulate is markedly increased during pregnancy. The adaptive significance of this change may be to protect the mother from blood loss.

Changes in Metabolism

As noted repeatedly in this book, successful reproduction depends on metabolic energy. Pregnancy increases the maternal requirements for energy and other nutrients. In order to complete pregnancy successfully, the mother must make metabolic adjustments directed toward several processes including 1) providing oxygen and nutrients for fetal growth and development, 2) providing the fetus with sufficient energy reserves to survive periods of maternal feed restriction, and 3) providing the mother with sufficient energy reserves to survive periods of feed restriction during pregnancy and subsequent lactation. The specific energy requirements of pregnancy can be divided into four main categories:

- Metabolic costs associated with growth and development of products formed during gestation (i.e., fetus, placenta, extraembryonic membranes, amniotic fluid, and maternal tissue).
- Metabolic costs of the biosynthetic processes that form these tissues.
- Metabolic costs of maintaining these tissues.
- Metabolic costs of external work associated with moving a heavier body mass.

In order to appreciate how the maternal system adjusts to the increased metabolic demands associated with pregnancy it is necessary to consider the concept of energy balance. The following equation is a common way to

16

expresses such a relationship: $E_{Gross\,Intake} = E_{Metabolism} + E_{Accretion/Storage} + E_{External\,Work} + E_{Excreted}$. Keeping this model in mind, it now becomes clear that there are several strategies available to increase the amount of metabolic energy available to meet the demands of pregnancy: 1) increase energy intake, 2) repartition energy substrates to various tissues, 3) increase absorption and storage of energy, 4) reduce work (physical activity), and 5) reduce loss due to excretion. Females rely on the first three approaches to accommodate the metabolic demands of pregnancy. It is common for food intake to increase in pregnant females. In addition, the rate of passage of food through the digestive tract slows during pregnancy resulting in enhanced absorption of nutrients. Finally, there are profound metabolic changes associated with pregnancy. Initially, intermediary energy metabolism promotes accretion of body fat. Later in pregnancy, metabolism shifts to promote transfer of energy substrates to the fetus. These changes in intermediary energy metabolism are orchestrated by several metabolic hormones that act on a variety of tissues including the liver, adipose tissue, and muscle. These hormones are produced by the placenta and the mother. In addition, steroid hormones produced by the ovaries and placenta influence release and actions of metabolic hormones. The extent to which females reduce energy loss due to excretion is unknown.

PLACENTAL PHYSIOLOGY

The idea that the placenta is involved with fetal nutrition can be traced to Aristotle. However, details regarding the morphology of this organ didn't emerge until the seventeenth century. Today the placenta is viewed as an organ that mediates the transfer of various chemicals between the maternal and fetal circulations as well as an important endocrine gland that influences maternal and fetal physiology.

PLACENTAL TRANSPORT

In eutherian mammals, the placenta consists of the chorioallantoic membrane of the fetus and the endometrial tissue of the uterus. Terminology referring to the fetal component can be confusing because the chorion is also referred to as either the trophectoderm or the trophoblast. The latter term was first used by Hubrecht in the late nineteenth century and will be used throughout this chapter. The functional unit of placental transfer is the barrier between the fetal and maternal blood and consists of the endothelial cell of the fetus, the microvillus of the trophoblast, and the maternal tissue of the uterus (Figure 16-2).

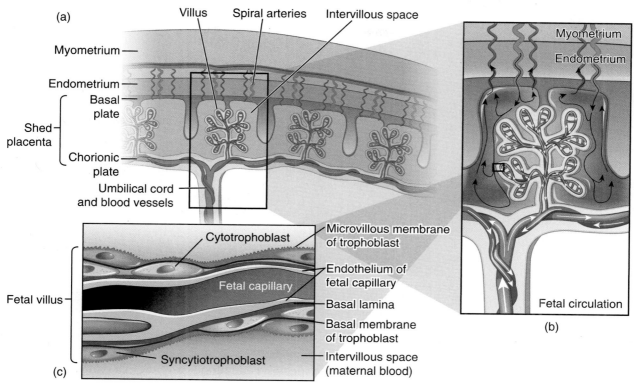

FIGURE 16-2

Schematic drawings depicting the interface between maternal and fetal tissue in the human placenta. Fetal blood flows from the umbilical artery to capillaries in the chorionic villi (composed of the syncytiotrophoblast and cytotrophoblast), whereas maternal blood flows into the intervillous space via spiral arteries (a). Thus maternal blood is in direct contact with the trophoblast (b and c).

The structure of this barrier varies considerably among mammals and affects the efficiency of transport between fetal and maternal circulations.

One source of variation in the maternal-fetal barrier is the structure of the trophoblast. The cells of this layer can function individually (cellular trophoblast) or as a syncytium (syncytial trophoblast) and can be arranged in either a single or multiple layers of cells. The barrier also varies with respect to the type of maternal cells that are apposed to the trophoblast. As noted in the previous chapter, the trophoblast makes contact with the uterine epithelium in ungulates, the endothelial cells of uterine capillaries in carnivores, and with blood in primates and rodents. The arrangement of fetal capillaries is a third source of variation. In primates and farm animals, they take on a villous (tree-like) arrangement, whereas in rabbits and rodents they form more complex labyrinths. Finally, the pattern of blood flow relative to the maternal

16

FIGURE 16-3

Three types of relationships between blood flow patterns in maternal and fetal blood vessels. The countercurrent arrangement provides the most efficient transfer of solutes, gases, and heat, whereas the concurrent flow provides the least efficient transfer.

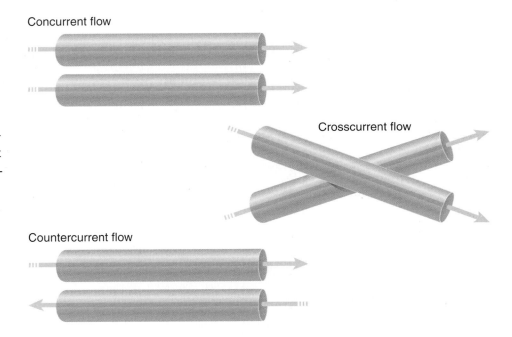

Concurrent flow

Crosscurrent flow

Countercurrent flow

circulation differs among species. Fetal blood can flow in a concurrent, crosscurrent, or countercurrent manner with respect to maternal blood flow (Figure 16-3). The arrangement of blood vessels affects the efficiency of transfer of solutes, gases, and heat between maternal and fetal systems. The countercurrent arrangement provides the most efficient transfer, whereas the concurrent provides the least efficient transfer.

The primate placenta has been the subject of numerous studies and therefore serves as a useful model for describing important principles of placental transport (Figure 16-4). The extent to which a substance (S) is transported into or out of the fetus can be expressed by the net transplacental flux (J_{net}) of that substance. This term can be expressed mathematically by the following formula: $J_{net} = J_{mf} - J_{fm}$, where J_{mf} is the maternal to fetal flux and J_{fm} is the fetal to maternal flux. A positive number indicates that there is a net flux of the substance from the mother to the fetus, whereas a negative number indicates a net flux from the fetus to the mother. J_{net} is a function of the effective concentration of a substance in the maternal blood, which is influenced by the following variables:

- Rate of maternal blood flow.
- Rate of dissociation from red blood cells or serum proteins (analogous to affinity between a hormone and receptor).

16

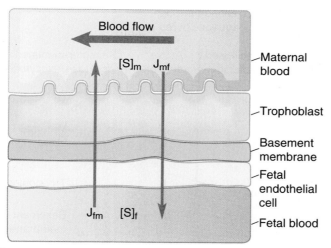

FIGURE 16-4

Schematic illustration depicting transport of a substance (S) between maternal and fetal blood supplies. The extent to which S is transported into or out of the fetus can be expressed by the net transplacental flux (J_{net}) of that substance that is simply the difference between the maternal to fetal flux (J_{mf}) and the fetal to maternal flux (J_{fm}), or $J_{net} = J_{mf} - J_{fm}$. A positive number indicates that there is a net flux of the substance from the mother to the fetus, whereas a negative number indicates a net flux from the fetus to the mother. The flow of a substance across the placenta is a function of $[S]_f$ and $[S]_m$ (concentration in fetal and maternal blood), which is determined by blood flow, affinity for red blood cells, and electrochemical gradients across the trophoblast.

- Potential difference across the trophoblast.
- Thickness of unstirred area of maternal blood (not flowing) near the microvillus border.

The cellular mechanisms that mediate the transfer of materials from the maternal blood or extracellular space across the trophoblast include (Figure 16-5):

- Simple diffusion of lipophilic molecules across trophoblast (e.g., oxygen).
- Restricted diffusion of hydrophilic molecules through aqueous channels (e.g., mannitol).
- Facilitated diffusion (e.g., D-glucose).
- Active transport (e.g., amino acids).
- Receptor-mediated endocytosis (e.g., immunoglobulins).

16

FIGURE 16-5

Schematic depiction of mechanisms for transporting substances across the human placenta. A substance can be transported across the trophoblast via one of several mechanisms. However, movement from the basal surface of the trophoblast into the fetal circulation occurs via diffusion.

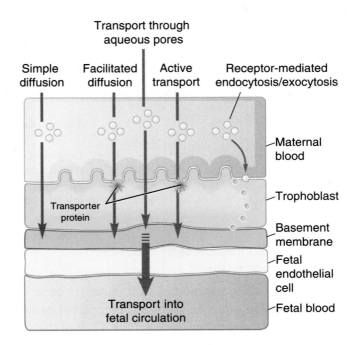

It is beyond the scope of this chapter to include detailed descriptions of which molecules are transported across the placental barrier. A detailed account of concentrations of various nutrients and wastes can be found in the references included at the end of this chapter. Some of the more important substances transported between the mother to the fetus include glucose, lactate, amino acids, lipids, ions, calcium, phosphorous, magnesium, trace minerals, proteins, respiratory gases, and various drugs.

The Placenta as an Endocrine Organ

In addition to mediating transfer of nutrients and wastes to and from the fetus, the placenta produces numerous hormones. Placental steroidogenesis is limited to progesterone and estrogens. The major polypeptide hormones produced by the placenta include placental lactogen, a variant of growth hormone, prolactin, relaxin, and chorionic gonadotropin.

Placental Steroidogenesis

Progesterone is the major steroid hormone produced by the placenta. With the exception of primates, the placenta does not produce estradiol until the final stages of pregnancy and is related to the accompanying drop in progesterone production. The significance of this will become clear when we consider the regulation of parturition. Primates are unique in the sense that the placenta produces appreciable amounts of estradiol throughout pregnancy. Circulating concentrations of estradiol in women follow the general pattern of progesterone. Both hormones increase throughout pregnancy and then drop at the time of birth.

One of the more important concepts associated with the maintenance of pregnancy is the extent to which the placenta contributes to the circulating pool of progesterone. Recall that progesterone is essential to maintain pregnancy. However, the source of progesterone isn't important. In some species the placenta becomes the major source of progesterone during pregnancy. In others, the corpus luteum is the sole producer of progesterone throughout pregnancy. The timing of the so-called luteal-placental shift in progesterone production varies among mammalian species (Table 16-1).

Placental Polypeptide Hormones

We have previously considered the importance of chorionic gonadotropin, a secretory product of the trophoblast that is important in maternal recognition of pregnancy in primates and horses. Although release of this hormone diminishes to baseline levels by 30 days of pregnancy, its production is sustained throughout pregnancy. The corpus luteum eventually becomes refractory to chorionic gonadotropin and the placenta becomes the major source of progesterone. Whether this hormone plays an important role in regulating other processes is unclear. Some researchers have suggested that chorionic gonadotropin might support growth of the placenta, regulate steroid hormone synthesis in the fetal testes and adrenal cortex, and regulate secretion of thyroid hormones in the mother.

TABLE 16-1 Length of gestation and timing of the luteal-placental shift in progesterone production in several species of mammals

Species	Length of Gestation (months)	Luteal-Placental Shift in Progesterone (day or month of gestation)
Canids	2	None
Felids	2	None
Sheep	5	50 days
Horses	11	70
Cattle	9	6–8 months
Swine	3.8	None
Human	9	60–70 days

16

The placenta also produces **placental lactogen** (or somatomammotropin), a polypeptide that expresses prolactin-like and growth hormone-like activities. The growth-promoting activity of this hormone is believed to be important in regulating growth of the fetus, whereas the lactogenic activity seems to be important in stimulating mammary gland activity in the mother.

In the human, some of the lactogenic and growth-enhancing activity of the placenta is also attributed to prolactin and an isoform of growth hormone, each of which are produced by the anterior pituitary gland.

In some species, the placenta produces relaxin, a hormone that plays an important role in parturition. This hormone is produced by the placentas of horses, primates, carnivores, swine, and rabbits, but not by ruminant placentas. In the latter case, the corpus luteum is the major source of this hormone.

PARTURITION

One of the more intriguing questions in reproductive physiology is what determines the length of gestation. The prevailing hypothesis is that some type of biological clock governs the timing of parturition. The precise nature of such a mechanism has not been elucidated for most mammals, but three general types seem possible. First, there may be a mechanism that tracks the number of cell divisions that occur from the time of fertilization. Second, the time elapsed since syngamy may be an important regulatory signal. This would involve some sort of endogenous rhythm in some biological variable within the mother, placenta, or fetus. Finally, the timing of parturition might be programmed within the genome of the fetus. In other words, at a particular stage of development fetus generates a signal that induces parturition. The third possibility clearly applies to the sheep. In this species, parturition appears to be dependent on the sequential development of the hypothalamic-pituitary-adrenal system. Work in other species such as humans suggests that alternative mechanisms exist. In spite of this variation there are important similarities among those mammals studied. In all cases, parturition involves changes in the contractility of the uterine myometrium. High levels of progesterone suppress motility of the myometrium resulting in weak and poorly coordinated contractions (known as Braxton-Hicks contractions in the human). In late pregnancy, progesterone concentrations decrease allowing the uterus to prepare for stimuli that can induce the type of strong and rhythmic contractions necessary to expel the fetus from the uterus and birth canal. These stimuli can be generated by the fetus, placenta, or mother. In almost all cases studies, $PGF_{2\alpha}$ appears to be the most important signal initiating the intense uterine contractility that is necessary for parturition.

Control of Parturition in the Ewe

One of the most important concepts regarding timing of parturition in sheep is that the signal that induces birth originates within the fetus (Figure 16-6). This idea arises from the fact that disruption or removal of the anterior pituitary or adrenal gland delays or prevents parturition. Unless these altered fetuses are aborted, they will continue to enlarge and distend the uterus to the point where it interferes with the mother's ability to eat.

The signal that triggers parturition in the sheep is cortisol. At a particular point in development (the last few days of pregnancy), the fetal hypothalamic-pituitary-adrenal axis awakens resulting in an increase in production of cortisol by the fetal adrenal gland. This increase in cortisol release is due to several changes. First there is an increase in release of corticotrophin-releasing hormone (CRH) from the hypothalamus. This together with an increased responsiveness of the anterior pituitary gland to CRH causes release of adrenocorticotrophin (ACTH), which in turn acts on the adrenal cortex to stimulate synthesis and release of cortisol, a steroid hormone. With respect to

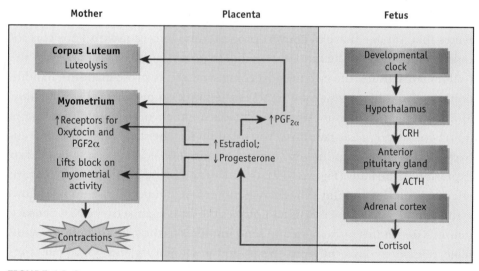

FIGURE 16-6

Summary of mechanism whereby the fetus initiates parturition in the ewe. The process begins with activation of the hypothalamic-pituitary-adrenal axis. An increase in release of corticotropin releasing hormone (CRH) from the hypothalamus causes an increase in adrenocorticotrophin (ACTH), which acts on the adrenal cortex to stimulate synthesis and secretion of cortisol. This adrenal steroid then acts on the placenta to alter biosynthesis of estradiol, progesterone, and $PGF_{2\alpha}$. The resultant changes in placental hormone production trigger changes in the maternal reproductive system that promote uterine contractions and eventually parturition.

16

FIGURE 16-7

Schematic illustration depicting the role of PGF$_{2\alpha}$ in establishing the positive feedback loop that generates intense, rhythmic contractions that are necessary for expulsion of the fetus in the ewe.

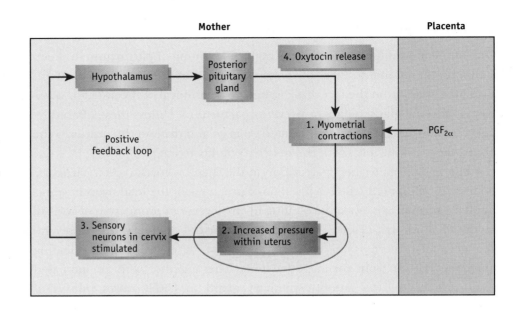

parturition, the major target tissue of cortisol is the placenta. Recall that in sheep the bulk of progesterone is produced by the placenta after day 50 of pregnancy. Fetal cortisol acts on this tissue to increase the expression of enzymes that are necessary to convert progesterone into estradiol. Thus cortisol causes an increase in the ratio of estradiol to progesterone. This change in steroid milieu activates the enzyme-governing synthesis of prostaglandins. In addition, the drop in progesterone removes the block on myometrial contractility. Moreover, the increase in estradiol increases numbers of myometrial receptors for oxytocin and prostaglandins. Prostaglandin sets the stage for a positive feedback loop that leads to intense, rhythmic contractions that eventually lead to expulsion of the fetus (Figure 16-7). Oxytocin and prostaglandins exert positive effects on uterine motility. The increase in prostaglandin release initiates powerful contractions of the uterus and pushes the fetus against the cervix. Sensory neurons detect this movement and send impulses to the hypothalamus to stimulate release of oxytocin from the posterior pituitary gland. Oxytocin is released in a pulsatile manner, which results in rhythmic uterine contractions. This sets into motion a positive feedback relationship between cervical stimulation and release of oxytocin (called the Ferguson reflex in humans). The feedback loop is broken once the fetus is pushed out of the birth canal.

In addition to pushing the fetus through the birth canal, cortisol triggers two other events that are necessary for successful birth. First, the increase in estradiol release from the placenta acts directly on the female reproductive tract to enhance secretions, which serve to lubricate the birth canal. Second, prostaglandin causes luteolysis and release of relaxin from the corpus luteum.

16

Relaxin causes the ligaments that support the pelvis to relax thereby enlarging the pelvic opening.

Control of Parturition in Other Mammals

Cortisol appears to play an important role in timing parturition in swine, goats, and cattle. The mechanism in the cow appears to be similar, if not identical, to that in the ewe. However, the regulation of parturition in the doe and sow are slightly different from that in the ewe. The major difference between the sheep and these other ungulates is the source of progesterone during pregnancy. In the sow and doe, the corpus luteum, rather than the placenta, is the primary source of progesterone during pregnancy. In these cases, cortisol induces synthesis of prostaglandin by the placenta, and this raises the estradiol:progesterone ratio by inducing regression of the corpus luteum. The importance of cortisol in control of parturition in horses is equivocal, but an increase in prostaglandin plays a pivotal role in the cascade of events leading to increased myometrial activity.

The role of the fetal adrenal gland in humans appears to be much less important than in sheep, swine, goats, and cattle. Unlike these animals, the signal that triggers the cascade leading to parturition is generated by the placenta, not the fetus (Figure 16-8), suggesting that this event may be regulated by a placental clock. The human placenta produces appreciable amounts of CRH late in pregnancy. This hormone then establishes two positive feedback loops; that is, one with the fetal pituitary-adrenal system and the other involving the fetal membranes. With respect to the first loop, CRH acts on the fetal pituitary gland to enhance ACTH release, which then acts on the fetal adrenal to elevate production of cortisol. Cortisol then feeds back positively on the placenta to further enhance CRH release. The resulting rise in cortisol alters placental steroidogenesis to favor production of estradiol, which then acts on the myometrium to enhance its motility. A second positive feedback loop involves CRH and prostaglandins. CRH produced by the placenta induces synthesis of prostaglandins by the amnion and these hormones feedback on amniotic tissue to induce production of CRH. This positive feedback loop causes a rapid increase in prostaglandins, which then act on the myometrium to induce contractions. Once contractions begin, the Ferguson reflex ensues and parturition soon follows.

16

Stages of Parturition

The cascade of endocrine events leading to parturition corresponds to visible changes in the mother during labor. For convenience these changes can be

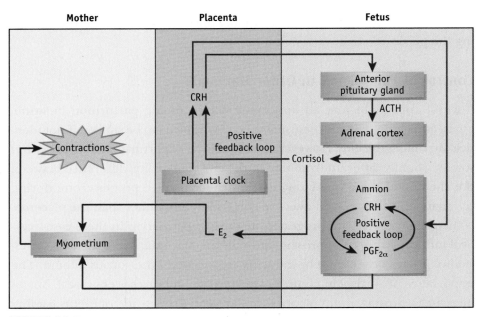

| Mother | Placenta | Fetus |

FIGURE 16-8

Schematic illustration summarizing the mechanism whereby the placenta initiates parturition. Unlike species such as the sheep, in humans the placenta is the source of the signal that initiates parturition. Corticotropin-releasing hormone (CRH) produced by the placenta acts on the fetal pituitary gland to stimulate release of adrenocorticotropin (ACTH), which acts on the adrenal gland to induce release of cortisol. Cortisol then acts on the placenta to promote release of E_2 and additional cortisol as well as on the amnion to stimulate release of $PGF_{2\alpha}$. E_2 and $PGF_{2\alpha}$ interact to promote contractions of the myometrium thereby causing the onset of labor.

grouped into three distinct phases. The first stage marks the beginning of labor and is brought on by an increasing estradiol:progesterone ratio. The two major events that characterize this stage are 1) a progressive relaxation and dilation of the cervix and 2) increased motility of the myometrium resulting in distinct and noticeable contractions of the uterus. These contractions increase the pressure on the cervix, which establishes the positive feedback loop between cervical stimulation and oxytocin release. The end of this phase is marked by movement of the fetus into the birth canal. The duration of this stage of parturition averages 1 to 12 hours in most mammals.

The second stage of parturition is characterized by strong and rhythmic contractions of the uterus, which culminate in the delivery of the infant (Figure 16-9). It is difficult to interrupt this stage because of the positive feedback loop, which generates the contractions. These contractions,

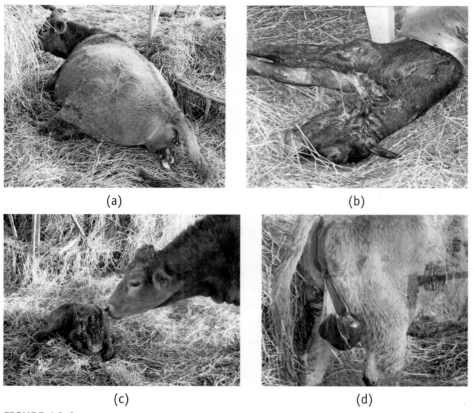

(a) (b)

(c) (d)

FIGURE 16-9

Photographs taken of a cow during the second and third stages of parturition. This stage is characterized by strong, rhythmic contractions of the uterus that move the calf into the birth canal. When this occurs the fetal membranes rupture and fluids are expressed from the reproductive tract. During this stage the cow assumes a recumbent position and begins abdominal straining to expel the calf (a). If the calf is situated properly (head first with front legs fully extended), movement through the birth canal is rapid (b). Once the calf is expelled the cow immediately stands up and begins licking and nuzzling the newborn (c). Following birth, the fetal membranes can be seen protruding from the cow's reproductive tract (d). This marks the third and final stage of parturition. Eventually these will be dislodged and expelled from the birth canal. Photographs provided by Mr. Michael Meyer, undergraduate student at the University of Kentucky.

16

together with application of intense abdominal pressure by the mother, lead to expulsion of the fetus. This stage progresses quickly as long as the fetus is situated properly in the birth canal. In most cases, this amounts to a head-first presentation with the front legs fully extended. Once the fetus is in the birth canal, the allantochorion ruptures and releases fluid ("breaking water"). At this point the amniotic membrane appears as a fluid-filled sac protruding from the vulva. This soon ruptures and the mother begins regular bouts of abdominal straining until the infant is expelled. This stage can last as little as 5 minutes or as long as 12 hours.

The third and final stage of labor involves detachment and expulsion of the placenta. These processes are largely the result of continued myometrial contractions that may continue for several days after birth. The duration of this stage varies greatly. In some cases (cats) the placenta is expelled with the neonate. In other cases, it may take several hours for the placenta to be expelled.

BOX 16-1   Focus on Fertility: Preventing Birth

Although research in reproductive biology has provided the basis for numerous technologies that enhance reproduction, results of such work have also been applied to developing methods for preventing birth. Birth control methods have been sought and valued by all human societies. People have always expressed the desire to control if and when they have children. It is also important to prevent birth in livestock. For example, it is sometimes necessary to terminate a pregnancy in heifers or cows in the case of an accidental breeding or during certain pathologic conditions. Birth control has also been considered as a means to manage reproduction in pets as well as feral animals that

have become pests. Figure 16-10 lists the major categories of birth control methods currently used in humans. Similar methods have been developed and adopted for use in other species of mammals.

The earliest methods of birth control most likely involved management of sexual behaviors. Most of these are still practiced in many cultures. Behavioral methods of birth control include 1) abstinence; 2) outercourse (masturbation and alternatives to penile-vaginal intercourse), 3) *coitus interruptus* (withdrawal of the penis from the vagina prior to ejaculation), 4) fertility awareness (avoiding sexual intercourse during the fertile period of the ovarian cycle), and 5) extended

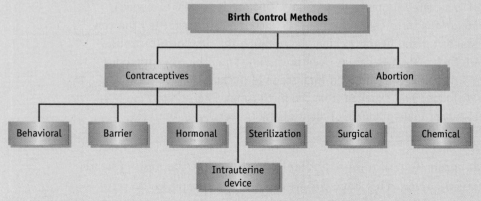

Figure 16-10 Major approaches to birth control in humans. Note that methods fall into one of two major categories: those that prevent fertilization (contraceptives) and those that disrupt pregnancy following fertilization, usually after implantation (abortion).

16

breast-feeding (prolongs lactational amenorrhea). With the exception of abstinence and outercourse, none of these methods have proven to be highly reliable as a means to prevent birth.

The second major category of birth control includes the barrier methods. The most common devices include 1) male condoms (a polyurethane sheath that fits over the penis), 2) female condoms (a polyurethane sheath that fits into the vagina), 3) vaginal sponge (disk-shaped polyurethane device that contains spermicide), 4) spermicide alone (foam, cream, film, or suppositories with spermicidal properties), and 5) diaphragms and cervical caps (dome-shaped rubber disks the cover the anterior vagina creating a barrier to prevent sperm from entering the cervix). The effectiveness of these barrier methods is highly variable, ranging between 10 and 50 percent (i.e., percentage of pregnancies occurring when used).

Hormonal methods have been shown to be the most effective method of birth control other than complete abstinence (less than one birth per 100 women per year). All of the hormonal methods currently available are administered to women. Although the method of delivery varies among methods, the modes of action are essentially the same; that is, induction of a progestational state (pseudopregnancy). Hormonal methods differ in terms of delivery method and whether or not they consist of progestin alone or progestin plus estrogen. The "combined" birth control pill, the first oral contraceptive, consists of both types of steroids and acts by preventing ovulation. The so-called minipill is also an oral contraceptive, but contains only progestin. Its primary mode of action is to thicken the cervical mucus, thereby preventing sperm from reaching the oocyte. Contraceptive hormones can also be provided by means other than pills. Alternative methods include: 1) skin patch, 2) vaginal ring (flexible ring inserted into the vagina and releases progestins and estrogens),

3) injection (progestin alone or progestin plus estrogen), and 4) subcutaneous implant (six matchstick-sized rubber rods filled with progestin).

A fourth approach to human birth control is the intrauterine device. This is a T-shaped object that contains copper. It is inserted into the uterus where it prevents fertilization, but the mode of action is poorly understood.

The fifth method of birth control is surgical sterilization. This approach includes tubal ligation in women (occlusion of the fallopian tubes) and vasectomy in men (transaction of the ductus deferens). Both of these procedures disrupt transport of gametes, thereby preventing the interaction between sperm and oocyte.

Each of the aforementioned birth control methods is classified as contraceptive; i.e., preventing conception. These approaches are fundamentally different from the sixth method of birth control: induced abortion. Induced abortion can be defined as the intentional termination of pregnancy. There are two major types of abortion: surgical and chemical. Surgical abortion involves the use of special instruments to remove the fetus and membranes from the uterus. A chemical abortion is induced by administering drugs (abortifacients) that terminate pregnancy. In humans this procedure is known as medical abortion. There are two major approaches to medical abortions: methotrexate and mifepristone (RU-486) in combination with a prostaglandin analogue. Methotrexate inhibits the metabolism of folic acid, which interferes with implantation of the embryo into the uterus. Mifepristone acts as a progestin receptor antagonist. In other words, it prevents progesterone from binding to its receptor. The resulting lack of progesterone support causes menstruation and softening of the cervix, conditions that are disruptive to pregnancy. Prostaglandin analogues are given within a few days after RU-486 to induce uterine contractions and promote emptying of the womb.

16

POSTPARTUM PERIOD

As noted earlier, parturition can be understood as a sudden disruption of the maternal-fetal exchange system. Viewed in this way, it is not difficult to imagine that the maternal system may not be equipped to establish a new pregnancy soon after the birthing process. Indeed, in most mammals, females enter a period of infertility during the postpartum period. The period following parturition has been referred to as the puerperium (from the Latin *puerpera*, which means woman in childbirth). This term refers specifically to the period during which the uterus involutes, or returns to its normal size. Involution involves expulsion of the fluid remains of the placenta (lochia), repair of the endometrium, and reduction in size of the uterus accompanied by an increase in its tone. Although it is true that the uterus is resistant to implantation before the completion of uterine involution, other mechanisms contribute to the infertility that is characteristic of the postpartum period. In general, the time required for females to resume fertile ovarian cycles is not highly correlated with the time required for the uterus to recover from pregnancy. For example, uterine involution is completed by 30 days in beef cows, but these animals typically resume ovarian cycles 50 to 60 days after parturition. In dairy cows, the uterus is completely involuted by 45 to 50 days, whereas ovarian activity resumes much earlier (10 to 25 days postpartum).

Lactation appears to be the major factor responsible for the postpartum hiatus in reproductive activity in cows, ewes, sows, and most primates, but not in the bitch, queen, or mare. Part of this effect is due to a negative energy balance that results from the heavy metabolic demands associated with milk production. However, the suckling stimulus itself also appears to suppress reproductive activity of the mother. In all species studied so far, the suckling-induced inhibition of reproductive activity involves suppression of pulsatile luteinizing hormone (LH) release, which prevents development of ovarian follicles to the preovulatory stage. Eventually, the maternal system overcomes these inhibitory effects and the female resumes ovarian cyclicity. The first postpartum ovarian cycle is typically characterized by a lack of estrus and a short luteal phase. The second cycle is usually normal, but may be less fertile than subsequent ones.

The Rat

Rats are prolific creatures. Females have the ability to give birth to one litter of pups immediately after weaning a previous litter. This reproductive strategy reflects the fact that the female rat expresses estrus and ovulates within

24 hours of birth. Gestation length in rats averages 21 days and pups are usually weaned by 30 days of age. This means that females could give birth to one litter while they are taking care of another. How does the female avoid nurturing two litters of pups? The answer lies in a phenomenon known as **delayed implantation** (Figure 16-11). If conception occurs while the female is nursing, the blastocyst will form, but then enter a quiescent state and not implant until 5 to 7 days later than normal. This strategy is used by other species to synchronize birth with environmental conditions that favor rearing of offspring. The delay in implantation extends the length of pregnancy to 30 days, a time by which weaning of the previous litter occurs.

If the female rat conceives during lactation, the corpora lutea of pregnancy do not fully regress. Therefore, the ovaries contain two sets of corpora luteus: those remaining from the previous pregnancy and the newly formed ones. This is due to the high circulating levels of prolactin which are caused

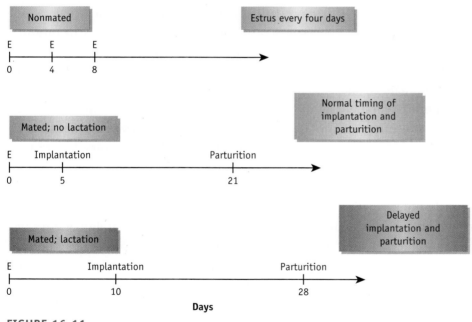

FIGURE 16-11

The effects of lactation on timing of implantation and birth in female rats. The rat has a 4-day estrous cycle (top panel). If a female rat conceives, and is not lactating, implantation occurs by the fifth day after mating and parturition will occur 21 days after estrus (middle panel). When a female rat conceives while lactating, implantation of embryos is delayed for 5 days, allowing the mother to wean one litter before giving birth to another (bottom panel). When this occurs, parturition will also be delayed.

16

by the suckling stimulus. During the early portion of lactation, large antral follicles regress, and the ovaries contain only those of small and medium sizes. This is due to the low-frequency pattern of pulsatile LH secretion that characterizes this period. The suppression of LH pulse frequency is due to a direct effect of suckling on gonadotropin-releasing hormone (GnRH) release as well as the aforementioned hyperprolactinemia. It is also noteworthy that suckling attenuates the positive feedback actions of estradiol on LH release during this period. Toward the end of lactation, pulsatile LH resumes and the estradiol-LH positive feedback system is restored. This leads to the resumption of follicle growth, estrus, and ovulation.

Domestic Ungulates

The sow, ewe, and cow express distinct periods of anestrus and infertility following parturition. The duration of these periods is an important determinant of lifetime reproductive rate. In domestic livestock this can be translated to mean lifetime production efficiency. The length of time a female is not reproductively active affects how much meat and/or milk she can produce each year. The economic importance of this trait has provided incentive for numerous studies regarding the control of the postpartum anestrus in domestic animals. In all cases, lactation suppresses pulsatile release of LH, which prevents development of follicles to the preovulatory stage and therefore estrus and ovulation. The average length of the postpartum anestrus is 3 to 6 weeks for sows and ewes and between 30 and 150 days in cows. In all cases, frequency of LH pulses increases gradually throughout the postpartum period (Figure 16-12). During this time, the ovaries will exhibit waves of follicle growth, but formation of an ovulatory follicle does not occur until LH pulse frequency reaches some threshold level. As in the rat, the ability of estradiol to induce an LH surge is compromised until late in the postpartum period. One of the major differences in mechanisms regulating postpartum anestrus is the nature of the stimulus that suppresses LH secretion. In sows and ewes, it is clear that suckling exerts a direct effect on hypothalamic release of GnRH. In cows, the presence of calves rather than suckling per se may be the mediating signal. In none of these cases is hyperprolactinemia a factor in suppressing LH secretion. Although circulating levels of prolactin are elevated in each case during lactation, there is no evidence to support the idea that prolactin inhibits LH secretion in these species. Finally, it should be noted that other environmental variables interact with suckling to affect length of the postpartum interval in livestock. A negative energy balance due to poor nutrition and/or parasite infestation prolongs the postpartum anestrus. Photoperiod

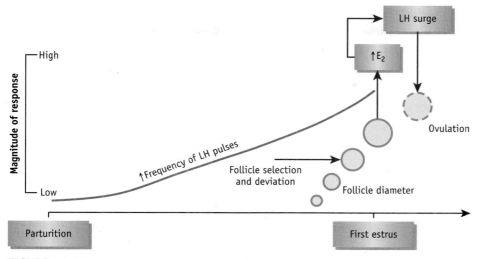

FIGURE 16-12

Major physiological events lead to the return of ovarian cycles in female mammals during the postpartum period. Following parturition, LH release from the pituitary gland is extremely low due to a low frequency of LH pulses. As time passes, the frequency of LH pulses gradually increases. At some time during the postpartum period, LH secretion is sufficient to drive a dominant follicle to the preovulatory stage, resulting in estrus, ovulation, and resumption of ovarian cycles. The rate at which the pulsatile LH secretion recovers depends on the presence of the offspring, nutritional status, and other environmental variables.

is another confounding variable. For example, ewes normally lamb in the spring, which is the time of onset for seasonal anestrus. Thus it is difficult, if not impossible, to distinguish between a seasonal and postpartum anestrus in this case. Studies of postpartum anestrus in sheep have been done with ewes mated during the late breeding season such that they lamb during the early- to mid-breeding season, long before onset of seasonal anestrus. Although domestic cattle are not facultative seasonal breeders, photoperiod has been shown to influence the length of the postpartum anestrous period. In general, long day lengths shorten the interval between calving and first estrus.

Primates

The apes (including humans), Old World Monkeys, and most New World monkeys exhibit a lactation-induced infertility following parturition. In all cases, the early portion of the postpartum period is characterized by a low frequency of LH pulses and an absence of ovulatory follicles. Eventually, the frequency of LH pulses increases, resulting in follicle maturation, ovulation,

16

and resumption of menses. The time required for resumption of cycles is directly proportional to the intensity of suckling by the infant. Although suckling causes hyperprolactinemia, there is no support for the idea that high prolactin levels are responsible for the low pulsatile secretion of LH. However, the elevated concentrations of prolactin provide support for the corpus luteum of pregnancy, which delays luteolysis, causing the structure to remain and produces some progesterone during lactation.

SUMMARY OF MAJOR CONCEPTS

- Hormones associated with pregnancy alter maternal physiology in ways that prepare the mother for pregnancy and lactation and allow her to successfully complete the pregnancy.
- The major functions of the placenta are to provide a means for transferring nutrients, gases, and wastes between the fetal and maternal systems.
- Parturition is induced by a prostaglandin-induced increase in motility of the myometrium. The signal that triggers this response varies among species and can originate in either the fetus or placenta.
- The reproductive strategies of mammals include a postpartum period of infertility that is regulated to a large extent by suckling and involves suppression of pulsatile release of LH.

DISCUSSION

1. Pregnant women frequently suffer from edema (fluid accumulation), especially in the lower extremities. Explain how this occurs.

2. If you had the choice between using either oxytocin or cortisol to induce parturition in a ewe, which would you prefer? Explain your answer.

3. RU-486 is a progesterone receptor antagonist. In other words, it binds to the progesterone receptor but does not elicit a biological response. This drug is also known as "the abortion pill." Explain how such drug induces abortion. Would you expect the drug to be more effective at a particular time in pregnancy? Why or why not?

4. The postpartum anestrus is shorter in the milked dairy cow than the suckled beef cow. Propose a hypothesis to explain this difference. How might you go about testing this hypothesis?

16

REFERENCES

Jenkin, G. and I.R. Young. Mechanisms responsible for parturition; the use of experimental models. *Animal Reproduction Science* 82–83:567–581.

McLean, M. and R. Smith. 2001. Corticotrophin-releasing hormone and human parturition. *Reproduction* 121:493–501.

McNeilly, A.S. 1994. Suckling and the control of gonadotropin secretion. In: E. Knobil and J.D. Neill, *The Physiology of Reproduction Vol. 2.*, second Edition. New York: Raven Press, pp. 1179–1212.

Morriss, Jr., F.H., R.D.H. Boyd, and D. Mahendran. 1994. Placental transport. In: E. Knobil and J.D. Neill, *The Physiology of Reproduction Vol. 2.*, second Edition. New York: Raven Press, pp. 813–861.

Norman, S. and R.S. Youngquist. 2007. Parturition and dystocia. In: R.S. Youngquist and W.R. Threfall, *Current Therapy in Large Animal Theriogenology*, second Edition. St. Louis: Saunders Elsevier, pp. 310–335.

Ogren, L. and F. Talamantes. 1994. The Placenta as an endocrine organ: polypeptides. In: E. Knobil and J.D. Neill, *The Physiology of Reproduction Vol. 2.*, Second Edition. New York: Raven Press, pp. 875–945.

Planned Parenthood. 2006. *Report: A History of Birth Control Methods*. Katharine Dexter McCormick Library, New York.

Rivera, R., I. Yacobson, D. Grimes. 1999. The mechanism of action of hormonal contraceptives and intrauterine contraceptive devices. *American Journal of Obstetrics and Gynecology* 181:1263–1269.

Solomon, S. 1994. The primate placenta as an endocrine organ: steroids. In: E. Knobil and J.D. Neill, *The Physiology of Reproduction Vol. 2.*, Second Edition. New York: Raven Press, pp. 863–873.

Spencer, T.E. and F.W. Bazer. 2004. Conceptus signals for establishment and maintenance of pregnancy. *Reproductive Biology and Endocrinology* 2:49–63.

Stewart, C.L. and E.B. Cullinan. 1997. Review—Preimplantation development of the mammalian embryo and its regulation by growth factors. *Developmental Genetics* 21:91–101.

Stock, M.K. and J. Metcalfe. 1994. Maternal physiology during gestation. In: E. Knobil and J.D. Neill, *The Physiology of Reproduction Vol. 2.*, Second Edition. New York: Raven Press, pp. 947–983.

Wiebe, E., S. Dunn, E. Guilbert, F. Jacot, and L. Lugtig. 2002. Comparison of abortions induced by methotrexate or mifepriston followed by misoprostol. *Obstetrics and Gynecology* 99:813–819.

16

Seasonal Regulation of Reproduction

BACKGROUND

As noted in Chapter 1, the success of a particular species depends on the ability of individual members of the species to survive and reproduce viable offspring. In order to accomplish this, an organism must have the ability to cope with its environment. Most mammals live in environments that change with the season. Under these circumstances, natural selection favors individuals that can cope with seasonal changes in variables such as ambient temperature and food availability. With respect to reproduction, natural selection promotes adaptations that restrict reproductive activity to times when environmental conditions favor pregnancy and rearing of young. Therefore, it should come as no surprise that most mammals exhibit some degree of seasonal variation in their reproductive activities. In the most extreme cases, the so-called seasonal breeders, reproduction is restricted to only part of the year. In less extreme cases, animals will show reproductive activity throughout the year, but the intensity of activity varies with season. In females that breed seasonally, expression of ovarian cycles is confined to a particular time of year. At other times these animals experience a complete cessation of estrus and ovulatory cycles. In seasonally breeding males, testicular size, testosterone production, and spermatogenesis decrease at certain times of the year. The degree of this decrease in testicular function can vary between complete infertility to reduced fertility.

REPRODUCTIVE STRATEGIES AND THE ENVIRONMENT

The reproductive characteristics of individual members of a species constitute the species' **reproductive strategy**. A reproductive strategy is a function of an individual animal's genotype, but the particular genes responsible for these traits are expressed because of natural selection. When considering how an animal's environment causes changes in its reproductive activity, it is useful to think in terms of ultimate and proximate causes. You have already encountered these concepts in reference to the causes of sexual behavior. Ultimate causes of environment-induced changes in reproductive activity are evolutionary processes that shape the annual reproductive cycle of a *species*. In contrast, proximate causes refer to the physiologic mechanisms that mediate the effects of various environmental stimuli on *individuals*. The relationship between food availability and reproductive activity in seasonal breeders such as sheep illustrates the difference between ultimate and proximate factors. Over the eons, seasonal changes in availability of food imposed selection pressure that favored lambing and suckling of lambs during a time of year when food is abundant (i.e., during spring and summer). It is not the availability of food per se that drives seasonal patterns of reproduction in sheep. Rather these animals use annual fluctuations in photoperiod as cues to predict when food availability is generally greatest. In this case, food availability is an ultimate cause of seasonality. This is not to say that food availability doesn't exert direct effects on reproduction. The restriction of food intake resulting from a lack of adequate food supplies suppresses fertility in all mammals by depriving them of the amount of calories required to support reproductive processes. This direct effect of food intake on the physiologic mechanisms regulating reproduction is an example of how food availability acts as a proximate cause of variations in reproductive activity.

The focus of this chapter will be on the ultimate causes of environmental effects on reproduction in mammals; that is, mechanisms regulating seasonal breeding. It is not possible to provide detailed accounts of the numerous ways mammals have adapted to seasonal fluctuations in their environments. Therefore, we will examine only a few of the most thoroughly documented examples. As we study these cases, it is important to keep in mind that the seasonal components of a species' reproductive strategy are shaped by both intrinsic and extrinsic variables. Intrinsic variables include: life span, ultimate body size, length of the female's reproductive cycle (from puberty to weaning of offspring), feeding strategy, and the presence of some seasonal survival mechanisms, such as hibernation. Extrinsic variables include: the nature, severities and timing of climatologic changes, dietary challenges, competition for resources, and predator pressure.

Strategies for Seasonal Breeding

Mammals cope with seasonal changes in environment via two basic strategies: 1) reacting directly to variations in an environmental variable and 2) reacting to environmental cues that predict periods that are favorable to reproduction.

17

In the first case, reproduction is not linked to seasonal cues such as annual patterns of day length. Mammals that rely on this strategy are facultative seasonal breeders, meaning that seasonal fluctuations in reproductive activity occur only when there are seasonal fluctuations in some variable that is necessary for reproduction (e.g., adequate food). The reproductive strategies of facultative seasonal breeders can be viewed as opportunistic. In other words, they will breed at opportune times. A good example is the house mouse (*Mus musculus*). Under field conditions these animals reproduce between April and November in northern temperate climates. However, when living commensally with humans (i.e., in their dwellings with abundant food), they will reproduce throughout the year. In general, facultative seasonal breeders are small and have high reproductive rates due to rapid development, short gestation periods, and short postpartum periods. These characteristics allow these animals to produce large numbers of offspring for as long as conditions support the reproductive processes.

Larger mammals will also express seasonal fluctuations in reproductive activity when living in environments where there are marked fluctuations in food supply. For example, hunter-gatherer societies living in regions where there are marked fluctuations in food availability (e.g., the !Kung people of the Kalahari desert in Africa) exhibit seasonal fluctuations in pregnancy rate among women. In each of these examples, food availability is affecting reproductive activity via altering the number of calories available for reproduction; that is, it produces immediate beneficial or detrimental effects on the reproductive fitness of an individual.

Obligatory seasonal breeders rely on environmental cues to predict when conditions will be favorable for successful reproduction. These animals are typically larger, and live longer than the majority of facultative seasonal breeders. They also have lower reproductive rates due to slower rates of sexual maturation, longer gestation lengths, and longer postpartum periods. Because of their longer gestation periods, mating occurs at a time that precedes birthing by several months. These mammals rely on mechanisms to ensure that breeding occurs at a time that results in offspring being born when the mothers have adequate food to support lactation. The domestic sheep is arguably the most familiar and most thoroughly studied example of this type of seasonal breeder. In temperate climates, the ewes of most breeds express a period of seasonal anestrus. Although the length of this anestrous period varies among breeds, it typically occurs during long days (between April and September in the northern hemisphere). Figure 17-1 shows the average seasonal fluctuation in expression of estrous cycles for several popular breeds of sheep when maintained in temperate climates (e.g., latitude 45°N). Recall that the gestation

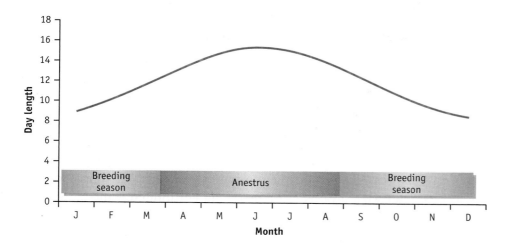

FIGURE 17-1

Average times of the breeding and anestrous periods relative to day length for sheep living in the northern hemisphere (latitude 45°N).

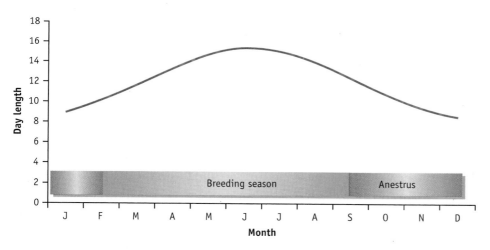

FIGURE 17-2

Average times of the breeding and anestrous periods relative to day length in mares living in the northern hemisphere (latitude 45°N).

length of sheep is approximately 5 months. Thus mating during the first part of the breeding season results in lambs being born in late February and early March when food availability is beginning to increase. Unlike facultative seasonal breeders, sheep and other obligatory seasonal breeders cannot take advantage of sudden increases in food supply during the anestrus season. Increased feeding does not overcome seasonal anestrus in these animals.

The sheep is an example of a short-day breeder. In other words, breeding is restricted to a time of year when photoperiod is short. Not all seasonal breeders are short-day breeders. For example, horses are long-day breeders. Figure 17-2 illustrates the major stages of seasonal reproduction in mares in northern, temperate climates. Mares begin to express estrous cycles as early as February, but ovulation is variable until May (spring transition period). Maximum fertility of mares occurs between May and September. Ovulation continues in a variable manner during the autumn transition to anestrus

17

(October through December). Because of their 12-month-gestation length, this breeding season ensures that foals are born in the spring and summer, a time when there is likely to be enough food to support lactation.

ENVIRONMENTAL FACTORS UNDERLYING SEASONAL BREEDING

The previous discussion has established that virtually all mammals have the capacity to express seasonal changes in reproductive activity. In this section, we will consider what aspects of the environment influence the patterns of reproduction in these animals. As noted in the previous section, food availability is a major driving force in this regard. Reproduction, like all other physiologic processes, is an energy-consuming process. Thus a deficit in energy consumption can suppress reproductive activity in an animal. Reduced reproductive activity can also result from an increase in use of energy for other vital processes, including cellular maintenance, thermoregulation, locomotion, and so on.

Food

As noted earlier, food can affect reproductive activity in both a proximate and ultimate way. Our concern in this section is with the proximate effects of food on reproduction. There is an abundance of information concerning the effects of food intake on reproductive activity of mammals. Most of this work has dealt with the effects of food restriction on sexual development and onset of puberty in females. Restriction of food intake delays age at first ovulation in all species studied. Figure 17-3 illustrates this effect in heifers. Other effects of food restriction on female reproductive activity include disruption of estrous cycles and reduced milk production. Pregnancy appears to be quite resistant to food deprivation, at least in large mammals that have large energy stores. In these cases, females draw on their own body stores of energy to maintain pregnancy. However, feed restriction during pregnancy can result in lower milk production and a longer postpartum anestrous period.

Restriction of food intake also disrupts reproductive activity in males. In adolescent males food restriction will impair steroidogenesis and delay sexual maturation. Severe and prolonged food restriction will disrupt both spermatogenesis and steroidogenesis in adult males.

Most studies have not been directed at identifying the particular nutrient deficiencies that are responsible for impaired reproduction during restricted feed intake. However, it is generally accepted that many of the detrimental

17

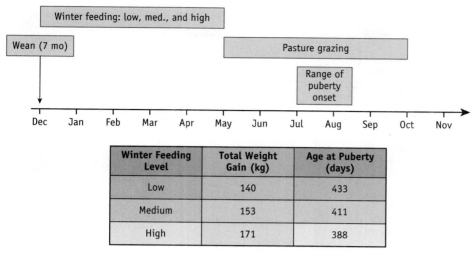

FIGURE 17-3

Summary of an experiment that describes the effects of level of nutrition on sexual maturation in beef heifers. A group of heifers was weaned in December when they averaged seven months of age. Between December and May, animals were fed low, medium, or high planes of nutrition. All groups were allowed to graze pasture for the remainder of the study. Age at puberty (age at first ovulation) was inversely proportional to level of nutrition. Data from Short and Bellows (1971).

effects of food restriction on reproduction can be attributed to a lack of dietary energy. Nevertheless a deficiency in particular nutrients (vitamins, minerals, amino acids, and fatty acids) can also disrupt reproduction in males and females.

Energy Metabolism and Reproduction

Food intake is only one of several variables that affect the amount of energy available for reproduction (Figure 17-4). The extent to which a particular amount of dietary energy can sustain reproductive activity depends on how much energy is consumed by other physiologic processes as well as how much energy the animal can mobilize from its storage depots. Major energy-consuming processes in adult mammals include cellular maintenance, thermoregulation, locomotion, growth, and lactation. During times of feed deprivation an animal offsets its deficit in dietary energy by mobilizing energy substrates from adipose tissue, muscle, and the liver. Whether or not an animal reproduces depends on a delicate balance among food intake, rate of energy consumption, and mobilization of energy substrates.

17

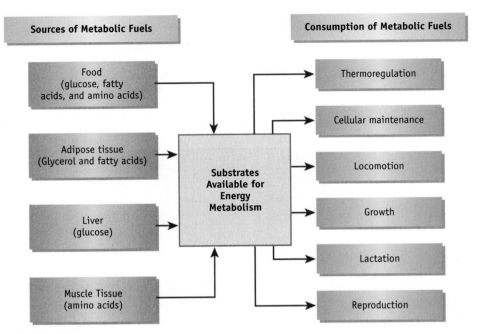

FIGURE 17-4

Schematic diagram illustrating the concept of energy balance. Metabolites for energy metabolism (metabolic fuels) can be derived from food as well as stores of fat, carbohydrate, and protein (yellow boxes). These energy substrates are used to support a variety of bodily functions (green boxes). When the rate of metabolic fuel production is the same as the rate of fuel consumption the animal is said to be in energy balance and neither gains nor loses body mass. When the rate of fuel production is greater than consumption the animal is in a positive energy balance and gains mass. Conversely, when rate of fuel production is less than rate of consumption a negative energy balance ensues and the animal loses mass.

Ambient Temperature

Mechanisms regulating body temperature consume a considerable amount of energy. Mammals are homeotherms; that is, they maintain their body temperatures within narrow ranges in spite of fluctuations in ambient temperature. An animal does not have to expend energy to maintain its body temperature when ambient temperatures are within its thermoneutral zone (Figure 17-5). When ambient temperatures rise above or fall below this zone, the animal makes metabolic and behavioral adjustments that allow it to maintain body temperature. The adjustments mammals make in order to maintain constant body temperature in high and low temperatures are widely different among species.

17

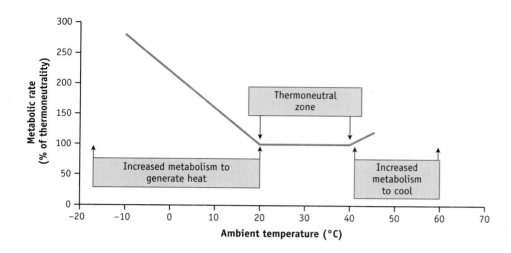

FIGURE 17-5

Relationship between ambient temperature and metabolic rate in mammals. Note that the range in ambient temperatures within which an animal does not have to expend energy to maintain its body temperature is called the thermoneutral zone.

Low Temperature

Coping with low ambient temperatures requires an increase in food intake, whereas adjusting to high ambient temperature does not. One of the most important ways a mammal prevents a drop in its body core temperature during low ambient temperatures is to generate heat by increasing its rate of metabolism. This raises the energy requirement of the animal, which it meets by increasing its food intake. The metabolic response to cold temperatures varies considerably among mammals, depending on how well they prevent heat loss. For example, small animals with large surface area-to-volume ratios and animals with poor insulation (due to lack of blubber or thick pelage) respond more robustly to a drop in temperature than larger animals that are well insulated. At any rate, when considering the effects of food intake on reproduction it is important to take into account the ambient temperatures that prevail during times of low food availability. Figure 17-6 illustrates this concept based on studies with rats. It is clear from this example that the effects of reduced food intake are exacerbated by low ambient temperatures. The important implication from these results is that the decrease in reproductive activity expressed by facultative seasonal breeders is probably attributed to the combined effects of a decrease in food intake as well as an increase in dietary energy requirements.

High Temperature

Ambient temperatures that exceed the thermoneutral zone of an animal can also suppress reproductive activity, but such effects are not entirely due to reduced availability of energy substrates. Elevated ambient temperatures disrupt reproductive processes via two major mechanisms; reduced appetite and

17

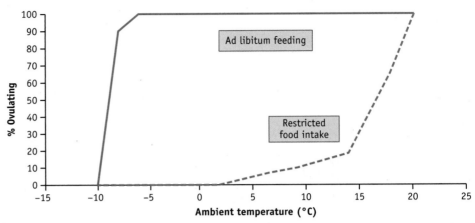

FIGURE 17-6

Effects of ambient temperature on ovulation rates in rats maintained on two planes of nutrition. Animals maintained on a restricted plane of nutrition do not have sufficient calories to maintain body temperature and reproductive activities and therefore give up reproduction in favor of thermoregulation. Data from Bronson and Heideman, 1994.

hyperthermia (elevated body temperature). Figure 17-7 summarizes these effects in the dairy cow, which has been the focus of study for many years. As noted in the previous chapter, females typically experience a period of negative energy balance early in the postpartum period. Heat stress during this period results in a reduction in feed intake thereby prolonging this period of metabolic insufficiency. In addition to reducing milk production, the reduced caloric intake suppresses pulsatile release of gonadotropin-releasing hormone (GnRH)/ luteinizing hormone (LH). This can disrupt follicular maturation (i.e., deviation), estrus, and ovulation resulting in ovulation of poor-quality oocytes, or under severe cases, extending the length of the postpartum anestrous period. Hot temperatures can also disrupt reproduction via behavioral mechanisms. Under such conditions males and females become lethargic and are less likely to express appropriate reproductive behaviors. This may prevent mating, or promote poor timing of insemination relative to ovulation.

Elevation of an animal's body temperature (hyperthermia) can exert direct (independent of reduced caloric intake) effects on reproduction. The physiologic mechanisms mediating the inhibitory effects of high temperature on reproduction have been studied extensively in livestock, but most of the work has focused on the female. In general, hyperthermia disrupts spermatogenesis in males and reduces conception rates in females. The major effect of heat stress in females is a compromised uterine environment.

17

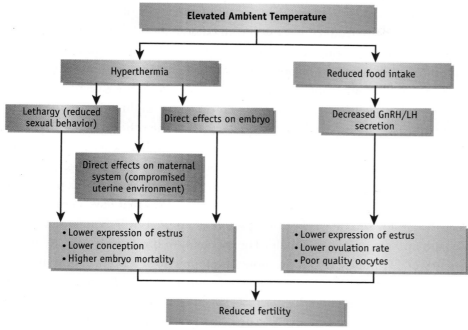

FIGURE 17-7

Mechanisms whereby high ambient temperatures can reduce fertility in dairy cows. Reduced fertility is due to the combined effects of hyperthermia and reduced food intake. With respect to hyperthermia, detrimental effects can be direct, affecting both the maternal and fetal systems, and indirect affecting the behavior of the animal.

However, there is also evidence that maternal hyperthermia exerts direct effects on the embryo. The combined effects on the uterus and embryo result in increases in embryo mortality.

Other Factors

Seasonal variations in the reproductive rates of mammals are influenced by variables other than food availability and ambient temperature. Our discussion of nutrition and reproduction has been narrowly focused on the importance of energy substrates. However, it is also likely that other nutrients can influence reproductive activity. Certainly deficiencies of vitamins, minerals, essential amino acids, and long-chain fatty acids can have negative effects on reproduction. Unfortunately, we lack detailed information concerning the roles of the substances in expression of seasonal patterns of reproduction. Three additional factors are likely to play a role in regulation of seasonal breeding are: 1) competition for limited resources, 2) predation pressure, and 3) social

17

interactions. Each of these can act as ultimate causes of seasonal breeding. Selection pressure from competition with or predation by other species could favor a breeding season that is either synchronized with or out of synchrony with those of competitors or predators. With respect to competition, the latter strategy would avoid over-consumption of resources. In the case of predation, synchrony between predator and prey reproduction might enhance the fitness of the prey species by satiating the predators. On the other hand, asynchrony between reproductive activities might provide some adaptive advantage to prey by reducing loss of very young offspring. Social interactions can act either as proximate or ultimate causes of seasonal breeding. An example of a proximate cause would be the so-called ram effect in ewes. Ewes that have been isolated from a ram for several weeks during seasonal anestrus will express estrus and ovulate upon exposure to a male. The mechanism involves an increase in frequency of LH pulses, which promote development of a preovulatory follicle. On the other hand, social cues can serve as a means to predict periods that are favorable to reproduction. In this case, social cues act as ultimate factors. Examples of this phenomenon will be described in the next section.

Before we leave this discussion of how the environment can cause seasonal changes in reproductive activity it is important to emphasize that none of the aforementioned factors acts alone. Rather, the pattern of reproduction expressed by members of a species reflects the combined actions of food availability and ambient temperature.

REGULATION OF SEASONAL BREEDING BY ENVIRONMENTAL CUES

As noted earlier, some mammals rely on seasonal cues to regulate the time of reproductive activity. These mechanisms are typically viewed as time-keeping mechanisms. In other words, they allow the animal to keep track of the time of year such that it can prepare metabolically for seasons that are either conducive or hostile to successful reproduction. Of course such mechanisms are only advantageous in climates where there are reliable annual cycles in climate and/or availability of nutrients. They would offer little advantage when climatic and dietary conditions are more variable. In this case opportunistic mating strategies offer an advantage.

Photoperiodic Control of Seasonal Breeding

Photoperiod appears to be the most important environmental cue regulating seasonal reproduction in mammals. In latitudes north and south of the

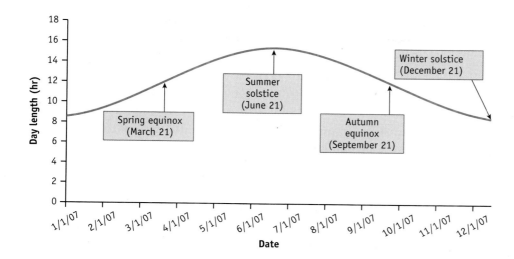

FIGURE 17-8

Annual fluctuations in day length in the northern hemisphere.

equator there are annual cycles in day length that do not vary from year to year. Figure 17-8 illustrates the annual cycle of photoperiod in the northern hemisphere (e.g., latitude 45°N). Twice during the year, the length of the light and dark periods is the same; i.e., the spring and autumn equinoxes. The longest day of the year occurs on the summer solstice, whereas the shortest day occurs on the winter solstice. Day length increases between the winter solstice and summer solstice, then decreases between the summer solstice and winter solstice. This pattern of photoperiod suggests three strategies whereby photoperiod can be used to regulate seasonal reproduction. First, a critical day length may both induce and terminate reproductive activity. For example, the spring equinox might induce gonadal activity, whereas the autumn equinox might terminate gonadal activity. Such a strategy would allow animals to reproduce more than one time each year. A second strategy is that a particular day length either stimulates or inhibits gonadal activity. For example, reproductive activity might begin at the spring equinox and continue for some predetermined length of time. This too would permit animals to reproduce more than once each year. Note, however, that this approach requires some type of internal timer that measures length of the breeding or nonbreeding season. The third strategy involves synchronization between an internal and external rhythm. In this case, the annual cycle of reproduction is driven by an endogenous **circannual rhythm,** which somehow becomes synchronized (entrained) with the external cycle of photoperiod. Although all of these strategies are possible, only the second and third have been documented in mammals. Syrian hamsters (*Mesocricetus areatus*) employ the second strategy, whereas the domestic sheep uses the third strategy. Each of these examples will now be considered.

17

Photoperiod and Reproduction in the Syrian Hamster

Gonadal activity in the Syrian hamster is at a minimum between November and January (Figure 17-9). In late winter (February) the gonads begin to **recrudesce** (become active again) and reach full activity around the time of March. Gonadal activity remains high throughout the spring and summer, but then diminishes beginning in late summer. This seasonal pattern of gonadal activity is produced via the following mechanisms. First, gonadal regression is induced when day length becomes less than 12.5 hours. The gonads remain inactive for 4 to 5 months, but then begin to recrudesce because the animal becomes refractory to the short day lengths. They will remain active throughout the summer, but then regress due to decreasing day length. It is clear that in the Syrian hamster, short days signal the end of the breeding season and that an internal interval timer, involving refractoriness to the inhibitory effects of short days, times the duration of the breeding season.

Photoperiod and Reproduction in the Sheep

The photoperiodic regulation of reproductive activity in the ewe has been studied extensively and intensively for decades. As noted earlier, sheep are short-day breeders. In temperate zones, ewes begin exhibiting estrous cycles in late summer and will continue to do so until early spring, if they are not mated

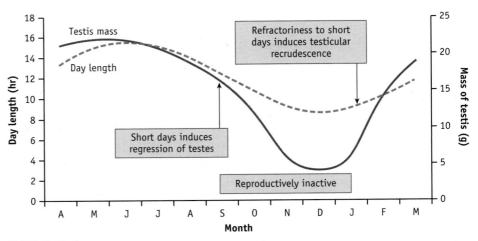

FIGURE 17-9

Regulation of testicular activity by photoperiod in the Syrian hamster. Exposure to short day lengths causes the testes to regress, causing cessation in reproductive activity during the winter months. However, after several months the hamster becomes refractory to short-day lengths and the testes begin to recrudesce. The animal gains full reproductive competence during the spring and summer when day length is long.

(Figure 17-10A). If ewes are maintained under a constant photoperiod for several years, they will continue to show recurring cycles of reproductive activity (Figure 17-10B). However, these cycles will eventually become unsynchronized (i.e., not all ewes at the same stages at the same time) and will not coincide with

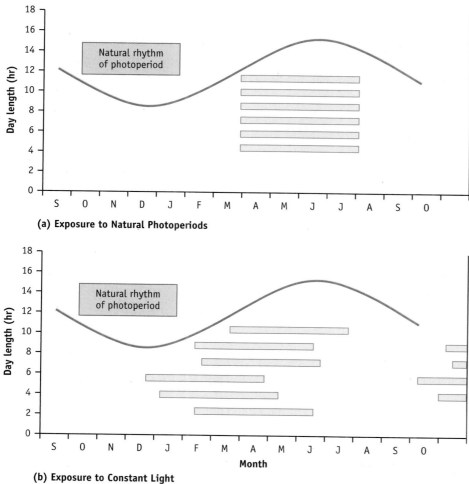

(a) Exposure to Natural Photoperiods

(b) Exposure to Constant Light

FIGURE 17-10

Expression of endogenous annual rhythms in reproductive activity (blue bars) in ewes exposed to constant day length for more than 2 years. When ewes are maintained under natural conditions where photoperiod fluctuates in an annual cycle, the onset and cessation of reproductive activity is approximately the same for all animals. In contrast, when ewes are exposed to constant day length (e.g., 12L:12D), each ewe continues to show distinct anestrous periods, but there is tremendous individual variation in the timing of these periods and they gradually become out of phase with the natural rhythm of photoperiod. This response has been interpreted to mean that sheep have an endogenous rhythm in reproductive activity that becomes entrained to the natural rhythm in photoperiod.

17

natural cycles of day length (reproductive activity does not necessarily occur during short-day lengths) because the periods of these cycles are less than 1 year. This response illustrates the existence of an endogenous cycle of reproductive activity (circannual rhythm). When sheep are exposed to natural fluctuations in photoperiod, their internal rhythms in reproductive activity become entrained by the external rhythm of day length.

The synchrony between the ewe's annual cycle of reproduction and the annual cycle of day length is evolutionarily significant in two ways. First, this strategy synchronizes the estrous cycles of ewes. Second, it causes lambing to occur at a time when food availability favors lactation and survival of young lambs.

Other Environmental Cues

How might animals predict opportune times for breeding when photoperiod is not a reliable indicator of environmental conditions? For example, due to dynamic weather conditions, photoperiod is not a reliable predictor of the onset and termination of plant growth in high-altitude environments such as the one found in the American Rocky Mountains. The montane vole appears to rely on the emergence of some types of grass as a predictor of availability of green grass. The stimulus appears to be a compound that is present in high concentrations in fresh green shoots. Other plant compounds may act as predictors of environmental conditions in other species.

NEUROENDOCRINE MECHANISMS MEDIATING THE EFFECTS OF ENVIRONMENT ON REPRODUCTION

The interface between an animal's external environment and its reproductive activity involves neuroendocrine mechanisms. In other words, the ability of an external stimulus to influence gonadal activity requires detection of the stimulus by the central nervous system, generation of a neuronal signal, and transformation of the neuronal signal into an endocrine signal that effect changes in reproductive activity. Our understanding of interfaces between the environment and the reproductive system is most complete with respect to how changes in photoperiod and feed intake affect gonadal activity. There is also a fair amount of information concerning how social cues influence reproduction.

Food

The importance of adequate nutrition in developing and maintaining reproductive activity is well documented. One of the more familiar examples is the relationship between nutrition, growth and onset of puberty. As noted earlier, growth rate during the prepubertal period is inversely related to age at puberty. It is also well known that infertility occurs when adults lose significant amounts of body weight. For example, amenorrhea and anovulation are common in women who engage in intensive athletic training or who suffer from various types of anorexia. A nutritional anestrus is also well documented in livestock. The vast majority of studies dealing with the relationship between food intake and reproduction have involved females. However, there is sufficient evidence to support the idea that food restriction reduces fertility in both males and females and involve essentially the same neuroendocrine mechanisms.

Several hypotheses have been developed to explain how nutrition influences gonadal activity. According to the first hypothesis, a so-called critical amount of body fat is required for normal reproductive activity. In spite of the popularity of this hypothesis, there is no empirical data to support it. There is support for modifications of this hypothesis. There is a growing consensus that some sort of metabolic signal, rather than body fat, mediates the effects of food intake on reproduction (Figure 17-11). According to the first version of this hypothesis, metabolic hormones (insulin, growth hormone, etc.) that reflect different metabolic states act as signals that regulate the reproductive system. Alternatively, the availability of metabolic fuels (e.g., glucose, fatty acids, and amino acids) acts might act as signals mediating the effects of nutritional status on reproductive activity. In either case, metabolic signals that reflect feed restriction might act to suppress pulsatile LH secretion, whereas signals that reflect the well-fed state might enhance pulsatile LH secretion. There are ample data showing that feed restriction reduces LH pulse frequency by suppressing the pulsatile release of GnRH. Such decreases in LH secretion impair gonadal function by disrupting steroidogenesis and gametogenesis.

Photoperiod

In order to understand how photoperiod regulates seasonal breeding, it is important to consider the following issues: 1) How do animals keep track of seasonal changes in day length? and 2) How are signals that provide information about these changes transformed into signals that affect the reproductive system? The most complete answers to these questions come from work with the ewe. The answer to the first question requires an understanding of the neuronal pathways that monitor photoperiod. The retina is the only photoreceptor in mammals and is therefore the most likely candidate for

17

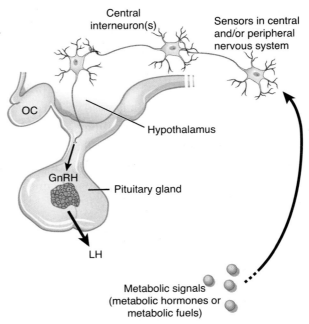

FIGURE 17-11

Hypothesis explaining the relationship between nutritional status and reproductive activity. According to this hypothesis, various signals that reflect an animal's metabolic status (i.e., metabolic signals) are detected by chemoreceptors located within the central and(or) peripheral nervous systems. These sensors monitor the metabolic status of the animal and regulate GnRH release via various afferent neuropathways.

detection of photic stimuli. Information about daylight is transmitted from the retina into the central nervous system via the retino-hypothalamic tract (Figure 17-12). Neurons from the retinal photoreceptors innervate the supra-chiasmatic nuclei and impinge upon other neurons that project to the superior cervical ganglion via the accessory optic tract. Neurons of this tract interact with ganglionic neurons, which re-enter the brain and terminate at the pineal gland and endocrine gland that produces melatonin. This hormone is released into the blood (the pineal gland is a circumventricular organ and is not protected by a blood-brain barrier) as well as the cerebrospinal fluid. Melatonin then acts on the hypothalamus to regulate seasonal changes in the pattern of GnRH release, which effects changes in gonadal function.

The role of the pineal gland is to provide information about daily photoperiod. Melatonin is secreted in a circadian pattern; that is, secretion is minimal during light and maximal during darkness. Thus the pattern of

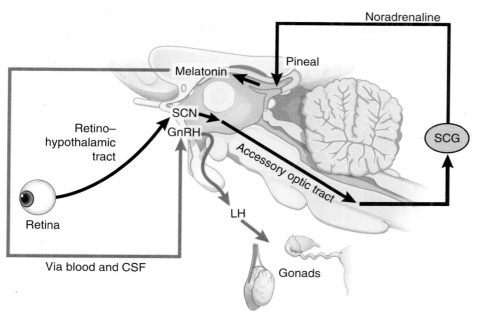

FIGURE 17-12

Schematic illustration of the neuroendocrine pathway mediating the effects of photoperiod on reproductive activity in mammals. This particular illustration corresponds to the rodent, but a similar mechanism appears to exist in other species. Information about day length is detected by the retina and transmitted to the hypothalamus via a retino-hypothalamic neuropathway. Direct neural inputs from the optic nerve enter and terminate in the suprachiasmatic nucleus (SCN) of the hypothalamus. Information from the SCN is ultimately transmitted to the superior cervical ganglion (SCG) via a series of interneurons that make up the accessory optic tract. Sympathetic neurons from the SCG project to the pineal gland and regulate release of melatonin. The pattern of melatonin depicts the length of the dark period. Somehow this signal influences neurons controlling GnRH release. Note that the pathway between the retina and pineal gland is neural (black arrows), whereas the pathway between the pineal gland and gonad is endocrine (red arrows).

melatonin release provides an accurate depiction of the daily light-dark cycle (Figure 17-13).

The precise role of melatonin in regulation of seasonal breeding patterns in sheep is unclear. However, reproductive biologists agree that melatonin itself does not stimulate or inhibit reproductive activity. Its role appears to be more of a time-keeping signal than a driver of the seasonal transitions between gonadal activity and quiescence. One well-accepted hypothesis is that patterns

17

FIGURE 17-13

Circulating patterns of melatonin during the light-dark cycles of long and short day lengths. Note that melatonin is elevated only during the dark period.

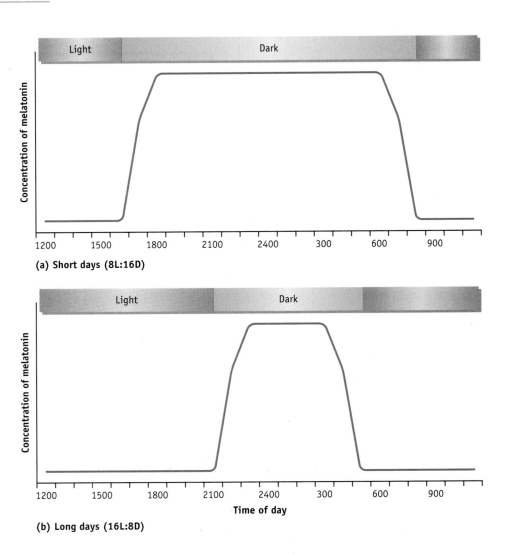

(a) Short days (8L:16D)

(b) Long days (16L:8D)

of melatonin mediate the synchronization between an endogenous circannual rhythm in reproductive activity and the external annual rhythm in day length.

Although our understanding of how melatonin regulates hypothalamic release of GnRH is limited, there is a good deal of information regarding the effects of photoperiod release of gonadotropins. The pulsatile pattern of LH release appears to play a central role in mediating the effects of photoperiod on gonadal activity. Much of our understanding of these effects comes from work with ewes. We have already considered the regulation of the ovarian cycle of ewes in Chapter 10. Recall that estrus and ovulation require development of a pre-ovulatory follicle, which emerges from a wave of follicle growth. During seasonal anestrus, ewes exhibit waves of follicle growth, but dominant follicles fail to attain the pre-ovulatory stage of development. This is due to

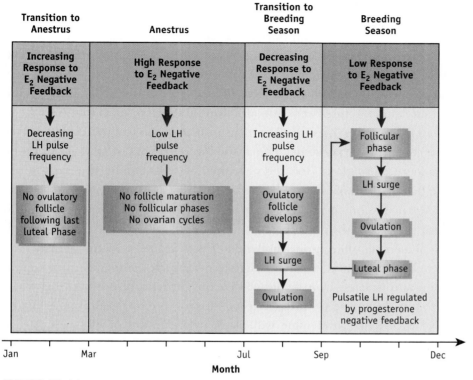

FIGURE 17-14

Effects of day length on the hypothalamic-pituitary-ovarian axis in ewes. During the transition from the breeding season to anestrus, the responsiveness of the hypothalamic pituitary axis to the negative feedback action of E_2 increases, thereby reducing frequency of LH pulses. Due to the decrease in LH, follicles do not reach the pre-ovulatory stage and the ewe neither expresses heat nor an LH surge. During anestrus, the lack of high-frequency LH pulses prevents follicle maturation. When photoperiods decrease during the transition to the breeding season, response to E_2 negative feedback decreases, permitting high-frequency LH pulses which stimulate follicles to the preovulatory stage. This leads to the first estrous cycle. Estrous cycles are sustained so long as pulsatile LH secretion increases following regression of corpora lutea.

the fact that LH is released in a low-frequency pattern during anestrus. In sheep, exposure to long days results in a decrease in LH pulse frequency. This is the result of steroid-dependent and steroid-independent mechanisms. The steroid-dependent mechanism involves an increase in responsiveness to estradiol negative feedback, a state that resembles the pre-pubertal period. Figure 17-14 summarizes the current theory for the hormonal control of seasonal breeding in the ewe.

17

Social Factors

There are no studies to support the idea that social factors can serve as cues to regulate seasonal breeding in mammals. However, such cues can and do affect reproductive activity in ways that can dramatically alter seasonal patterns of reproduction. One of the more dramatic examples is the so-called ram effect in ewes. Introduction of a ram to ewes that have been isolated from males for several months induces high-frequency patterns of LH release during anestrus. The response is attributed to a pheromone that is secreted by glands located near the horn pits and becomes dispersed in the fleece. Because such an increase in LH secretion is necessary for estrus and ovulation, it appears that introduction of a ram to ewes might disrupt the anestrous period.

Similar male pheromone-mediated effects on reproductive activity have been well documented in rodents. The estrous cycles of female mice kept in a group and isolated from males will become synchronized following exposure to a male mouse **(Whitten effect)**. Moreover, exposure of newly-mated female mice to a novel male will block implantation of embryos **(Bruce effect)**. These effects are caused by pheromones present in the male's urine that presumably disrupt the reproductive activity of females by acting on the hypothalamic-pituitary system.

BOX 17-1   Focus on Fertility: Effects of Space Travel on Reproduction

When biologists speak of environmental effects on the reproductive physiology of animals it is generally assumed that the environment being considered is the one here on Earth. What about the extraterrestrial environment? During the past 40 years space travel has become common. The International Space Station is continually inhabited by humans and is viewed by some as a step toward the colonization of space. The colonization of space will require that humans, and possibly other animals, spend long periods of time away from Earth. Under these circumstances it is inevitable that procreation will be attempted. What are the chances that life forms that evolved within the Earth's $1 \times g$ gravitational field can reproduce successfully in hypogravity ($<1 \times g$)? No one really knows, but there is a growing number of studies to address this question. At this time there is only one report documenting the mating of a mammalian species during a space-flight. In this case a male rat was reported to mate with several female rats, but no pregnancies resulted. Investigators speculated that the embryos were resorbed, but it is also possible that the lack of pregnancies was attributed to infertility of the males and/or females.

Very few studies have involved assessment of fertility during or after actual space flights. Due to the expense and difficulty of conducting long-term studies in space, many investigators rely on models that simulate spaceflight. One model involves confining humans to bed rest with a 6-degree head-down tilt. The resultant shift in body fluid and reduced resistance placed on muscles reproduces major symptoms of hypogravity. A similar approach has been developed for rats. In this case the animal is suspended from the base of its tail to achieve a 30-degree head-down position.

Based the results of experiments conducted during actual or simulated spaceflights it is reasonable to conclude that hypogravity affects reproductive traits in both males and female.

There is general agreement that actual and simulated spaceflight have negative effects on testicular function in males. Male rats experience a reduction in testes weight and a decrease in circulating concentrations of testosterone regardless of the experimental model. In most, but not all, cases these effects are accompanied by reduced spermatogenesis. Men subjected to 60 to 120 days of bed rest with head-down tilt express altered sperm morphology and a reduced number of motile sperm. The physiologic basis for these results is unclear. However, it is worth noting that unlike the human, the inguinal canals of rats does not close. Therefore, the reduced sperm count in rats exposed to actual or simulated hypogravity may be attributed to retraction of the testes into the abdominal cavity. Recall that body temperature is detrimental to spermatogenesis.

More recent studies have focused on the effects of hypogravity on sperm activity. Bovine sperm subjected to 360 seconds of freefall and sea urchin sperm sent on space shuttle missions exhibited higher velocities than sperm kept at $1 \times g$. The impact of these effects on fertility remain unclear at this time.

Our knowledge of the effects of hypogravity on female reproduction is even less than that for males. There is no information on the effects of low gravity on the ovaries of nonpregnant females, but weights of ovaries and numbers of preovulatory and atretic follicles in postpartum rats that were in space between days 9 and 20 of pregnancy were no different from those of rats that remained on Earth during the same time period. Finally, of the three experiments involving female rats carried on spaceflight mission, only one showed detrimental effects of hypogravity on pregnancy. In this case, rats exposed to spaceflight during gestation had reduced weight gain, prolonged parturition and gave birth to lighter pups that showed lower perinatal survival rates. There is no information concerning the effects of actual spaceflight on the menstrual cycle of women because female astronauts are required to suppress ovarian cycles with the birth control pill. Studies involving women exposed to simulated spaceflight conditions are limited and inconclusive.

Whether or not humans can eventually colonize space depends on whether the physiologic processes that regulate our reproductive processes will function properly under the conditions that prevail in extra-terrestrial environments. Microgravity is one such condition. However, there are other environmental variables that might also pose barriers to mammalian reproduction. In addition to hypogravity, people who travel in space will be exposed to various types of electromagnetic radiation as well as periods of hypergravity, each of which can exert their own effects on reproductive processes.

17

SUMMARY OF MAJOR CONCEPTS

- Reproductive success depends on the ability of an animal to coordinate its reproductive activity with environmental conditions that are conducive to successful reproduction.

- Harmony between an animal's reproductive state and favorable environmental conditions can be achieved by responding directly to food availability, or by using environmental cues to predict when food availability is sufficient to support reproduction.

- The extent to which sufficient calories will be available to support reproduction depends on food availability, ambient temperature, and energy stores.

- The annual rhythm in photoperiod is highly repeatable and therefore serves as the most common environmental cue to synchronize reproductive cycles with food availability.

- Photoperiod and metabolic status influence reproductive activity via neuroendocrine mechanisms that regulate pulsatile LH secretion.

DISCUSSION

1. It is generally agreed that changes in food availability cause changes in reproductive activity. Differentiate between food as a proximate factor and food acting as an ultimate factor affecting reproductive activity. Give an example of each case.

2. Imagine that you are observing two groups of ewes. One group is allowed to consume feed on an ad libitum basis during the anestrous season, whereas the other group is maintained on a diet that is below maintenance (i.e., the animals lose approximately 20 percent of their body weight during anestrus). Would you expect each group to begin showing estrous cycles at the same time in the late summer/early autumn? Explain your answer.

3. *Bos taurus* (domestic cattle) are not seasonal breeders in the sense that cows will express regular estrous cycles throughout the year if they are provided with adequate nutrition. What would you expect the reproductive patterns of cows to be when they are maintained strictly under range conditions in temperate climates? Explain your answer.

4. When female rats with good body condition (ample fat stores) are given a drug that blocks the metabolism of glucose, they continue to show

estrus and ovulate. However, if the same treatment is administered to thin females, they will fail to express estrus and ovulation. Explain these results.

REFERENCES

Bronson, F.H. and P.D. Heideman. 1994. Seasonal regulation of reproduction in mammals. In: E. Knobil and J.D. Neill, *The Physiology of Reproduction Vol. 2.,* Second Edition. New York: Raven Press, pp. 541–585.

De Rensis, F. and R.J. Scaramuzzi. 2003. Heat stress and seasonal effects on reproduction in the dairy cow—a review. *Theriogenology* 60:1139–1151.

Engelmann, U., F. Krassnigg, and W-B Schill. 1992. Sperm motility under conditions of weightlessness. *Journal of Andrology* 13:433–436.

Foster, D.L. and S. Nagatani. 1999. Physiological perspectives on leptin as a regulator of reproduction: Role in timing puberty. *Biology of Reproduction* 60:205–215.

Gangrade, B.K. and C.J. Dominic. 1984. Studies of the male-originating pheromones involved in the Whitten effect and Bruce effect in mice. *Biology of Reproduction* 31:89–96.

Gerlach, T. and J.E. Aurich. 2000. Regulation of seasonal reproductive activity in the stallion, ram, and hamster. *Animal Reproduction Science* 58:197–213.

Goodman, R.L. 1994. Neuroendocrine control of the ovine estrous cycle. In: E. Knobil and J.D. Neill, *The Physiology of Reproduction Vol. 2.,* second Edition. New York: Raven Press, pp. 659–711.

Short, R.E. and R.A. Bellows. 1971. Relationships among weight gains, age at puberty, and reproductive performance in heifers. *Journal of Animal Science* 32:127–131.

Schillo, K.K. 1992. Effects of dietary energy on control of luteinizing hormone secretion in cattle and sheep. *Journal of Animal Science* 70:1271–1282.

Tash, J.S. and G.E. Bracho. 1999. Micogravity alters protein phosphorylation changes during initiation of sea urchin sperm motility. *FASEB Journal* 13(Suppl):S43–S54.

Tou, J., A. Ronca, R. Grindeland and C. Wade. 2002. Models to study gravitational biology of mammalian reproduction. *Biology of Reproduction* 67:1681–1687.

17

Ethical Issues in Reproductive Physiology

- Define ethics.

- Explore the relationship between science and ethics.

- Discuss approaches to addressing ethical issues in reproductive physiology.

INTRODUCTION

Our desire to satisfy our curiosity about the world may be sufficient justification for studying the reproductive physiology of mammals. Nevertheless, from the perspective of society, such endeavors require more pragmatic reasons. As noted in Chapter 1, the major reason industrialized societies allocate resources to studying reproductive physiology is directly related to a desire to achieve a sustainable balance between human populations and vital resources such as food. Thus far, our emphasis has been on developing a scientific understanding of the physiologic processes controlling reproduction in mammals. However, our appreciation for this knowledge would be incomplete without also considering the social implications of putting this knowledge to use. In this chapter we will explore ideas concerning how we should use this knowledge. In other words, we will consider some ethical perspectives of reproductive physiology.

WHAT IS ETHICS?

One way to gain appreciation for what **ethics,** also known as moral philosophy, is all about is to consider a popular issue such as whether or not we should regulate human birth rates with methods that terminate rather than prevent pregnancy; that is, induced abortion. There is no question that we have the technological means to do this. Our knowledge of human reproduction has given rise to surgical and nonsurgical methods that effectively terminate pregnancy. The question of whether or not we *ought* to do this is more controversial. Ethics seeks to provide answers to how people ought to behave. So, the question of whether or not we should use abortion as a means of birth control is an ethical question.

The reason such questions are so controversial is that there are usually more than one answer to them and there are typically good reasons to support opposing answers. Who is to say which view is the correct one? Is there only one correct view? Can there ever be agreement on such controversial issues? In this chapter we will explore how ethics might help us understand and possibly address issues such as these.

Normative Ethics

Each of us is confronted with a multitude of ethical decisions each day. Should I attend class? Should I check on my friend who is ill? Should I purchase clothing made by people who work in sweatshops? Should I donate my time or money to a charity? When you choose a course of action for these types of questions you have made an ethical judgment. When you tell a friend that she should return the credit cards she found, you are giving ethical advice. When you tell a teacher that his grading policy is unfair, you are making an ethical evaluation. This type of ethical activity is typically referred to as normative ethics. This involves making and/or evaluating judgments about what ought to be done or giving advice about such ideas based on prescribed norms and standards. It is important to note that each day we make many of these types of judgments almost automatically, without much critical reflection.

Philosophical Ethics

At times we encounter situations where it is difficult to make snap decisions about what should be done. For example, you might have to decide between studying for an exam and having lunch with a good friend. You may be torn between the obligations you have to yourself and the ones you have to your friend. A decision regarding what you should do may require some thought. In other words, you weigh the pros and cons of each course of action and then make a decision as to which obligations are stronger in this case. When you start analyzing and evaluating your options in order to make an ethical judgment, you are engaged in philosophical ethics. Philosophical ethics involves analyzing (taking apart) and critiquing normative judgments along with the reasons that support these judgments. An ethical judgment along with its supporting reasons (premises) is known as an ethical argument. Philosophical ethics deals with the critical analysis of ethical arguments.

Ethical Theory

Philosophers who deal with the theoretical aspects of ethics are said to be ethical theorists. Work in ethical theory involves developing ideas regarding the nature of good and right as well as how we should go about promoting these values. Ethical theories develop from our cultural notions of what is

18

good and right. Ethical theorists attempt to clarify these ideas and transform them into law-like formulations. For example, it was long recognized that people seek happiness and try to avoid suffering and that actions that increased happiness and/or minimized suffering are considered good, whereas actions that reduce happiness and/or increase suffering are considered bad. Some ethical theorists took these ideas and formulated what is known as the utilitarian theory of ethics. Briefly, the theory embraces the notion that we should judge our actions as good or bad based on the extent to which they promote happiness and/or minimize suffering. Another theory that is quite influential in our society is deontological ethics. According to this view, each of us has a duty to respect the individual rights of others. We will discuss these ideas in more detail later in this chapter.

Ethics and Social Issues

We can now return to our original question of how ethics might help us address controversial issues associated with reproductive physiology. At the very least, ethics offers us a means to understand the nature of these issues. Analysis of opposing arguments illuminates the sources of disagreement between two perspectives. Opposing views are often grounded in different ethical theories, or different interpretations of a particular ethical theory. Consider the issue of abortion, for example. One of the major sources of disagreement between pro-abortion and anti-abortion groups deals with the question of whether or not a human conceptus has a right to life. In this case, there is disagreement over how rights-based ethics should apply to this issue.

Can ethics be more than just a tool to understand disagreements over what should be done? In theory, the answer to this question is yes. One of the fundamental assumptions of Western ethics is that humans should choose the most rational course of action because it is in our nature to do so. In other words, we should choose the course of action that is supported by the best reasons. Thus ethical truth (i.e., what is truly the right thing to do) is generally understood to be the best ethical argument. Many people are skeptical about this view. The issues with which we struggle most are extremely difficult to resolve because each of the opposing groups is convinced that they are right and the other groups are wrong. Compromise seems out of the question over issues such as abortion. Ethical analysis can reveal whether or not an argument is valid, but in cases where each of the competing arguments is valid, we may have to accept that there might be more than one right answer (i.e., an ethical pluralism). Nevertheless, societies are confronted with the task of addressing issues, while at the same time dealing with the reality

18

of ethical pluralism. Frequently such issues are settled politically rather than ethically. Because of this problem, some ethicists (pragmatists) focus their attention more on how ethical issues should be addressed politically rather than the analysis and critique of ethical theories and ethical arguments.

ETHICS AND SCIENCE

Having described the nature of ethics, we can now consider how science and ethics are related. The general goal of science is to explain how the world is. Scientific inquiries are aimed at developing detailed, meticulous, well-documented, and verifiable accounts of the world. In order to eliminate bias, scientific investigators work toward being objective in their descriptions and explanations. In science, objectivity requires that accounts of the world be impartial (not favoring a particular perspective), accurate (without error), and rational (interpretations of data are limited to only what the results support).

Although objectivity is the goal of science, it is not an outcome that occurs automatically from our scientific methods. Science is ultimately a human activity and humans, even scientists, can be biased, make errors, and behave irrationally. In fact, objectivity may be more of an ideal that scientists strive to achieve knowing full well that is never fully achieved. It may be more appropriate to say that a scientist can develop an objective attitude than to say that a scientist can be completely unbiased, accurate, and rational. In this sense, objectivity is a virtue much like honesty or bravery.

Why can't science be truly objective? The main reason is that all scientific theories are value laden to some extent. In other words, scientific theories embody the personal preferences of scientists as well as the values of society. Values, including moral values, influence what scientific questions are posed, how scientific research programs are developed, how scientific experiments are conducted, and how scientific data are interpreted. To put it simply, the scientific answers we obtain depend to a large extent on the questions posed by scientists, and there is no reason to assume that such questions are not biased in some way. This becomes clear when we consider who sets the agendas for scientific research. To a large extent, the questions scientists address are those of interest to the people willing to pay for scientific research; that is, government agencies and private industry. These forces shape not only which issues will be addressed but also how they will be addressed. Consider the issue of starvation in nonindustrialized regions of the world. This problem can be viewed as a question of resource production, resource consumption, or overpopulation. If the issue is viewed as one of production, then research, on production of food makes sense. On the other hand, if the issue is viewed as one

18

of consumption, then it makes sense to support research on the efficiency of resource use. Finally, if the issue is viewed as one of overpopulation, it seems logical to direct research toward human reproduction. There is evidence to support each view, but the perspective one chooses determines how one will address the issue. This illustrates how three different sets of scientific facts can serve three different approaches to research, each of which advocates a different set of perspectives and values. Perpetuating one particular approach (e.g., increasing food production) to scientific research promotes only one concept of what the world should be like (e.g., using particular technologies that increase crop yields); that is, it promotes a particular concept of ethics.

Reasoning from Facts to Values

Our faith in the objectivity of science often compels us to conclude that an ethical judgment can be derived from scientific facts. As noted earlier, the opposite is more likely true: that is, an ethical perspective leads to construction of scientific facts. Arguments that draw ethical conclusions strictly from empirical observations are irrational. The following example illustrates this. Suppose the prevailing view is that starvation is the result of overpopulation and this prompts international agencies to provide huge sums of money for scientists to engage in research on developing and distributing new birth control methods to nations with high birth rates. Population data clearly show that the populations of some nations either do or will surpass the ability of these countries to feed all of their people. Clearly, a reduction in birth rates will help alleviate this problem. This might lead you to conclude that measures to reduce birth rate are good and therefore should be imposed (an ethical conclusion). The argument to support this conclusion can be written as follows:

- Conclusion: Birth control policies should be imposed in nations with large populations and limited resources in order to reduce birth rates and ultimately create a sustainable balance between population and food.

- Premise 1: The populations of these countries will soon be too large to support with available resources.

- Premise 2: Reducing birth rates will bring populations down to sizes that can be sustained with available resources.

- Premise 3: Policies that promote birth control reduce birth rates.

At first glance the argument may appear to be both valid (logical) and sound (based on true premises). However, closer examination reveals that the argument is not at all valid. In fact it is an example of jumping to conclusions.

18

Notice that the conclusion is an ethical conclusion; that is, it prescribes what ought to be done about the issue of starvation. Also notice that each of the premises is an empirical claim. Descriptions of how the world *is* do not necessarily lead us to conclusions regarding how the world *ought* to be. There seems to be a missing premise; that is, one that makes some general ethical claim. Consider the following modification of the previous argument in support of birth control:

- Conclusion: Birth control measures should be imposed in nations with large populations and limited resources in order to reduce birth rates and ultimately create a sustainable balance between population and food.

- Premise 1: The populations of these countries will soon be too large to support with available resources.

- Premise 2: Reducing birth rates will bring populations down to sizes that can be sustained with available resources.

- Premise 3: Policies that promote birth control reduce birth rates.

- Premise 4: A society should take whatever actions are necessary to create a sustainable balance between population and food.

By including premise 4, the argument becomes valid. In other words, the conclusion is now a direct consequence of the premises. Unlike premises 1 through 3, premise 4 cannot be proven to be true or false simply by making empirical observations. This premise is an example of a **general moral premise;** that is, a widely accepted, law-like claim regarding what is right. The truth value of this type of premise is based on our judgment of what is appropriate based on existing norms and standards as well as our experiences. Often these premises are left out of arguments concerning public policy. Nevertheless, when arguing in favor of a particular course of action, it is just as important to be clear about the general moral premise as it is to be accurate in using factual premises. As noted earlier, society embraces different values and there is often disagreement regarding which value is appropriate for a particular issue. Some people may argue that societies should not use any measures to develop a sustainable society if such measures infringe upon the rights of individuals.

Integrating Science and Ethics

The main point to be gleaned from the previous discussion is that scientific accounts of the world are often biased and that such bias is not necessarily detected and/or eliminated by conventional scientific methods as is frequently assumed. Science alone cannot discern, let alone evaluate, the ethical perspectives it embraces. Scientific research aimed at developing methods for reducing birth rate cannot address the question of whether or not birth rate ought to be reduced. This is an appropriate role for ethical analysis. Thus, if society expects to have a science that is responsible in the sense that it perpetuates a desired ethical perspective, then it has to find a way to integrate science and ethics. This can be accomplished if we reject the belief that science is valuefree and insist that scientific research be evaluated from both ethical and scientific perspectives. In other words, each of us, scientist and nonscientist alike, should ask not only if it can be done, but also whether or not it should it be done.

Based on the previous chapters it is clear that our knowledge of the reproductive physiology of mammals is extensive. This knowledge has been used to manipulate the reproductive activity of many mammalian species, including humans. What sorts of ethical issues are associated with this understanding? In general we can divide these issues into two categories: 1) issues arising from how reproductive physiology research is conducted and 2) issues arising from the way our knowledge of reproductive physiology is put to use. The former area includes consideration of research ethics; that is, how research should be done. This would include important questions concerning if humans and nonhuman animals should be used in research and, if so, how they should be treated when serving as research subjects.

The second type of ethical issue deals with reproduction technologies. A general question is whether or not we should attempt to manipulate the reproductive activity of humans and nonhumans. More specific questions deal with whether or not particular technologies should be developed and used. As mentioned earlier, the dominant ethical concerns of society involve humans. With respect to the use of reproduction technologies, an important issue is whether or not human embryos and fetuses should be destroyed as a means to prevent birth, or to provide stem cells for research and medical treatments. This is not to say that there are no ethical issues concerning the use of these technologies in nonhuman animals. If these animals are to be included in the so-called moral community, then it is appropriate to consider whether or not such technologies affect their well-being as well. This of course raises a more general question; should animals be raised and killed for human use?

It is not feasible to provide comprehensive ethical analyses for all of the ethical issues related to reproductive physiology. The purpose of this chapter is to describe how ethics can be used to better understand controversial issues associated with the field of reproductive physiology, with the hope of finding reasonable and fair solutions to these issues. The following sections will provide ethical analyses of two important issues: 1) Should abortion be allowed? and 2) Should nonhuman animals be used for research and food? The controversial question underlying both of these issues is, who or what should be considered part of the moral community? Disagreement about the nature of the moral community provides the basis for other ethical issues in reproductive physiology; for example, whether or not human embryos should be cloned in order to provide stem cells (see Box 18-1).

ETHICS AND THE HUMAN FETUS (ABORTION)

Abortions were permitted in the United States until the beginning of the twentieth century. The motivation for making abortion illegal stemmed from a desire to discourage illicit sexual activity, a belief that abortion was not a safe medical procedure and the concern among some groups that it was morally wrong to kill a fetus. However, in January 1973, the Supreme Court ruled on the now famous case of *Roe v. Wade* thereby legalizing the practice of abortion in this country. The rationale for striking down laws that prohibited abortion was that they violated a woman's constitutionally-protected right to privacy. It is noteworthy that the Court did not make a judgment concerning whether or not a fetus should be considered a person in the ordinary sense of the word. It did, however, conclude that a fetus is not a person in the *legal* sense. The majority opinion of the court includes the following statement:

> …[N]o case could be cited that holds that a fetus is a person within the meaning of the Fourteenth Amendment . . . the word "person," as used in the Fourteenth Amendment, does not include the unborn . . .

The important implication of this conclusion is that a human fetus has no rights that are protected by the constitution of the United States. But does this make abortion right? What is legal is not necessarily ethical. Recall that it was once illegal for women to own property or to vote. Laws that granted these rights to women were enacted only after there was growing consensus that it was unethical to deny a woman such rights. Whether or not abortion is a practice we ought to allow is an ethical issue as long as there is disagreement over whether or not it should be allowed. Although resolution of this issue seems elusive, the following discussion of ethical arguments concerning the moral status of abortion is offered in the spirit of achieving a deeper understanding of this important issue.

18

Is a Human Fetus a Person?

Before delving into a detailed discussion of this issue, it is important to clarify some of the language used in the ethical and legal literature concerning abortion. As described in an earlier chapter, development of the conceptus is typically divided into embryonic and fetal phases. In the discourse concerning abortion, the term embryo and fetus are used interchangeably. For convenience and consistency, the term fetus will be used in this chapter as a general term to refer to the conceptus.

Much of the debate concerning the morality of abortion hinges on whether or not the human fetus should be considered to be a person in the moral sense. This aspect of the debate reflects the traditional assumption that only persons have moral standing. One of the earliest and most influential ethical arguments against abortion was developed by John Noonan, a law professor. According to Noonan a human being is created at the time of conception. Therefore a human fetus has moral standing that confers on it the right to be protected from harm. Noonan concludes that abortion is rarely justified.

Noonan's argument rests on the premise that a human fetus is a person. In order to verify this premise, he first discusses how the fetus can be classified as a human being, and then focuses on refuting various arguments that make a distinction between a fetus and a person. Noonan refers to the Christian theological concept of "ensoulment" to demonstrate that a human conceptus is a person and insists that this idea does not require a theological basis and can be understood in secular terms. What theologians call a human soul can be translated to the term "rational soul" or, in more scientific terms, the human genome. Basically Noonan asserts that "a being with a human genetic code is a [hu]man," or to put it more simply, a human is anyone conceived by human parents. From this line of reasoning, Noonan concludes that a human being forms at the time of conception.

After arguing that personhood is established at conception, Noonan turns his attention to refuting claims that a human fetus can be distinguished from a person in morally significant ways. He notes that several major distinctions have received attention in the debate over abortion. These include viability (whether the fetus can survive on its own), experience (whether the fetus can have experiences), sentiments (how parents and other persons feel about the fetus), sensation (whether the parents can feel the fetus), and social visibility (whether the fetus is socially perceived as human). Noonan challenges these ideas by providing counterexamples that suggest these distinctions are unclear at best. For example, he argues that the ability to survive independently (i.e., viability) fails to provide a clear distinction between a fetus and a person because children between 3 and 5 years of age are absolutely dependent on another person's care. The problem with using such distinctions to establish a particular point in development is that development is a continuum without demarcations. For Noonan the clearest distinction between the human and nonhuman occurs at syngamy.

Mary Anne Warren offers one of the strongest challenges to Noonan's argument against abortion. A clearer understanding of the disagreement between Noonan and Warren can be gained by outlining the general structure of Noonan's argument.

- Conclusion: It is wrong to abort a human fetus.
- Premise 1 (General Moral Premise): It is wrong to kill innocent human beings.
- Premise 2: A human being is any living thing possessing a full complement of the human genome.
- Premise 3: Human fetuses possess the human genome.
- Premise 4: Human fetuses are innocent.
- Premise 5: Abortion kills human fetuses.

Warren's critique of Noonan's argument focuses on Premise 2. Her major criticism is that he fails to distinguish between two commonsense definitions of the term human; that is, one that has ethical significance (the moral sense) and one that does not (the genetic sense). Warren notes that Noonan defines the fetus as human in the genetic sense, but does not provide justification for assuming that this makes it human in the moral sense. Based on this criticism, she sets out to establish the boundaries of the moral community and argues that human fetuses fall outside of these boundaries.

"Genetic humanity," according to Warren, is not a sufficient condition for moral humanity. In other words, possessing the genes that give rise to human traits does not necessarily make something a person. What characteristics are required for personhood? Warren provides a "rough and approximate list of the most basic criteria": 1) consciousness, 2) the ability to reason, 3) expression of self-motivated activity, 4) the ability to communicate, and 5) self awareness. She asserts that these traits are consistent with commonsense notions of what it means to be a person and notes that human fetuses fail to meet any of these criteria. In other words, human fetuses are human beings who are not people. Warren goes on to argue that only persons have full moral rights, a position that is consistent with Western traditions of ethical theory. According to this tradition human fetuses do not have full moral rights because they do not meet any of the well-accepted criteria for personhood.

18

An obvious question might be, at what stage of development do human life forms become people? Noonan points out that after conception there is a high probability (80 percent) that a zygote will develop into a person. How does Warren address this challenge? She argues that at no time during pregnancy does a fetus meet the criteria necessary for it to have a right to life. As far as the fetus' potential to become a person is concerned, Warren acknowledges that there "may well be something immoral . . . about wantonly destroying potential people when doing so isn't necessary to protect anyone's rights," but asserts that any right to life a potential person might have "could not possibly outweigh the right of a woman to obtain an abortion, since the rights of any actual person invariably outweigh those of any potential person." It is important to note that this portion of Warren's argument deals with the issue of **moral significance**. Warren's main disagreement with Noonan involves the issue of **moral considerability**; that is, what criteria define what counts when making moral judgments (is the fetus a person?). This is a different question than asking which members of the moral community have the greatest value (whose rights carry the most weight?).

Moral Significance and a Fetus' Right to Life

Philosopher Judith Jarvis Thomson focuses on the issue of moral significance to develop a moral defense of abortion. To begin, Thomson takes issue with Noonan's assertion that a fetus becomes a person at the moment of conception as well as Warren's view that a fetus is not a person at any time during pregnancy. The problem with both views is that fetal development is a continuum of events. Thomson compares this phenomenon to the development of an acorn into an oak tree. It is unreasonable to conclude that an acorn is an oak tree simply because acorns develop into oak trees. Likewise, it is unreasonable to view the human zygote, embryo, or fetus as a person simply because these things develop into people. Both are examples of a slippery slope argument, which is a logical fallacy. In other words, it is not impossible to make reasonable distinctions.

Thomson is inclined to accept the assertion that a fetus becomes a person "well before birth" and is willing to accept, for the sake of argument, Noonan's view that this occurs at the time of conception. This means that Thomson also accepts the premise that the human fetus has a right to life. However she deviates from Noonan's view by arguing that a person's right to life does not outweigh a mother's "right to decide what shall happen in and to her body." Thomson relies on a clever analogy to make her point.

Imagine that you awaken one morning and find yourself in a hospital bed with an unconscious, renowned violinist who has a fatal kidney disease and requires continuous blood transfusions from a donor with a compatible blood type. The Society of Music Lovers searched medical records and found that you are an ideal donor. They arranged for you to be kidnapped and taken to a hospital where your circulatory system is surgically connected to the musician's. When you awaken the director of the hospital explains that he doesn't approve of what the Society of Music Lovers did to you. Nevertheless a disconnection at this point would kill the violinist. Therefore, he cannot unplug you because this would violate the musician's right to life. The director tries to console you by telling you that this won't be a permanent situation because the treatment has to last for only 9 months. Thomson asks if you have a moral obligation to accept this relationship with the violinist.

By constructing this scenario and asking if you have a moral duty to remain connected to the musician, Thomson sets the stage for exploring what it means to exercise respect for another person's right to life. Does this mean that we have a duty to provide everything someone requires to stay alive under any circumstances? How much of a moral burden must we take on in order to fulfill our duties to other people? In other words, how much does our own right to life matter in terms of our duties to another person's right to life? To address these types of questions, Thomson develops a detailed account of what it means to have a right to life.

To illustrate the nature of a person's right to life, Thomson develops a second thought experiment involving a popular celebrity. Suppose you are suffering from a life-threatening disease and the touch of the celebrity's hand will save your life. Although it would be nice if your friends traveled to Hollywood and convinced the celebrity to return with them and heal you, Thomson asserts that you have no right that demands that your friends do this. Thus as in the case of the violinist, "having a right to life does not guarantee having either a right to be given the use of or a right to be allowed continued use of another person's body—even if one needs it for life itself." Thus the notion that it is wrong to abort a fetus does not follow from the assumption that the fetus has a right to life.

Finally Thomson addresses the issue of whether or not there might be any circumstances during which it is unjust to kill a fetus. The assumption underlying this view is that there are conditions under which a woman ought to allow a fetus to use her body. According to Thomson, this may be the case, but this doesn't mean the fetus has a right to the mother's body. Consider the situation where an uncle gives a box of chocolates to two boys. If one of the boys decides not to share, we would consider his actions unjust and therefore

18

wrong. However, his decision not to share would be ethically acceptable if the chocolates were given only to him; that is, if he were the owner of the chocolates. It would be nice if the boy shared the candy, but he is not obligated to do so. Likewise, it might be generous for a woman to allow an unwanted pregnancy to continue to term, but in no way is the mother obligated to let the fetus use her body in this way. The reasonableness of this position is illustrated by the fact that no country in the world legally requires a person to be a so-called Good Samaritan; that is, requiring a person to make large sacrifices in order to sustain the life of another who has no right to demand them. Thus, although we may be disturbed when people witness a murder without attempting to deter the murderer, there are no laws for bringing charges against such witnesses. It should be noted that some nations have enacted so-called Good Samaritan laws. The intent of these laws is to protect from blame those people who offer aid to others who are injured or ill. In some cases, such laws require people to come to the aid of people in distress, but only when doing so does not place the aid-giver in harms way. The main point is that our general sense of moral obligation to others is not consistent with the idea that helping others does not mean that we have to suffer undue distress.

It may be unclear why the right to life of human beings warrants a moral responsibility to them following birth, but not before. Birth marks the time when the person is no longer occupying or using the mother's body. Thomson argues that we have no "special responsibility for a person unless we have assumed it." When parents decide to let a pregnancy continue to term and take the infant home with them instead of offering it for adoption, they have assumed responsibility for it thereby conferring rights upon it. Thus Thomson makes a distinction between the act of a removing fetus from a mother's body and the act of a mother securing the death of her child. She has the right to do the former, but not the latter. Thomson is quick to point out that her pro-abortion argument should not be construed to mean she is arguing in favor of a woman's right to secure the death of her "unborn child," and concedes that at some point in pregnancy the fetus has the ability to survive outside the mother's womb. At this point there might very well be a moral obligation to the fetus. Returning to the case of the violinist, if by some miracle he survives after unplugging yourself from him, it would be wrong to kill him. Likewise, it would be wrong to kill a fetus if it survived after detachment from the womb. Thomson seems to be advocating some constraints on abortion (restricting the practice to stages where the fetus cannot survive on its own) rather than advocating abortion on demand under any circumstances.

Why is it Wrong to Kill a Person?

The previous discussions of abortion clearly illustrate that debate on this issue hinges upon the concept of moral considerability and moral significance. Don Marquis finds these questions almost unsolvable and favors a more theoretical account of why abortion is immoral. He notes that although we accept that it is wrong to kill an existing human being, it is unclear why this is so. If we can articulate why it is wrong to kill an actual person, then we might be able to understand why it is wrong to kill a potential person. Marquis' answer is that it is wrong to kill someone because it deprives him or her from a future life. In other words, the wrong is done to the victim not to the murderer or the friends and relatives of the one who is murdered. Marquis views the loss of the victim's life as a loss of "all those activities, projects, experiences, and enjoyments which would otherwise have constituted [the] future personal life [of the victim]." Such a loss applies to all people whether they have been born or have yet to be born. According to Marquis, a future life has moral value and any act which deprives someone of a future life should be considered immoral.

A major concern with Marquis' argument deals with moral significance. Are all future lives likely to have the same value? First, it seems that one could argue that the future life of the mother has greater moral value than the life of her fetus. Second, it also seems reasonable to assume that some future lives are likely to be so miserable (e.g., due to an extremely painful and(or) debilitating illness) that termination before birth is justifiable.

MORALITY AND NONHUMAN ANIMALS

Our analysis of the abortion issue reveals that one of the major points of contention between pro-abortionists and anti-abortionists is disagreement over the nature of the moral community; that is, what things should count when making moral judgments. The same type of disagreement arises when considering how we should treat nonhuman animals. Extending moral status to nonhuman animals is no less controversial than extending this status to human fetuses.

Historical Perspective

Throughout the history of Western civilization humans have relied heavily on the use of nonhuman animals for a variety of purposes including food, fiber, power, entertainment, and more recently, scientific research. Many of these practices require killing and/or inflicting pain on these creatures. Much of

18

the treatment we impose on nonhuman animals would be considered unethical if applied to humans. The reason we humans seem so willing to inflict harm on nonhuman animals can be attributed to the idea that we have no moral obligations to them. Although this has been the dominant view for much of our history, there has always been disagreement regarding the moral status of nonhuman animals. During the past 30 years, a growing number of people reject the notion that humans have no moral obligations to nonhuman animals.

The ancient Greek philosopher Aristotle (384–322 BCE) argued that nonhuman animals have value only insofar as they serve the needs of humans; that is, that they have only instrumental moral value. The concept of instrumental value is illustrated by our use of nonhuman animals for food. According to this perspective, the reason to refrain from harming a cow is that it might make it less useful (e.g., produce less milk for human use), not because it causes the cow to suffer or is a violation of its moral rights. This theme is prevalent in the ideas of some of the most influential philosophers in Western culture. Thomas Aquinas (1224–1274), relying on biblical scriptures, professes that nature, including nonhuman animals, exists only for human benefit. Rene Descartes (1596–1650), viewed by many as "the father of modern philosophy," seems to take this idea to an extreme, asserting that nonhuman animals are merely machines, like clocks, and lack souls, free will, or consciousness. Thus it is not immoral to kill them, eat them, or harm them in any way because they aren't aware of what happens to them. Francis Bacon (1561–1626), often referred to as the "father of modern science," accepted the view that nonhuman animals have no moral value and like Descartes advocated the dissection of live animals (vivisection) to advance our scientific understanding of the world. Finally, Immanuel Kant (1724–1804), a German philosopher whose ethical theory on moral duty has shaped much of current views on morality, argued that humans have no direct duties to nonhuman animals because they lack rationality.

Not everyone shared the aforementioned views regarding the moral status of nonhuman animals. In ancient Greece, followers of Pythagoras (c. 550–c. 500 BCE) advocated vegetarianism on the grounds that they believed nonhuman animals should be treated with respect. Perhaps the most well-known advocate of animal well being is Jeremy Bentham, an English philosopher, who formalized the **utilitarian theory** of ethics. Bentham argued that it is not reason, but the capacity to suffer, that makes nonhuman animals morally considerable. These ideas were developed further in the late twentieth century in response to a change in society's views regarding the relationships between humans and nonhuman components of nature.

18

During the 1960s the so-called environmental movement was ignited by Rachel Carson's *Silent Spring*, a critique of agricultural insecticide use, and the Sierra Club's national campaign against damming the Colorado River inside the Grand Canyon National Park. These developments encouraged a critical questioning of traditional ethical theory. Western ethical theory was developed to clarify how humans should interact with each other. As noted earlier, these theories assume that we have direct ethical obligations to only human beings. Our duties to nonhuman elements of nature are viewed as indirect; that is, our actions toward these things matter only insofar as they impact other human beings. For example, it would be wrong to kill a person's pet dog because it is considered unethical to destroy the personal property of a human being or because it might perpetuate violence against humans, not because it harms the dog. As the environmental movement took shape, ethical theorists began questioning traditional views regarding the relationship between humans and the natural world and created the discipline of environmental ethics. One of the most significant issues addressed by this line of inquiry is whether or not nonhuman animals have moral standing. The first and perhaps most influential attempts to include nonhuman animals in the moral community came from applied ethicists; that is, ethical theorists who sought to apply traditional ethical standards to nonhumans.

Utilitarianism

As noted earlier, English philosopher Jeremy Bentham (1748–1832) is known as the founder of the reform-oriented perspective of utilitarianism. Bentham argued that the ultimate moral principle should be the "Principle of Utility."

> By the Principle of Utility is meant that principle which approves or disapproves of every action whatsoever, according to the tendency which it appears to have to augment or diminish the happiness of the party whose interest is in question; or what is the same thing in other words, to promote or to oppose happiness.

In other words, the interests of each being likely to be affected by a particular action should be taken into consideration when making a moral judgment; that is, the interests of one individual should not count any more than those of any other individual. Perhaps the most controversial aspect of Bentham's theory was the inclusion of nonhuman animals in the moral community. Bentham questioned the conventional view that the ability to reason was the necessary and sufficient condition that confers moral considerability on a being. According Bentham, "[t]he question is not, Can they *reason?* nor Can they *talk?* but, Can they *suffer?*

18

John Stuart Mill (1806–1873), a disciple of Bentham, refined these ideas in a way that utilitarianism is commonly understood today. However, he seems to place less emphasis on the interests of nonhuman animals. He implies that human interests carry more weight than those of nonhuman animals. Mill viewed an ideal state of affairs as one where everyone is as happy and well-off as they can be. His formulation of this theory is based on the "Greatest Happiness Principle:

> According to the Greatest Happiness Principle…the ultimate end, with reference to and for the sake of which all other things are desirable (whether we are considering our own good or that of other people), is an existence exempt as free as possible from pain, and as rich as possible in enjoyments.

According to Mill's view, each of us should act so as to bring about the greatest amount of happiness *for all those who will be affected*. It is important to emphasize that Mill and Bentham focus on individual happiness. The actions that produce the greatest amount of individual happiness are the right actions.

During the early 1970s, Peter Singer elaborated on earlier versions of utilitarian ethics to advocate an animal liberation movement. Singer argues that the social reform movements to end racism and sexism were based on a "basic principle of equality." In each of these cases, the interests of individuals who are part of a particular group are given less consideration than those who make up another group. In cases when the interests of two races conflict, racists violate the principle of equality by placing greater significance on the interests of members of their own race than on the interests of members of other races. Likewise, when the interests of males and females clash, sexists violate the principle by placing greater emphasis on the interests of members their own sex than on the interests of members of the other sex. Singer believes that such a principle can also be applied to nonhuman animals. Recall that in utilitarian theory the morally relevant trait is having an interest in avoiding pain. This basic interest applies both to human and nonhuman animals. Extending the basic principle of equality to nonhuman animals does not mean that they should be treated in exactly the same way as humans. Recall that Singer's principle entails equal *consideration* of interests. For example, both humans and horses have interests in avoiding pain, but this does not mean that they require the same conditions or treatments to address these interests. As Singer notes, a slap on the rear of a horse to move it forward may be insignificant, but a slap of the same force applied to an infant would be too harsh a treatment.

Singer portrays the use of nonhuman animals for research and food production as violations of the basic equality principle and refers to the attitude that justifies such practices as "speciesism." According to Singer, to be a speciesist (i.e., giving greater weight to the interests of humans than to those of nonhumans) is just as unethical as being a racist or sexist. Singer's view does not necessarily exclude the possibility of using animals for research or raising them and killing them for food or research. According to utilitarian theory, harm can be inflicted as long such harm doesn't diminish overall happiness *and* if there are no alternatives to inflicting the harm. This idea underlies current animal care guidelines that advocate use of pain relieving measures in animal research. However, Singer condemns most animal experimentation and livestock production because he believes that such actions do not enhance happiness and are unnecessary. With respect to the use of animals for research, Singer argues that the vast majority of medical research involving animals does not contribute to the overall well being of humans or animals. With respect to using animals for food, Singer argues that modern livestock production systems are cruel and that eating meat is unnecessary from a nutritional standpoint. Eating animals just because they taste good is not justified when other nonhuman foods can replace animal flesh.

Animal Rights Theory

American philosopher Tom Regan began writing about the moral status of nonhuman animals shortly after Singer's work appeared. Although Regan agrees with Singer's assertion that nonhuman animals have moral standing, he believes that Singer's approach was inadequate to prevent many of the harms suffered by animals.

Regan views the utilitarian principle advocated by Singer as a consequential (emphasizing results) ethic aimed at "bringing about the greatest possible balance of good over evil . . . the greatest possible balance of satisfaction over dissatisfaction, taking the interests of everyone affected into account *and* counting interests equally." Regan's concern with the utilitarian approach is that it permits too much harm to nonhuman animals. According to Regan, Singer assumes, but does not demonstrate, that a failure to follow the equality of interests principle will result in worse consequences than following it. In fact, Regan asserts, giving the interests of animals less weight than those of humans can actually *enhance* overall satisfaction. This is precisely the argument animal researchers and animal agriculturalists use to justify the use of animals for experimentation and food; that is, such pain is outweighed by the good it generates. Regan argues that, in theory, Singer's principle could be used to justify more radical or egregious forms of differential treatment of

18

animals (and humans for that matter). Is there a way to ensure that animals are protected from harm regardless of the consequences of our actions?

Regan's approach to including nonhuman animals in the moral community is grounded in the ideas of Kant, who is noted for his ethical theory based on the concept of moral duties. Kant embraced the idea that humans are self-conscious individuals capable of making rational decisions. From this concept of human nature Kant argued that morality should be about doing what is rational. His notion of moral duty is based on the idea of a categorical imperative. The following example illustrates this concept. Suppose you want to become an accomplished guitarist. If you want to become an accomplished guitarist, you must practice. This is a "hypothetical imperative." In other words, what you should or ought to do is based your desire, in this case your desire to become an accomplished guitarist. Another way of looking at this is to say that if you want a particular outcome A, you ought to do B. In such cases, you can escape the imperative simply by denying the desired outcome. If you don't want to be an accomplished guitarist, you don't have to practice playing the guitar. Kant's view of morality deals with a categorical imperative, obligations that do not depend on outcomes. According to Kant, a categorical imperative is a rational prescription of what one should do in *all* circumstances. Humans are bound to it because they are rational beings. A failure to act morally is a failure to act rationally. Perhaps the most familiar formulation of Kant's categorical imperative is:

> Act only according to that maxim by which you can at the same time will that it should become a universal law.

This is a familiar moral perspective in our society and has often been compared to the "Golden Rule." Unless you are willing to be treated in a particular way, then you shouldn't treat others in this way. According to this view, you ought not to steal from another person because you wouldn't want to be the victim of theft. To will the universal maxim that everyone should steal from each other would be illogical. In doing so you would advocate actions that would be harmful to yourself.

An important implication of Kant's categorical imperative is that humans should be treated as ends and never as means. This is related to the notion that a categorical imperative is not dependent on outcomes. Thus, the way you treat another person ought not to depend on what you would like to happen. For example, it would be wrong to harm another person based on the justification that doing so will benefit you in some way. According to Kant, such actions are wrong because that person matters morally. A human being is a moral agent and is therefore deserving of respect. Using another person for the benefit of oneself or others is disrespectful.

18

An important concept in Kant's ethical theory is inherent or **intrinsic value.** The moral value that we have a duty to respect others is not dependent on whether or not the overall good is served, as in utilitarian theory. This latter type of value is often called instrumental value. Kant is adamant that humans have moral value irrespective of their usefulness. In other words, it is value inherent in us. It is the value each human has in and of itself.

Kant's ideas offer a means for understanding the idea of individual **rights.** It is useful to think about rights in terms of what purpose they serve. The purpose of rights is to place limits on what one individual should do to another individual. For example, the right to speak freely means that a person should not interfere with another person's speech. Kant's ideas can be understood to mean that each of us has a moral duty to respect the individual rights of others. Rights prevent us from treating each other as means.

Regan argues that the best way to ensure that the interests of animals are protected is to extend to them certain individual rights. According to Regan, individual moral rights "place a justifiable limit on what the group can do to the individual." Clearly this view is related to Kant's advocacy of respect for individual humans. What does Regan mean by a justifiable limit? This means that an individual can suffer harm, but if and only if:

- There is good reason to believe that overriding an individual's right will prevent and is the only realistic way to prevent a vastly greater harm to other innocent individuals.

- There is good reason to believe that allowing the individual harm is a necessary link in a chain of events that collectively will prevent vastly greater harm to innocent individuals and there is good reason to believe this chain of events is the only way to prevent such harm.

- There is good reason to believe that only by overriding an individual's rights will there be reasonable hope of preventing vastly greater harm to other innocent individuals.

Regan's concept of individual rights should seem familiar. After all it is the basis of our social and political system. However, should it apply to nonhuman animals? To answer this question, Regan returns to traditional notions of rights, particularly ideas regarding the basis of such rights. Kant argued that individual humans have inherent value that is worthy of respect (protection). Inherent value is the value of something in and of itself; not because that something is useful (i.e., instrumental value). In other words, humans have value because of some particular trait or set of traits, distinct from their utility or skill. According to Kant this means that humans should be treated as ends and not as means.

18

Do nonhuman animals have inherent value? We typically regard animals in terms of their instrumental value. Laboratory rats are valuable as research tools. Farm animals are valuable to us as sources of food and fiber. Companion animals are valuable to us because they entertain us, provide company, and so on. Is this the only type of value these creatures have? Are there traits unique to animals that have value irrespective of their use to us? Regan thinks that animals have inherent value. Each individual dog, cat, rat, gopher, cow, and so on has a life independent of its relationship with humans (or independent of the value a human places on them). Moreover, according to Regan, an individual animal and an individual human is a "subject of a life," meaning that what is better or worse for each is not dependent logically on what others do or do not do. If we accept Kant's notion that human nature is worthy of respect and that human rights are a means of ensuring that respect is upheld, it is logical to assume that nonhuman animals have natures that are also worthy of respect and postulate the existence of animal rights to ensure that such respect is upheld.

Against Moral Obligations to Nonhuman Animals

Singer and Regan rely on traditional ethical theories to argue that humans have direct moral duties to nonhuman animals, and that such duties require that we not use them for research and food. Not all contemporary ethicists agree with the view that these traditional theories can or should be applied in this way. Those who disagree with Regan and Singer point out that traditional ethics were developed for use in the world of humans, and that any attempt to apply these theories beyond this realm requires special scrutiny. The analyses of some ethicists lead them to conclude that extensions of traditional ethics to nonhuman animals are misguided, meaning that these animals have no moral standing. We shall explore this perspective in the next section.

Moral Justification for Using Nonhuman Animals for Food and Research

Jan Narveson examines both utilitarian and rights-based approaches to extending moral considerability to nonhuman animals, and reaches a conclusion that it markedly different from those of Singer and Regan. Briefly, Narveson concludes that "[a]lthough it may be unfortunate for animals that we make meals out of them [and perform research on them], we are morally justified in doing so."

Peter Singer concludes that the utility gained by raising and killing animals for food does not outweigh the overall costs and thus advocates vegetarianism. Narveson concludes that it is not at all clear that vegetarianism is the outcome supported by a utilitarian analysis of animal agriculture. His skepticism is based on difficulties in assigning utility to human and nonhuman animals. Assuming, as does Singer, that the interests of one individual count the same as any other individual, Narveson demonstrates that support of vegetarianism is questionable. Such confusion is illustrated by the following analysis. Within the utilitarian framework the fact that several people benefit from eating one chicken or pig seems to support the view that it is right to kill and consume these animals. However, on a global basis, it seems likely that the costs to the millions of chickens and pigs killed and eaten each year far outweigh any utility humans might gain from eating them, and thus vegetarianism would appear to ensure best outcome. But wait a minute! Recall from Chapter 1 that on a global basis more people can be fed on grains than with meat from grain-fattened livestock. Thus it is reasonable to conclude that raising large numbers of animals for food helps keep human populations smaller which certainly boosts the utility for those of us who can survive. Clearly utilitarianism doesn't provide a clear-cut answer to the question of whether or not we should eat nonhuman animals.

The concerns Narveson raises about utilitarian approaches to extending moral status to nonhuman animals are similar to those raised by Regan. However, Narveson finds Regan's approach to be even less helpful. According to Narveson, animals do not have rights. His argument relies on the so-called **social contract** view of morality.

According to social contract theorists such as Thomas Hobbes (1588–1679), John Locke (1632–1704), and most recently John Rawls, humans have unlimited freedoms in the absence of society (i.e., a "state of nature"), but life under such conditions is difficult and inefficient. Social contract theorists hypothesize that in order to make life more bearable, rational beings (humans) come together and form civil society that is intended to improve living conditions. Thus, formation of society is a trade-off concerning individual rights. In order to live peacefully, individuals give up some rights, but the social contract also preserves certain fundamental rights that allow each individual to express his/her nature. According to Locke these rights include life, liberty, and property. Of course there is no tangible social contract. Rather, the contract is a moral attitude that is required to sustain civil society, and is reflected in the norms and laws of society. The motivation for accepting the contract is rational self-interest. In other words, those who enter into the contract give up total freedom, but in return are guaranteed that their basic rights will be protected, thus making them better off.

The popularity of the social contract theory is likely related to the fact that it provides practical accounts of what morality entails, who has it, and who is bound by it. From this perspective, morality is a solution to resolving

18

conflicts that arise when self-interested human beings interact. Narveson emphasizes that according to this theory, morality is restricted to individuals who meet two important criteria: 1) having the ability to derive more benefit from abiding by the contract than from not doing so and 2) being capable of entering into and keeping an agreement. In other words, the social contract recognizes as having rights only those beings who meet these criteria. Nonhuman animals meet neither of these criteria. Therefore, they can not be considered to be part of the moral community and as such have no rights.

Carl Cohen shares Narveson's concerns about extending rights to nonhuman animals, and provides a detailed critique of Regan's argument. The focus of Cohen's critique hinges on Regan's distinction between moral patients and moral agents as well as Regan's claim that nonhuman animals have the same type of inherent value that humans have. Cohen and Regan agree that nonhuman animals are moral patients that lack the ability to formulate and use moral principles, whereas humans are moral agents who can formulate and use moral principles. However, Cohen questions how Regan can hold this view and then conclude that nonhuman animals have an inherent value that requires moral respect. According to Kant it is the ability to reason that constitutes such inherent value. Because nonhuman animals lack the ability to formulate and use moral principles (i.e., reason), how can Regan assign them moral value? Cohen seems willing to concede that humans might have some obligations to nonhuman animals, but that this doesn't mean that they have rights. According to Cohen, rights are only part of the human world. The world of nonhuman animals is amoral.

One of the obvious problems with arguments that seek to deny nonhuman animals moral standing on the grounds that they lack moral abilities is that such criteria also exclude certain types of human beings to whom we typically grant rights; for example, very young infants, comatose people, and so on. If an ability to enter a contract or do right or wrong are prerequisites for moral status, then shouldn't we deny these people moral standing? The fact that we don't suggests that moral standing involves more than being a rational, self-interested being.

It may very well be, as Narveson and Cohen suggest, that traditional ethics cannot be successfully applied to nonhuman animals. However this doesn't necessarily mean that these animals lack moral status. Alternatively, our difficulty in assigning moral status to nonhuman animals might be due to the fact that our moral theories are flawed. Viewed in this light, the arguments of Narveson and Cohen seem more conceptual (advocating certain definitions of terms) than moral. According to some environmental ethicists, our traditional ethical theories are anthropocentric; that is, they emphasize human traits as the basis of morality. In this case, it would be helpful to reject traditional assumptions about ethics and turn to the task of developing a new ethical framework from which new theories can be developed. One such approach is ecological ethics.

ECOLOGICALLY BASED ETHICS

One of the major features of traditional ethical theories is that they are human-centered (anthropocentric) and focus on individuals (individualistic). However, individual humans live in a world in which they are related to each other, to other organisms as well as to nonliving portions of the environment. Consider the loner who lives alone in a large city. Even this person exists in a relationship with his immediate surroundings and human neighbors. This person and his neighbors also exist in a larger set of relationships that make up society and the environment. In other words, this person's actions have direct and indirect impacts on other people, nonhuman animals, water, air, and so on, and these other components of the environment affect him/her in both direct and indirect ways. The critical question is whether or not these are morally significant interactions. The following thought experiment might help us develop an answer. Suppose you are the last person left on Earth. Now suppose that you set out to use all the remaining technology to destroy all the remaining wilderness areas, simply because you find it stimulating. Would your actions be considered unethical? According to traditional ethical theory, there would be nothing ethically wrong with doing this because you would not be violating the interests or rights of anything that has moral status. On the other hand, something seems terribly wrong about this conclusion. Our intuitions might indicate that there is something ethically wrong with destroying nonhuman nature. Now ask yourself, what aspects of these wilderness areas have moral value. Is it only the individual animals, individual animals and plants, entire species of organisms, the entire ecosystem, or all of these things? Some environmental philosophers have worked to develop an ethical theory that is based on the principles of ecology, the branch of biology that considers the interrelationships among organisms and their environment. Charles Taylor, an early proponent of ecological ethics, argues that individualistic ethics is based on a "delusion of self-sufficiency," and therefore fails to account for the value of relationships. Such relationships include social (among humans) as well as ecological (among living and nonliving components of the environment).

Philosophers who find fault with the individualistic and anthropocentric approach to ethics typically turn to the ideas of Aldo Leopold (1887–1949), a forest ecologist and author of *A Sand County Almanac*. In this book,

18

published in 1949, Leopold describes the natural history of a region of Central Wisconsin near the Wisconsin River and from his experiences in this ecosystem he advocates a new ethic, which he calls "the Land Ethic." Unlike traditional ethical approaches, which value individual interests or rights, Leopold's ethic values the health of the community of life, which he refers to as "the land."

> A thing is right when it tends to preserve the integrity, stability, and beauty of the biotic community. It is wrong when it tends otherwise.

Leopold's idea is radical in the sense that it challenges foundational assumptions of traditional ethics. Specifically, it assumes that humans are part of the so-called biotic community and have no more and no less value than any other component. In Leopold's ethic, the unit of moral considerability is the web of relationships among living and nonliving components of the environment, and the primary moral value is well being of wholes such as ecosystems, species, and entire planets. Interestingly, Leopold doesn't view his ideas as replacing traditional ethical theory. Rather he views it as a principle that should be added on to traditional ethical principles. The major implication of an ecological ethic such as Leopold's is that individuals diminish in importance with respect to moral decisions. In this sense, Leopold's ethic is holistic rather than individualistic.

Ecofeminism

One of the most thorough critiques of traditional ethics comes from feminist theorists. Some feminists argue that an emphasis on individual rights reflects the dominant perspectives of privileged white males in Western societies and ignores other important values that are associated more with the traditional lives of women. In other words, the dominant conceptual framework of Western societies, including its moral values, is patriarchal or dominated by the experiences and values of men.

Ecofeminism is one type of feminism that focuses on intersections between gender and environmental issues. Ecofeminsists point out that the traditional social roles of women (maintaining households and rearing and caring for children), make them more reliant on social and ecological relationships than men and as a result, their experiences contribute to a conceptual framework that differs from the one associated with the traditional social roles of men. This alternative conceptual framework gives rise to different value systems which have been overlooked by traditional liberal theory. Karen Warren describes this in the following way:

An ecofeminist ethic provides a central place for values typically unnoticed, underplayed, or misrepresented in traditional ethics (e.g., values of care, love, friendship, and appropriate trust). These are values that presuppose that our relationships to others are central to an understanding of who we are.

Warren views the dominant conceptual framework of Western societies as oppressive in the sense that it "functions to explain, maintain, and justify relationships of unjustified domination and subordination." She and other ecofeminists argue that this conceptual framework is patriarchal because it justifies the subordination of women by men. They also note that this oppressive conceptual framework also justifies the subordination of the nonhuman others in nature. Thus there is a connection between the subordination of women and nature by men and the two types of subordination are mutually reinforcing.

How might ecofeminist ideas about ethics apply to evaluating issues associated with reproductive physiology? Let us consider the issue of raising and killing pigs for food. As noted by Chris Cuomo, traditional ethical theories (including those of Regan and Singer) portray ethical issues as conflicts between individuals. In our example this would be a conflict between saving pigs and saving humans. Cuomo argues that this view oversimplifies the issue. What if we view this issue through a different lens, a "wider and more sensitive lens," an ecological lens? In other words, what if we view pigs and humans as "relational beings—richly enmeshed sets of interests, matter, and meaning?" This view reveals ethical issues to be more complex than simple conflicts between individuals. The discomfort inflicted on pigs that are raised and slaughtered for food and the discomfort experienced by humans who are deprived of pork are certainly important factors in making moral judgments about this practice. However, there are other relevant factors to consider. When one considers the social, political, and ecological contexts in which the consumption of pigs takes place, one begins to recognize that both pigs and humans can be victims of this practice. For example, the hardships suffered by the humans working on pig farms, meat packing plants, and related industries as well as the suffering experienced by pigs can all be traced to the thoughts and practices underlying our economic system and all the institutions that support it. Examples such as this support Cuomo's claim that "suffering is social." Suffering is also ecological. Certain practices kill individuals, destroy healthy relationships, and create unhealthy relationships. Viewing pig consumption, or any other practice, in this way extends the view of eating pork beyond the farm or slaughterhouse. What impact do the wastes

18

generated by pig consumption have on the environment? What moral, social, and political attitudes are reinforced and perpetuated by engaging in this type of activity? Are they attitudes that promote the values of love and caring, or ones that promote arrogance and dominance? What are the implications of promoting such attitudes on the quality of human and nonhuman life?

BOX 18-1   Focus on Fertility: Cloning

The possibility of cloning a human being has been a popular theme in science fiction. Recent advances in reproductive physiology suggest that the idea may not be so far fetched. In 1997, Dolly the sheep became a household name because she was the first adult mammal to be successfully cloned. Clones are not as unusual as one might first think. For example, identical twins are natural clones. Before Dolly, clones derived from totipotent cells of early bovine and rodent embryos were used to produce several genetically identical individuals. The cloning of Dolly was noteworthy because it was the first time an identical twin was produced from an adult (Dolly was 6 years old at the time). The technique used to produce such a clone is called **somatic cell nuclear transfer** (Figure 18-1). This involves replacing the haploid nucleus of an oocyte with the diploid nucleus of a somatic cell from the animal to be cloned, and then transferring it to the reproductive tract of a recipient for development. Following the success with Dolly, this technique has been used to produce clones of sheep, cattle, pigs, goats, deer, cats, dogs, mules, rats, and mice. It should be noted that the success rate of this technique is extremely low. Most clones die during pregnancy or soon after birth. Even the clones that survive experience serious health problems, one of which is accelerated aging.

Two applications of cloning have emerged from research in this area. The first, known as **reproductive cloning,** is to produce offspring. The second application **(therapeutic cloning)** is for the production of embryonic stem cells for research and possibly medical applications. Compared to other assisted reproduction technologies cloning has generated the most concern. Much of this concern arises from the possibility of human cloning. It is important to point out that there has not been a scientifically documented case of a successful cloning of human embryos, and it appears that the cloning of primate embryos will be much more difficult than that of other mammalian species. Nevertheless there is considerable debate over the possibility of cloning a human. These concerns are primarily ethical. With respect to reproductive cloning, the most common criticism is based on the idea that cloning is wrong because it interferes with the natural order of life. Some view this as "playing God." A more secular concern deals with the rights of the human embryo, or the future person developing from it. Assuming a child can be cloned from an adult cell, it is likely that certain expectations will be imposed on the individual and that these somehow infringe upon the person's right to autonomy, or to a unique identity.

Ethical concerns regarding therapeutic cloning are slightly different. Human embryos have to be destroyed when they are used to generate stems cells to treat illnesses. Of course this is

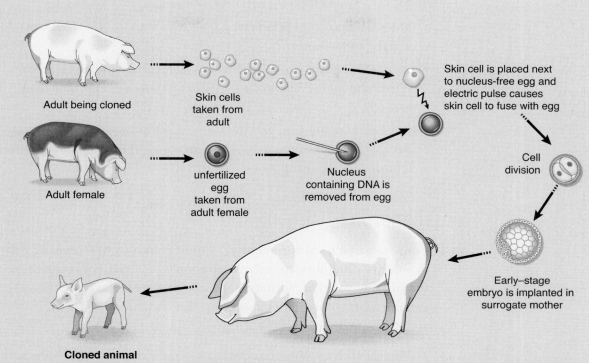

Cloned animal

Figure 18-1 Summary of major steps in cloning a pig by the method of somatic cell nuclear transfer. An unfertilized oocyte (egg) is recovered from a donor female and somatic cells are collected from the animal that is to be cloned. The haploid nucleus from the oocyte is removed from the oocyte and replaced by the diploid nucleus taken from a somatic cell. The diploid egg is allowed to divide *in vitro* until it reaches the blastocyst stage. The embryo is then transferred to a recipient female and allowed to develop to term. Illustration from the National Institute of Health Information on Stem Cells.

unacceptable to those who embrace the notion that human embryos are entitled to the full compliment of human rights, including the right to life. The same arguments used to oppose induced abortion also apply to the issue of therapeutic cloning of human embryos. Although human embryos have not been cloned, the idea of harvesting stem cells from "unwanted" embryos has been promoted by some scientists as a means to develop methods for using stem cells to treat certain diseases. The main argument opposing a ban on use of embryonic stem cells is that the loss of these embryos is justified by the potential benefits derived from such research (curing lethal diseases). Nevertheless, opposition to the use of embryonic stem cells has led to a federal ban on research involving these types of cells. This prohibition has prompted some scientists to explore ways of harvesting and using stem cells from adults.

The issue of human cloning is likely to remain a contentious issue in society. Whether or not we will face the reality of human clones depends on a complicated mixture of scientific, sociologic, moral, and political variables.

18

SUMMARY OF MAJOR CONCEPTS

- Ethics is a branch of philosophy that deals with what is considered to be right and wrong.

- Science and ethics are related by the fact that scientific research reflects the moral values of scientists and those who sponsor their research.

- Issues such as abortion and the use of nonhuman animals for food and research can be understood by analyzing the ethical arguments that support opposing views.

DISCUSSION

1. Construct an ethical argument to support your own view regarding the practice of using human embryos to supply stem cells for research. Be sure to state your conclusion clearly and provide factual principle and a general moral principle that support your conclusion.

2. Discuss what a utilitarian might address the issue of abortion.

3. Ecological ethics is often referred to as a radical approach to ethics. In what way(s) is this approach radical?

4. Discuss what Peter Singer and Aldo Leopold would conclude about the practice of killing deer in order to reduce herd sizes to a manageable size. Would they agree or disagree? Explain.

REFERENCES

Cohen, C. 1997. Do Animals Have Righs? *Ethics and Behavior* 7:103–111.

Cuomo, C. 2005. Ethics and the Eco/Feminist Self. In *Environmental Philosophy: From Animal Rights to Radical Ecology,* Fourth Edition. Upper Saddle River: Prentice Hall, pp. 194–207.

Leopold, A. 1949. *A Sand County Almanac.* New York: Oxford University Press, pp. 201–226.

Marquis, D. 1989. Why Abortion is Immoral. *The Journal of Philosophy* 86:183–202.

National Human Genome Research Institute. Accessed January 7, 2008, from: http://www.genome.gov/10004765.

National Institutes of Health: Stem Cell Basics. In *Stem Cell Information:* Accesed January 7, 2008, from: http://www.nih. gov/info/basics/defaultpage.

Narveson, J. 1977. Animal Rights. *Canadian Journal of Philosophy* 7:161–178.

Noonan, J. 1968. Deciding Who is Human. *Natural Law Forum 13*:134–138.

Regan, T. 2005. Animal Rights, Human Wrongs. In *Environmental Philosophy: From Animal Rights to Radical Ecology,* Fourth Edition. Upper Saddle River: Prentice Hall, pp. 39–52.

Rachels, J. 2003. *The Elements of Moral Philosophy.* Boston: McGraw Hill.

Singer, P. 2005. All animals are equal. In: *Environmental Philosophy: From Animal Rights to Radical Ecology,* Fourth Edition. Upper Saddle River: Prentice Hall, pp. 25–38.

The Oxford Companion to Philosophy. 1995. Oxford: Oxford University Press.

Thomson, J. J. 1971. A Defense of Abortion. *Philosophy and Public Affairs* 1:47–66.

Warren, K. J. 2005. The Power and the Promise of Ecofeminism, Revisited. In: *Environmental Philosophy: From Animal Rights to Radical Ecology,* Fourth Edition. Upper Saddle River: Prentice Hall, pp. 252–280.

Warren, M. 1973. On the Moral and Legal Status of Abortion. *The Monist* 57:43–61.

18

GLOSSARY

A

Acrosome The Golgi-derived cap that covers the top two-thirds of the sperm nucleus and contains enzymes that facilitate entry of the sperm through the zona pellucida of the oocyte.

Acrosome reaction Changes in the ultrastructure of the sperm head resulting from binding of the outer acrosomal membrane to the zona pellucida which allows dispersal of acrosomal contents.

Activational actions Effects of steroid hormones that stimulate or inhibit activity of neurons in the central nervous system.

Acrosin A trypsin-like enzyme produced by sperm.

Adenohypophysis The anterior lobe of the pituitary gland, consisting of the distal, intermediate, and infundibular portions.

Adnexa The connected parts of an organ or tract.

Adventitia The outer layer of connective tissue covering an organ or blood vessel. It is formed from outside the organ and is not an integral part of the organ as is the serosa.

Allantois The fetal membrane that develops from the hindgut and forms the umbilical cord and placenta in eutherian mammals.

Allantochorion The extra-embryonic membrane that forms when the allantois and chorion fuse.

Alpha-fetoprotein A serum protein produced during the prenatal and early postnatal period in rats. The protein is believed to sequester estradiol, thereby protecting the female rat from the masculinizing effects of this hormone on the brain.

Amnion The extra-embryonic membrane that envelopes the embryo suspended in amniotic fluid.

Ampulla A sac-like dilation of a duct, such as the ductus deferens and oviduct.

Anaphase The stage of mitosis and meiosis during which the chromosomes migrate from the equatorial region toward the poles.

Androgen A class of steroid hormones that stimulate the male accessory sex glands and promotes development of male secondary sex traits (e.g., testosterone).

Androgen-binding protein A protein produced by Sertoli cells that binds and sequesters testosterone in the seminiferous tubule.

Anestrus A prolonged period during which a female is sexually quiescent and does not express estrus.

Anovulatory wave A follicular wave that does not culminate with ovulation.

Anti-Müllerian hormone A peptide produced by developing Sertoli cells that causes regression of the mullerian ducts in males during differentiation of the genital ducts.

Antrum An enclosed or nearly enclosed cavity such as that found in tertiary follicles.

Apoptosis The genetically programmed self-destruction of a cell.

Appetitive behavior Behaviors that bring an individual into contact with the goal; for example, searching for a sexual partner.

Asexual reproduction Reproduction without sex; that is, without the re-combination of genetic material from separate sources.

Assisted reproduction technologies Any technique that assists sexual reproduction; e.g., artificial insemination, in vitro fertilization, or embryo transfer.

Atresia Death of an oocyte accompanied by degeneration of the surrounding follicle.

Autocrine The process whereby a hormone acts on the cell that produces and releases it.

Autopoiesis Literally, the formation of self. The metabolic processes that allow an organism to sustain itself in the presence of a continually changing environment.

Autosomes Any chromosome that is not a sex chromosome.

B

Basement membrane A layer of connective tissue underlying the base of a layer of epithelial cells. Also known as the lamina propria.

Binding protein Any serum protein that binds a hormone.

Blastocoele The central cavity of the blastocyst.

Blastocyst An early form of the embryo consisting of an inner cell mass with a thin layer of trophoblasts forming the wall of the blastocoele.

Blastomere A cell formed after cleavage of the egg following fertilization. Also known as a cleavage or embryonic cell.

Blood-brain barrier A barrier that restricts passage of solutes from the blood to the cerebrospinal fluid.

Blood-testis barrier A barrier between the interstitial capillaries and the lumen of the seminiferous tubules that restricts passage of solutes from blood to the tubular compartment.

Broad ligament The peritoneal fold that envelopes and suspends the uterus, uterine tubes (oviducts), and ovaries from the walls of the pelvis.

Bruce effect The phenomenon whereby exposure to a novel male induces abortion in newly-mated female rats.

Bulbospongiosus muscle In males, the skeletal muscle located at the bulb of the penis which constricts the pelvic urethra to promote excretion of urine or enhance expulsion of semen after ejaculation.

Bulbourethral gland An accessory sex gland located near the bulbous penis which secretes ejaculatory fluids into the urethra during ejaculation.

C

Capacitation The process whereby egg-binding proteins on the surface of sperm are unmasked thereby rendering the cell capable of binding to the zona pellucida and undergoing the acrosome reaction.

Capsule A layer of dense connective tissue that envelopes an organ.

Carrying capacity The population that a particular area will sustain without depleting resources.

Caudal Toward the tale.

Centrioles Tubular structures consisting of a wall of nine triplets of microtubules that exist in pairs near the nucleus of a cell.

Centromere The constricted point of a chromosome that serves as the point of attachment to the spindle fiber.

Cerebrospinal fluid The filtrate of blood that is comparable to serum secreted into the ventricles of the brain and which circulates throughout the ventricles and subarachnoid space within the central nervous system.

Chiasmata An intersection between two fibrous bundles. With respect to cytogenetics, the chiasmata is the point at which the chromatids of homologous chromosomes come into contact with each other. Also known as chiasm or chiasma; for example, optic chiasm.

Chorion The outermost fetal membrane consisting of the extraembryonic mesoderm, trophoblast, and villi.

Chromatid One of the two arms (strands) of a chromosome visible when the chromosome condenses during prophase of mitosis and meiosis. The chromatids of a chromosome are joined at the centromere.

Chromosomal sex The complement of sex chromosomes exhibited by males (XY) and females (XX).

Cilia Motile extensions of the surface of a cell.

Circannual rhythm A rhythm with a period of approximately 1 year.

Circadian rhythm A rhythm with a period of approximately 24 hours.

Circumventricular organs Chemoreceptors located near the ventricles of the brain and which are outside the blood-brain barrier.

Clitoris The erectile body located at the anterior border of the vulva between the folds of the labia minora.

Compaction The process whereby cells of the morula converge to form a compact morula. At this stage, cells of the embryo are differentiated into those of the inner cell mass and the outer trophectoderm.

Consummatory behavior Behaviors related to completion of a goal.

Corona radiata The single layer of cells (usually columnar) arising from the cumulus oophorus that anchor the zona pellucida in a follicle.

Corpus albicans Literally means white body. The scar tissue remaining on the surface of the ovary following regression of the corpus luteum.

Corpus hemorrhagicum Literally means bloody body. The blood clot that forms on the ruptured follicle following ovulation.

Corpus luteum Literally means yellow body. The endocrine body that develops from the tissue of the ruptured follicle following ovulation.

Cortex The outer portion of an organ separate from the inner medulla.

Cortical reaction Accumulation of secretory vesicles on the surface of the oocyte in response to the fusion of the sperm membrane with the vitelline membrane of the oocyte.

Cotyledon A structure of the fetal placenta supplied with blood vessels and consisting of chorionic villi which attach to a uterine endometrial structure known as the caruncle.

Cranial Toward the head.

Cremaster muscle The skeletal muscle that suspends and retracts the testis.

Cryptorchidism The failure of the testes to descend from the abdominal cavity into the scrotum.

Cumulus oophorus The mass of granulose cells that surround the oocyte.

Cytokinesis The extranuclear changes in the cytosol of a cell during division.

Cytotrophoblast The inner layer of the trophoblast of an embryo. This layer together with the embryonic mesenchyme forms the chorion.

D

Decidualization The process whereby the mucosal epithelium of the endometrium is transformed into the maternal tissue of the placenta.

Defeminization The loss of feminine features during sexual differentiation.

Delayed implantation Postponement of placentation. This is one of several mechanisms whereby the time of parturition is synchronized with a time of year that is favorable to survival of offspring.

Deviation The process whereby the largest follicle continues to grow and smaller follicles undergo atresia.

Diestrus The postovulatory phase of the estrous cycle during which the corpus luteum is fully developed and functional.

Diploid The normal complement of chromosomes in somatic cells for a given species.

Distal Away from the center of a body or point of origin of a structure.

Dominant follicle The largest follicle of a follicular wave.

Dorsal Referring to the back or upper surface of an animal.

Ductus deferens The duct running from the epididymis of the testis to the pelvic urethra.

E

Ectoderm The outer layer of cells in an embryo at the stage when the three germ layers (ectoderm, mesoderm, and endoderm) are established.

Efferent ducts Microscopic ducts draining the rete testis and emptying into the head of the epididymis.

Egg-binding protein The protein on the sperm membrane that serves as a ligand for a sperm receptor located on the zona pellucida.

Embryo An organism at an early stage of development. Usually the embryonic stage ends when the species of the organism is recognizable.

Emergence The stage of follicle development during which secondary follicles become tertiary follicles.

Endocrine Referring to secretions internally into the blood or extracellular fluid.

Endocrine gland Ductless glands that secrete products into the extracellular fluid or blood.

Endoderm The innermost layer of cells in an embryo at the stage when the three germ layers (ectoderm, mesoderm, and endoderm) are established.

Endometrial cups Clusters of tightly packed cells from the trophoblast that become embedded deeply in the endometrium and produce eCG.

Endothelium The layer of squamous cells that line blood vessels and lymphatic vessels.

Entropy The fraction of energy that cannot be used to perform work.

Epiblast The cells that give rise to the three germ layers of the embryo (ectoderm, mesoderm, and endoderm).

Epididymis The coiled tubule attached to the testis connecting the efferent ducts with the ductus deferens. It consists of the head (caput), body (corpus), and tail (cauda).

Epithelium Avascular layer of cells that reside on a basement membrane and line all free surfaces.

Equine chorionic gonadotropin A glycoprotein hormone produced by the endometrial cups with biological activity similar to LH.

Estrus The phase of the estrous cycle during which the female is sexually receptive.

Ethics The academic discipline that deals with what is good and bad as well as what constitutes moral duty.

Excurrent ducts The system of ducts that carry fluids away from the testis.

Exocrine gland Glands that secrete materials into ducts.

External genitalia Reproductive organs located on the exterior of the body (penis and scrotum in males and clitoris and vulva in females).

Extra-embryonic membranes Membranes outside the body of the embryo involved with protection and nutrition.

F

Fertile Having the ability to conceive and give birth to offspring.

Fetus The unborn offspring of a viviparous animal. Typically this term is used to refer to the stage of development following implantation.

Fibroblast A star- or spindle-shaped cell with cytoplasmic processes found in connective tissue and capable of producing collagen.

Follicle-stimulating hormone A pituitary glycoprotein hormone that causes ovarian follicles to develop from the secondary to the tertiary stage.

Follicular wave The pattern of growth of tertiary follicles characterized by a rapid increase in diameter followed by ovulation or atresia.

Folliculogenesis The development, growth, and maturation of ovarian follicles.

Freemartin A sterile female born as a twin to a male with which it exchanges blood. Both the male and female twins are chimeras having a mixture of cells with XX and XY sex chromosomes.

G

Gametes Specialized sex cells that contain a haploid complement of chromosomes.

General moral premise A commonly held postulate regarding what constitutes moral duty or obligation.

Genital ducts The ducts of the genitals (epididymis and ductus deferens in males, and vagina, uterus, and oviducts in females).

Genital fold The bilateral creases on either side of the embryonic genital tubercle that fuse to form the shaft of the penis in males and remain separated to a cleft or vestibule of the female vagina.

Genital swelling The mound of tissue surrounding the embryonic genital folds that differentiates to form the scrotum in males and the vulva in females.

Genital tubercle The knobby prominence of the undifferentiated external genitalia that gives rise to the glans penis in males or the clitoris in females.

Germinal epithelium The layer of cuboidal epithelial cells lining the surface of the ovaries.

Gonadal dysgenesis A defect in the development of the gonads.

Gonadal ridges (Genital ridges) Knots of connective tissue superficial and medial to the embryonic kidney that will differentiate into the testes or ovaries.

Gonadotropin A hormone that promotes the growth and function of the gonads (e.g., LH and FSH).

Gonadotropin-releasing hormone (GnRH) A neuro-hormone consisting of 10 amino acids, produced by neurosecretory cells of the hypothalamus and released into the hypothalamic-hypophysial portal system. Its major function is to stimulate release of luteinizing hormone and follicle-stimulating hormone.

Gonads The organs that produce gametes; that is, the testes and ovaries.

Granulosa cells Epithelial cells located within the lumen of the ovarian follicles.

Gubernaculum The cord of connective tissue that develops from the embryonic mesenchyme and connects the fetal testis to the scrotum. Shortening of this structure is involved with descent of the testis into the scrotum.

H

Haploid Half the number of chromosomes found in somatic cells.

Hatching of the embryo Escape of the developing blastocyst from the zona pellucida.

Hermaphroditism The condition characterized by the presence of both testicular and ovarian tissues.

Hilus (hilum) Point at which nerves and vessels enter and leave an organ.

Homeostasis The set of physiologic mechanisms that act to sustain a steady-state condition in an organism.

Hormone Any biochemical that serves as an intercellular messenger.

Hypersex The process by which different organisms merge and develop a permanent symbiotic relationship.

Hypoblast The layer of cells that is situated next to the yolk sac and below the epiblast of the embryo.

Hypophysis An endocrine gland suspended from the ventral surface of the hypothalamus and embedded in the hypophysial fossa, which is a depression in the dorsal surface of the sphenoid bone. Also known as the pituitary gland.

Hypospadia Developmental defect of the penis characterized by a failure of the genital folds to fuse and which leaves a urethral opening on the ventral surface of the penis.

Hypothalamus The ventral and medial portion of the diencephalons, which forming the walls of the third ventricle and lying between the optic chiasm (anterior limit) and mammillary bodies (posterior limit).

I

Infertile Reversible condition of a reduced ability to reproduce.

Infundibulum Any funnel-shaped structure including the prominence at the base of the hypothalamus or enlarged portion of the anterior portion of the oviduct.

Inguinal canals Openings through the abdominal oblique muscles through which the testicular spermatic cord extends to support the testes.

Inguinal hernia Protrusion of intestine through the inguinal canal due to failure of the inguinal ring to close.

Inguinal rings The arrangement of abdominal muscle that forms the inguinal canals.

Inner cell mass A collection of compacted cells in the blastocyst, which is differentiated from the outer trophoblast. This structure will develop into the embryo.

Interphase The phase between two successive cell divisions.

Intracellular messenger Any biochemical that is generated as a result of a hormone binding to its receptor which initiates a cascade of intracellular events that influence cell function.

Intrinsic value The value of an organism that is distinct from any instrumental value it might have.

Ischiocavernosus muscle A skeletal muscle extending from the ischium to the corpus cavernosum of the penis. Contraction of this muscle compresses the crus penis.

Isthmus The narrow part of the oviduct located between the ampulla and the uterus.

L

Labia The lip-shaped folds of the female external genitalia. They consist of an outer labia majora, and an inner labia minora.

Large luteal cells Endocrine cells of the corpus luteum derived from granulosa cells.

Lateral To the sides away from the median or midsagittal plane.

Leptotene The early stage of meiosis during which the chromosomes shorten and appear as long, separated filaments.

Leydig cells Interstitial testicular cells that occupy the space between the seminiferous tubules and produce testosterone.

Ligaments A fold of peritoneum that suspends organs from the abdominal wall.

Lobule Small lobes. Subdivisions of an organ created by septa that extend from the surface to the deep interior.

Low-density lipoprotein A complex of molecules consisting of lipid and protein. These compounds are internalized by cells and serve as the major source of testosterone.

Lumen The interior space of a tubular structure such as the intestine, vagina, or urethra.

Luteinization Morphologic and biochemical transformation of an ovulated ovarian follicle into a corpus luteum.

Luteinizing hormone A pituitary glycoprotein hormone that stimulates development of ovarian follicles to the preovulatory stage, induces ovulation, and causes the primary oocyte to complete the first meiotic division.

Luteolysis The dissolution of the corpus luteum.

M

Masculinization The acquisition of masculine traits during sexual differentiation.

Medial Toward the middle or near the median or mid-sagittal plane.

Median The center, middle, or midline.

Median eminence The ventral portion of the hypothalamus that forms a mound that is continuous with the infundibulum.

Mediastinum Central mass of connective tissue in the testis, joined to the tunica albuginea by numerous projections that divide the testis into lobules.

Medulla The center of an organ.

Meiosis A type of cell division in which a cell undergoes two divisions of the nucleus and gives rise to four gametes, each of which has the haploid number of chromosomes.

Meiotic sex Reproduction that requires the fusion of male and female gametes, each of which develops from meiosis.

Mesenchyme Connective tissue of the early embryo.

Mesentery A fan-shaped fold of peritoneum that suspends the bulk of the small intestine from the abdominal wall.

Mesoderm The middle layer of cells in an embryo at the stage when the three germ layers (ectoderm, mesoderm, and endoderm) are established.

Mesonephric duct The tubule that drains from the embryonic kidney (mesonephros) to the cloaca in the embryo.

Mesonephros The early embryonic kidney that eventually regresses as the metanephric kidneys develop.

Mesothelium A single layer of squamous cells that lines serous cavities, such as the peritoneum.

Metanephros The permanent kidney that develops from the uretic buds of the embryo.

Metaphase The stage of meiosis or mitosis during which chromosomes become aligned at the equator of cell.

Metestrus The phase of the estrus cycle between estrus and diestrus, during which the corpus luteum is developing.

Microcotyledons A structure of the fetal placenta supplied with blood vessels and forming chorionic villi, which attaches to the uterine endometrium. Similar to cotyledons, only these are not visible to the naked eye.

Mitosis The process of cell division whereby one cell gives rise to two offspring cells, each of which is identical to the parent cells.

Monoestrous Exhibiting one estrous cycle each year.

Moral considerability A trait or set of traits that confer(s) moral value to an entity (e.g. the ability to reason).

Moral significance The degree of moral value conferred to an entity. For example, a dog and human may each have moral value, but one is given a higher priority in moral decision making.

Morula The stage of embryo development during which the embryo consists of a compact mass of blastomeres resembling a mulberry.

Mucosa The innermost layer of tissue in a tubular organ consisting epithelial cells, basement membrane, and in some cases a layer of smooth muscle.

Mucous membrane The layer of epithelial tissue that lines the inner walls of tubular organs.

Mullerian ducts The pair of embryonic ducts, also known as paramesonephric ducts, that give rise to the upper vagina and uterus in females.

Muscularis The muscular layer of a tubular structure (usually smooth muscle) that lies beneath the serosa. In its most developed state this layer consists of two sublayers of circular and longitudinal muscle.

N

Negative feedback A control mechanism whereby a response inhibits the mechanism responsible for generating it.

Nerve fiber The axon of a nerve cell.

Nerve ganglion A grouping of nerve cell bodies located outside the central nervous system.

Nerve nucleus A grouping of nerve cell bodies located within the central nervous system.

Nerve tract A bundle of parallel nerve fibers within the central nervous system.

Neuroendocrine Referring to interactions and/or interfaces between the nervous and endocrine systems.

Neurohypophysis The posterior lobe of the pituitary gland.

Neuron A nerve cell.

Neurosecretory cells A nerve cell that releases a hormone into the extracellular space.

O

Omentum A fold of peritoneum passing from the stomach to other viscera.

Oocyte A primitive form of the female gamete.

Oogenesis Process by which the ovum is developed.

Oogonia Primitive germ cells of the female that undergo mitotic divisions to give rise to primary oocytes.

Organizational actions Actions of steroid hormones that induce permanent changes in structure of the central nervous system.

Ostium Entrance or small opening to a hollow organ.

Ovary The female gonad.

Oviducts Narrow extensions of the uterine horns consisting of the infundibulum, ampulla, and isthmus.

Ovulation The process whereby a mature ovarian follicle ruptures to release the oocyte.

Ovulatory wave A wave of follicle growth that ends with ovulation.

Oxytocin A polypeptide hormone produced in the hypothalamus and released from the posterior pituitary gland in response to suckling or stretching of the cervix to induce milk ejection from the mammary gland or contraction of the myometrium.

P

Pachytene The stage of meiotic prophase during which homologous chromosomes form pairs and divide longitudinally to form two identical chromatids. During this stage each homologous pair exists as a cluster of four chromatids.

Paracrine A mode of hormone action in which a hormone is released into the extracellular space and acts locally on adjacent target cells.

Parenchyma The cells of an organ that are responsible for its function and are supported by connective tissue.

Parietal Referring to the wall of a cavity.

Penis The organ of copulation in males.

Perineum The external region between the vulva (females) or penis (males) and the anus.

Peritubular myoid cell A smooth muscle cell located along the basement membrane of the seminiferous tubules.

Pheromone A chemical substance that is secreted by one individual and perceived by and inducing some physiologic and(or) behavioral response in another individual.

Pituitary gland An endocrine gland suspended from the ventral surface of the hypothalamus embedded in the hypophysial fossa, a depression in the dorsal surface of the sphenoid bone. Also known as the hypophysis.

Placenta The organ regulating metabolic exchange between the fetal and maternal systems and consisting of both fetal and maternal components.

Placental lactogen A lactation-promoting hormone produced by the placenta.

Placentation The process by which the placenta becomes organized and attaches to the endometrium.

Placentoma The placental unit in the cotyledonary placenta consisting of the fetal caruncle attached to the maternal caruncle.

Polyandry A mating strategy in which females have more than one male mating partner.

Polyestrous A pattern of reproductive activity in which females express more than one estrous cycle in a year.

Positive feedback A control mechanism in which a response enhances the signal that induces the response.

Prepuce A fold of skin that covers the glans penis.

Primary follicle An ovarian follicle consisting of an oocyte surrounded by a layer of cuboidal follicular (granulosa) cells.

Primary oocyte An intermediate cell type in oogenesis arising from an oogonium and remaining in the first meiotic prophase until ovulation.

Primitive sex cords Columns of cells formed by proliferation and inward migration of cells from the mesonephros and coelomic epithelium. Development of the sex cords causes the genital ridges to enlarge and grow into the mesonephros.

Primordial follicle The most primitive follicle consisting of the primary oocyte surrounded by a single layer of follicular cells with a thin basement membrane.

Primordial germ cells The progenitors of gamete cells that originate in the inner lining of the yolk sac near the developing allantois and migrate from the yolk sac to the genital ridges during early embryogenesis.

Proestrus The stage of the estrous cycle during which a follicle or follicles develop to the preovulatory stage leading to estrus and ovulation.

Progestin A generic term referring to any substance that exerts the same biological actions as progesterone.

Prolactin A polypeptide hormone produced by the anterior pituitary gland that acts on the mammary gland to initiate lactation in most mammals, and sustains the corpus luteum in rodents.

Prophase The initial phase of meiosis and mitosis characterized by: contraction of chromosomes and migration of centrioles and asters toward the poles of the cell. In meiosis the process is divided into six stages: preleptotene, leptotene, zygotene, pachytene, diplotene, and diakinesis.

Prostate gland One of the accessory sex glands located at the origin of the urethra at the site where the two ductus deferens converge.

Proximal Near the point of origin of a structure.

Pseudohermaphroditism A condition in which an individual possesses the gonads of one sex, but the external genitalia of the other sex.

Pseudopregnancy The condition where the pattern of progesterone is the same as that characterizing pregnancy, but without pregnancy.

Puberty The physiologic, morphologic, and behavioral changes that occur in association with developing the ability to reproduce.

R

Radioimmunoassay A method for quantifying the amount of a chemical substance, involving standard amounts of the substance, a highly purified form of the substance labeled with a radioactive isotope, and antiserum that specifically binds the substance.

Receptor A cellular protein (membrane or intracellular) that binds a substance with high affinity and specificity.

Recrudesce To become active.

Recruitment The process whereby a cohort of secondary follicles develops into tertiary follicles.

Reproductive strategy The reproductive characteristics of individual members of a particular species that are a function of natural selection.

Rete testis The network of microscopic tubules located in the mediastinum into which the straight segments of the seminferous tubules empty.

Rete tubules The microscopic tubules of the rete testis.

Retractor penis muscles A bilateral pair of smooth muscles that originate on the ventral surface of the caudal vertebrae, circumvent the rectum, and insert into the penis on the lateral and ventral surfaces.

Rights Claims or entitlements that place limits on what a group (society) can do to an individual.

Rostral Toward the head.

S

Sagittal The direction along the anterior-posterior plane.

Sterol Carrying Protein-2 (SCP-2) An intracellular protein that interacts with the cytoskeleton to facilitate transport of cholesterol across the cytoplasm.

Scrotum The sac that houses the testis that is composed of skin and an underlying layer of smooth muscle called the tunica dartos.

Secondary follicle An ovarian follicle that consists of the primary oocyte encased in a translucent membrane, called the zona pellucida, surrounded by two or more layers of follicular cells.

Secondary sex traits Sexually differentiated traits that may facilitate sexual contact, but are not required for the production, transport, or fusion of gametes.

Selection The process whereby a particular tertiary follicle emerges from the pool of secondary follicles a few hours before other members of the cohort.

Sella turcica Literally "Turkish saddle." The depression in the dorsal surface of the sphenoid bone which houses the pituitary gland.

Semen The fluid ejaculated through the penis following sexual stimulation. It consists of sperm, testicular fluid, and secretions from the vesicular, prostate, and bulbourethral glands.

Seminiferous tubules The microscopic tubules that make up a bulk of the testicular parenchyma.

Serosa The outermost tissue layer of a visceral organ located within the abdominal or thoracic body cavities, consisting of the surface mesothelium and irregular connective tissue.

Serous membrane A thin membrane (e.g., peritoneum) that produces a serum-like fluid.

Sertoli cells Elongated cells within the seminiferous tubules that ensheath and provide support for developing sperm. These cells are joined laterally by tight junctions to form the blood-testis barrier.

Sex The genetic and phenotypic traits that distinguish between males and females.

Sex chromosomes Chromosomes (X and Y in mammals) that contain genes that are particularly important in determining whether an individual will develop ovaries or testes.

Sex determination The mechanisms that establish the sex of an individual; in particular those that establish what type of gonad forms during early embryogenesis.

Sexual differentiation The process by which sexually dimorphic tissues take on the male or female characteristics.

Sexual reproduction A method of reproduction in which genes from separate sources are recombined.

Sexuality A characteristic describing the sexual feelings and behaviors of an individual.

Sigmoid flexure The S-shaped curve of the fibro-elastic penis when retracted.

Small luteal cells Hormone-producing cells of the corpus luteum that are derived from theca interna cells of the follicle.

Social contract A hypothetical agreement among individuals forming a society that defines the rights and duties of individuals.

Spermatic cord A band of connective tissue that extends from the abdominal cavity through the inguinal rings and supports the ductus deferens, lymphatics, nerves, and blood vessels that carry blood to and from the testes.

Spermatids Haploid cells that metamorphose into spermatozoa and are derived from secondary spermatocytes.

Spermatocytes Parent cells of spermatids derived from spermatogonia.

Spermatogenesis The process by which spermatozoa are produced from spermatogonia.

Spermatogonia Testicular stem cells that occupy the basal region of the seminiferous tubules and differentiate into spermatozoa.

Spermatozoon The male gamete cell (also called a sperm).

Spermiation The release of spermatozoa from the apical portion of the Sertoli cells into the lumen of the seminiferous tubules.

Spermiogenesis The process by which spermatids are transformed into spermatozoa.

Spindle The fusiform structure consisting of longitudinally-arranged microtubules that appears in a dividing cell and provides tracks along which chromosomes migrate to opposite poles of the cell.

Sex Determining Region of the Y Chromosome (SRY) A gene that acts as a switch to initiate expression of other genes that direct development of the male gonad.

Steroidogenic Acute Regulatory Protein (StAR) An intracellular protein believed to play a role in transporting cholesterol molecules from the outer mitochondrial membrane to the inner mitochondrial membrane where it encounters the side-chain cleavage enzyme.

Stigma A visible spot on the surface of a mature ovarian follicle marking the point at which the membrane will rupture causing ovulation.

Stroma The portion of an organ made up primarily of connective tissue (distinguished from the parenchyma).

Submucosa The layer of connective tissue directly beneath the mucosal epithelium of a tubular organ.

Superficial Near the surface.

Superovulation A hormone treatment inducing the ovulation of more than one follicle.

Syncytiotrophoblast The outer layer of the trophoblast.

Syngamy The process by which a sperm and oocyte conjugate.

T

Telophase The final stage of cell division during which the spindle disappears and two nuclei appear.

Tertiary follicle An ovarian follicle with a fluid-filled antrum consisting of an oocyte, distinct granulosal, and thecal layers of follicular cells.

Testes The male gonads.

Theca A layer of follicular cells lying outside the wall of the follicle. The layer consists of an inner theca interna and outer theca externa.

Totipotent Relating to the ability of a single cell to differentiate into any cell type, and therefore develop into part of or an entire organism.

Trabeculae A meshwork of bundles of connective tissue that originate in the capsule and traverse the body of an organ.

Transgenic sex The exchange of genetic material between organisms (e.g., conjugation in bacteria).

Transverse The plane which lies across the long axis of a body.

Trophectoderm Outermost layer of cells in the blastocysts that will give rise to the trophoblast.

Trophoblast The outermost layer of cells in the blastocyst that become the fetal portion of the placenta (i.e., the chorion).

Tuberohypophysial tract A bundle of axons emanating from neuron cell bodies in the hypothalamus and terminating in the posterior pituitary gland.

Tuberoinfundibular tract Bundle of axons emanating from neuron cell bodies in the hypothalamus and terminating in the infundibulum.

Tubulus contortus The convoluted portion of a seminiferous tubule.

Tubulus rectus The straight portion of a seminiferous tubule.

Tunica albuginea Collagenous capsule enveloping the testis and ovary.

Tunica vaginalis Serous membrane (derived from peritoneum) that envelopes the testis and epididymis.

U

Urachus The urinary canal of the embryo emptying into the allantois.

Urethra The tubular canal running from the bladder to the exterior.

Urogenital sinus An embryonic chamber common to the reproductive, urinary, and excretory systems. Also called the cloaca.

Uterus The womb. A hollow portion of the reproductive tract where the fetus implants and develops.

Utilitarian theory An ethical theory based on the assumption that right action is that which promotes the most pleasure.

V

Vagina The female organ of copulation extending from the uterus to the vulva.

Vaginal cavity The space formed between the visceral and parietal tunica vaginalis *once* the testes have descended.

Vaginal process The cavity that forms between the visceral and parietal tunica vaginalis *during* testicular descent.

Vaginal tunic The tunica vaginalis.

Ventral The undersurface of an animal.

Ventricle A cavity within the brain or heart.

Vesicular glands A pair of male accessory sex glands situated laterally to the ductus deferens and dorsal to the pelvic urethra.

Vestibule The small cavity located at the entry to the vagina.

Villi Folds of tissue in a tubular organ that project into the lumen.

Visceral Referring to the viscera; that is, internal organs.

Vitelline block Biochemical changes in the oocyte that prevent more than one sperm from binding to the vitelline (oocyte) membrane.

Vulva External genitalia of the female including the clitoris, labia majora and minora, and vestibule.

W

Whitten effect A phenomenon in which the estrous cycles of female mice kept in a group and isolated from males will become synchronized following exposure to a male mouse.

Wolffian ducts The embryonic ducts that develop into the epididymis and ductus deferens. Also called the mesonephric ducts.

Y

Yolk sac An extraembryonic membrane which is largely vestigial in eutherian mammals, serving only as a source of nutrients before the placenta develops.

Z

Zona bock Biochemical changes in the oocyte that prevent more than one sperm from binding to the zona pellucida, thereby preventing polyspermy.

Zona pellucida A glycoprotein membrane surrounding the oocyte.

Zygote The diploid cell that forms upon union of a sperm with an oocyte.

Zygotene The stage of meiotic prophase during which the pairing of homologous chromosomes begins.

INDEX

Note: Page numbers in bold type indicate pages on which terms appear in bold. Page numbers followed by *f* or *t* indicate figures and tables, respectively.